Signals and Communication Technology

This series is devoted to fundamentals and applications of modern methods of signal processing and cutting-edge communication technologies. The main topics are information and signal theory, acoustical signal processing, image processing and multimedia systems, mobile and wireless communications, and computer and communication networks. Volumes in the series address researchers in academia and industrial R&D departments. The series is application-oriented. The level of presentation of each individual volume, however, depends on the subject and can range from practical to scientific.

Indexing: All books in "Signals and Communication Technology" are indexed by Scopus and zbMATH

For general information about this book series, comments or suggestions, please contact Mary James at mary.james@springer.com or Ramesh Nath Premnath at ramesh.premnath@springer.com.

Sunday Cookey Ekpo
Editor

The Second International Adaptive and Sustainable Science, Engineering and Technology Conference

ASSET 2023 Proceedings

Springer

Editor
Sunday Cookey Ekpo
Manchester Metropolitan University
Manchester, UK

ISSN 1860-4862 ISSN 1860-4870 (electronic)
Signals and Communication Technology
ISBN 978-3-031-53934-3 ISBN 978-3-031-53935-0 (eBook)
https://doi.org/10.1007/978-3-031-53935-0

This Springer imprint is published by the registered company Springer Nature Switzerland AG
The registered company address is: Gewerbestrasse 11, 6330 Cham, Switzerland

Preface

The Adaptive and Sustainable Science, Engineering and Technology (ASSET) Council facilitates interactions amongst experts and professionals from communities of interest on developing and deploying adaptive and sustainable science, engineering and technology and applications for national security and socio-economic development. The council embraces the disciplines of applied science, engineering and technology, focusing on the total integrated complex systems effectiveness of national and global significance. The applications base is sustained through knowledge transfer partnerships and multidisciplinary research and development collaborations.

The ASSET conference serves as an interactive forum for the advancement of the practice of adaptive and sustainable systems across the multiple disciplines and specialty areas involved with the science, engineering and technology of integrated entities, complex systems and networks. The conference provides an avenue for ASSET practitioners, researchers, managers, developers, analysts, educators and users to exchange innovative ideas, concepts, applications and lessons learned in addressing domain-specific problems, applications-oriented topics, methodologies, standards and multidisciplinary research opportunities and findings relating to ASSET systems.

The theme of ASSET conference is "Developing ASSETs for the Present and the Future Generations."

ASSET 2.0 conference was held on 18–20 July 2023 in Ikot Akpaden, Nigeria and Manchester, UK, the same places and venues where the maiden edition, ASSET 2015, was held.

Development of ASSETs in Akwa Ibom State (Nigeria) and Manchester (UK)

Akwa Ibom State falls within the tropical rain forest zone with a dominant vegetation of green foliage of trees and shrubs. Hence, vegetables, seafood and domestic animals dominate the people's traditional economy. Akwa Ibom invests in and develops ASSETs that attract both local and international tourists and holiday makers. It is the highest oil-producing state in Nigeria and has large deposits of crude oil and natural gas that can be sustainably exploited.

Greater Manchester is a North-West County in England. Its contributions to the UK's ASSETs innovation and education span over the last 200 years and the Manchester city-region is a globally recognised centre of cultural, industrial and technological excellence.

The Second International Adaptive and Sustainable Science, Engineering and Technology (ASSET) Conference 2023

The ASSET 2023 conference programme included 1 Guest Speaker (Prof Peter Aaen); 11 Keynote Speakers (Prof Sang Choi, Prof Ifiok Otung, Prof Jaume Anguera, Dr Rafael Del Rey, Dr Aris Christos Alexoulis Chrysovergis, Prof Charles E. A. Uko, Prof Ismael Soto, Engr Paul Sheedy, Dr Kelvin Anoh, Prof Temitayo Shenkoya and Prof Keeley Crockett); 5 Invited Speakers (Dr Ayodeji Sunday, Stephen Muldowney, Dr Amirpasha Javid, Dr Eric Lou and Dr Hosam A. Aleem); 8 Short Courses (Prof Hans Degens, Dominic Heyhoe, Prof Liangxiu Han, Prof Faisal Riaz, Prof Peter Aaen, Prof Sang Choi, Prof William Holderbaum and Prof Samik Chakraborty); and 13 parallel technical sessions.

The success of the ASSET 2.0 conference was made possible by the collective efforts of many individuals. The Conference Organising and the Technical Programme Committees worked hard and smart in planning the conference logistics and the programme. They ensured that the hybrid (i.e., onsite and the online) format of the ASSET 2023 provided a comfortable experience for all the participants. The host institutions, Conference Track Chairs, Local Organising Committee, Programme Chairs, Publicity Committee and the external reviewers provided their timely and significant commendable contributions. The Platinum Sponsor, SmOp Cleantech Ltd and all the industry partners provided commendable assistance and noble support that made ASSET 2023 a rewarding, memorable and enriching experience that brought the global north and the global south together within the ASSETs ecosystem.

About This Volume

This book contains the presentations given in the various technical sessions of the ASSET 2023. A selection of 28 of the presentations spanning novel, significant, impactful and previously unpublished contributions to advances in adaptive and sustainable engineering, science and technologies have been written up and edited as full articles. These are presented as separate chapters of the present volume. The articles include a broad range of discipline-specific and multidisciplinary ASSETs topics and are grouped into eight sections, viz.:

- Mobile and Satellite Communications
- Adaptive and Complex Systems
- Internet of Things, Artificial Intelligence and Big Data
- Renewable Energy Generation, Transmission & Utilisation
- Sustainability, Environment and Climate Action
- Healthcare and Medical Applications
- Education, Entrepreneurship and Governance
- General and Multidisciplinary Topics

Short Courses

Short Course Talk (SCT) 1 Artificial Intelligence, Big Data Analytics, Internet of Things Sensors, Digital Twins and Biomedical Systems

1. *SCT-1A*: "Age-Related Changes of the Muscles and Human Body Systems" – Prof Hans Degens, Manchester Met University, UK
2. *SCT-1B*: "The Adoption of Artificial Intelligence-enabled Internet of Things Sensors" – Dominic Heyhoe, Software AG, UK
3. *SCT-1C*: "The Role of AI and Big Data in Sustainable Food Security for All" – Prof Liangxiu Han, Co-Director of Centre for Advanced Computational Science and Deputy Director of ManMet Crime and Wellbeing / Big Data Centre, Manchester Met University, UK
4. *SCT-1D*: "Basics of Designing Self-driving Vehicles" – Prof Faisal Riaz, ORIC Director, Mirpur University of Science & Technology, Pakistan

Short Course Talk 2 5G/6G Radio Access Technologies / Satellite-Cellular Convergence / Metamaterials & Metasurfaces / Holographic Beamforming Technologies / Smart Manufacturing / Sustainable Energy Systems

1. *SCT-2A*: "Quantitative Visualisation of Microwave Circuits Under Realistic Operating Conditions for 5G/6G Communications" – Prof Peter Aaen, Colorado School of Mines, USA

2. *SCT-2B*: "The Role of Artificial Intelligence in Space Missions: Existing and Emerging Capabilities and Applications" – Prof Sang Choi, NASA Inventor/Scientist, NASA Langley Research Centre, Virginia, USA
3. *SCT-2C*: "Strategy to Reduce Carbon Emission in Port Terminal" – Prof William Holderbaum, Manchester Met University, UK
4. *SCT-2D*: "Holographic Beamformer Metasurfaces Design for Mobile Communication Applications" – Prof Samik Chakraborty, Regent Education and Research Foundation, Kolkata, India

The second ASSETs research and development (R&D) results published in these proceedings vividly provide the way forward for a sustainable integration and utilisation of space and terrestrial assets for the present and the future generations. The innovative promising solutions established by following these R&D beacons will certainly provide world-leading advances for sharing at the future Adaptive and Sustainable Science, Engineering and Technology Conferences.

Manchester, UK

Sunday Cookey Ekpo

Technical Sessions Programme

ASSET 2.0 TECHNICAL PROGRAMME

The Second International Adaptive and Sustainable Science, Engineering and Technology (ASSET) Conference

Theme: Developing ASSETs for the Present and the Future Generations

Ikot Akpaden, Nigeria and Manchester, UK; Tuesday, 18 – Thursday, 20 July 2023; www.assetgrp.org; info@assetgrp.org

The Second International Adaptive and Sustainable Science, Engineering and Technology (ASSET) Conference

Theme: Developing ASSETs for the Present and the Future Generations

Ikot Akpaden, Nigeria and Manchester, UK; Tuesday, 18 – Thursday, 20 July 2023

www.assetgrp.org info@assetgrp.org

ASSET 2.0 TECHNICAL PROGRAMME

The Second International Adaptive and Sustainable Science, Engineering and Technology (ASSET) Conference

Theme: Developing ASSETs for the Present and the Future Generations

Ikot Akpaden, Nigeria and Manchester, UK; Tuesday, 18 – Thursday, 20 July 2023; www.assetgrp.org; info@assetgrp.org

Conference Abstract and Technical Programme

ASSET 2.0 Conference

18 – 20 July 2023

Ikot Akpaden, Nigeria and Manchester, UK

Scope of the Conference includes:

- Mobile and Satellite Communications
- Internet of Things, Artificial Intelligence and Big Data
- Sustainability, Environment and Climate Action
- Education, Entrepreneurship and Governance
- Adaptive and Complex Systems
- Renewable Energy Generation, Transmission & Utilisation
- Healthcare and Medical Applications
- General and Multidisciplinary Topics

ASSET 2.0 TECHNICAL PROGRAMME

The Second International Adaptive and Sustainable Science, Engineering and Technology (ASSET) Conference

Theme: Developing ASSETs for the Present and the Future Generations

Ikot Akpaden, Nigeria and Manchester, UK; Tuesday, 18 – Thursday, 20 July 2023; www.assetgrp.org; info@assetgrp.org

Organising Committee

General Chairs	Technical Programme Chairs	Short Courses Chairs	Exhibit Chairs	Local Organising Committee Chair	Best Paper Award Chair
Sunday Cookey Ekpo *Manchester Met University, UK*	**Isaac Oluwatayo** *University of Venda, South Africa*	**Umar Raza** *Manchester Met University, UK*	**Tshitso Mosolodi** *Snake Nation, South Africa*	**Imeh Okop** *Akwa Ibom State University, Nigeria*	**Ephraim Paul** *Akwa Ibom State University, Nigeria*
Imeh Okop *Akwa Ibom State University, Nigeria*	**Stephen Alabi,** *SmOp Cleantech, United Kingdom*	**Perpetual Eze-Idehen** *The University of Manchester, UK*	**Ubong Ukommi** *Akwa Ibom State University, Nigeria*	**Perpetual Eze-Idehen** *The University of Manchester, UK*	**Muhammad Ijaz** *Manchester Met University, UK*

Steering Committee

Peter Aaen, *Colorado School of Mines, Illinois, USA;* ***Itoro Esiet Udo,*** *University of Uyo, Nigeria;* ***Olawole Kuti,*** *Manchester Met University, UK;* **Daniel Nwandu,** *DCLM, Inc., UK;* **Joseph Essien,** *Lagos, Nigeria,* ***Kelvin Anoh,*** *University of Bolton, UK;* ***MfonObong Uko,*** *Manchester Met University, UK*

ASSET 2.0 TECHNICAL PROGRAMME

The Second International Adaptive and Sustainable Science, Engineering and Technology (ASSET) Conference

Theme: Developing ASSETs for the Present and the Future Generations

Ikot Akpaden, Nigeria and Manchester, UK; Tuesday, 18 – Thursday, 20 July 2023; www.assetgrp.org; info@assetgrp.org

Welcome Message

Welcome to the programme of the Second International Conference on Adaptive and Sustainable Science, Engineering and Technology (ASSET 2023).

We are all very glad to be able to meet onsite and online for this conference. This conference will enhance our collective global visibility for capacity-building and partnership.

Background: The ASSET Council facilitates interactions amongst experts and professionals from communities of interest on developing and deploying adaptive and sustainable science, engineering and technology and applications for national security and socio-economic development. The council embraces the disciplines of applied science, engineering and technology, focusing on the total integrated complex systems effectiveness of national and global significance. The applications base is sustained through knowledge transfer partnerships and multidisciplinary research and development collaborations.

Conference Objectives: This conference serves as an interactive forum for the advancement of the practice of adaptive and sustainable systems across the multiple disciplines and specialty areas involved with the science, engineering and technology of integrated entities, complex systems and networks. The conference will provide an avenue for ASSET practitioners, researchers, managers, developers, analysts, educators and users to exchange innovative ideas, concepts, applications and lessons learned in addressing domain-specific problems, applications-oriented topics, methodologies, standards and multidisciplinary research opportunities and findings relating to ASSET systems.

ASSET 2.0 conference will be held on 18 - 20 July 2023 in Ikot Akpaden, Nigeria and Manchester, UK, the same places and venues where the maiden edition, ASSET 2015, was held.

ASSET 2023 conference programme includes **1 Guest Speaker** (Prof Peter Aaen); **16 Keynote Speakers** (Prof Sang Choi, Prof Ifiok Otung, Dr Chris Clifton, Prof Jaume Anguera, Dr Rafael Del Rey, Dr Aris Christos Alexoulis Chrysovergis, Prof Charles E. A. Uko, Prof Ismael Soto, Engr Paul Sheedy, Dr Muhammad Ijaz, Dr Kelvin Anoh, Prof Bamidele Adebisi, Prof Temitayo Shenkoya and Prof Keeley Crockett); **5 Invited Speakers** (Dr Ayodeji Sunday, Stephen Muldowney, Dr Amirpasha Javid, Dr Eric Lou and Dr Hosam A. Aleem); **8 Short Courses** (Prof Hans Degens, Dominic Heyhoe, Prof Liangxiu Han, Prof Faisal Riaz, Prof Peter Aaen, Prof Sang Choi, Prof William Holderbaum and Prof Samik Chakraborty); and **13 parallel technical sessions.**

The success of the ASSET 2.0 conference is made possible by the collective efforts of many individuals. We appreciate our profound gratitude to the Conference Organising and the Technical Programme Committees. They worked hard and smart in planning the conference logistics and the programme. They ensured that the hybrid (i.e., onsite and the online) format of the ASSET 2023 provided a comfortable experience for all the participants. We extend our acknowledgement to the host institutions, Conference Track Chairs, Local Organising Committee, Programme Chairs, Publicity Committee and the external reviewers. Our heartfelt thanks go to the support and contributions of all the authors. We are eternally grateful to our Platinum Sponsor, SmOp Cleantech Ltd and profoundly thank all our industry partners for their commendable assistance and noble support. We hope you have a rewarding, memorable and enriching experience as we bring the global north and the global south together with the ASSETs ecosystem. Finally, we look forward to hosting you again in our future ASSET events.

ASSET 2.0 TECHNICAL PROGRAMME

The Second International Adaptive and Sustainable Science, Engineering and Technology (ASSET) Conference

Theme: Developing ASSETs for the Present and the Future Generations

Ikot Akpaden, Nigeria and Manchester, UK; Tuesday, 18 – Thursday, 20 July 2023; www.assetgrp.org; info@assetgrp.org

Attendees' Guideline

Time Zone: British Summer Time (BST) (GMT+1)

- The duration of oral presentation slot is 15 minutes (including 2-3 minutes Q&A time).
- We will appreciate your punctual arrival and active engagement in each session.
- Prepare and backup your presentation PPT and/or PDF files.
- The official oral presentation time schedule is for reference only. Where absence or some presentations finish in less than 15 minutes, kindly join your session earlier.
- A participation certificate will be awarded to each presenter.
- A best presentation certificate will be awarded to the best presentation selected from each session.

For Onsite Participants

- Location-specific instructions will be provided.

Attention

Please, take care of your belongings in public area. For your personal and property safety, delegates are suggested to wear representative card during conference and not to lend it to those unconcerned to enter event rooms. Conference does not assume any responsibility for loss of personal belongings of participants.

- Avoid staying too late in the city;
- Avoid being alone in the remote area;
- Be vigilant and careful with the strangers who offer you service, sign of charity and other appealing experiences at scenic spots; and
- You can search online for more Security and Tourist Information tips.

For Online Participants

Online Conference MS Teams Link:

ASSET 2.0 TECHNICAL PROGRAMME

The Second International Adaptive and Sustainable Science, Engineering and Technology (ASSET) Conference

Theme: Developing ASSETs for the Present and the Future Generations

Ikot Akpaden, Nigeria and Manchester, UK; Tuesday, 18 – Thursday, 20 July 2023; www.assetgrp.org; info@assetgrp.org

- https://teams.microsoft.com/l/meetup-join/19%3ameeting_ZDQ1YzQ1MTEtMzRlMS00MDU3LTk2M2QtZThhNTQ2NTlmMWI4%40thread.v2/0?context=%7b%22Tid%22%3a%22283ffb50-a30b-488c-90f4-cdae4f7ae6d1%22%2c%22Oid%22%3a%222d4ed2c9-af9e-40a9-a41b-080219478a98%22%7d
- Meeting ID: 391 180 210 642
- Passcode: MNsGMa

ASSET 2.0 TECHNICAL PROGRAMME

The Second International Adaptive and Sustainable Science, Engineering and Technology (ASSET) Conference

Theme: Developing ASSETs for the Present and the Future Generations

Ikot Akpaden, Nigeria and Manchester, UK; Tuesday, 18 – Thursday, 20 July 2023; www.assetgrp.org; info@assetgrp.org

Conference Schedule

DAY 1: TUESDAY, 18 July 2023		
8:45 – 09:00	Login via MS Teams Link sent and walk-in registration	
INAUGURAL CEREMONY		
Programme Director(s): **Dr Perpetual Eze-Idehen,** *The University of Manchester, UK* and **Dr Imeh Okop,** *Akwa Ibom State University, Nigeria*		
09:00 - 09:10	**INTRODUCTION**	**Dr Laurence Stamford,** *Head of Environmental Sustainability / Chair of the School of Engineering Board, The University of Manchester (UoM), UK*
09:15 - 09:25	**WELCOME ADDRESS**	**Nigeria: Prof Nse Essien,** *Vice Chancellor, Akwa Ibom State University, Nigeria*
		UK: Prof Steve Rothberg, *Provost / Deputy Vice Chancellor, Manchester Met University, UK*
09:30 - 09:45	**GUEST SPEAKER**	**"Building the Future Green World with ASSETs"** - *Prof Peter Aaen, Department Head, Colorado School of Mines, Illinois, USA*
09:50 – 10:20	**KEYNOTE SPEAKER 1**	**"Scientific Innovation and Challenges for Space Exploration"** – *Prof Sang Choi, NASA Inventor/Scientist, NASA Langley Research Centre, Virginia, USA*
10:25 - 10:30	**CONFERENCE THEME EXPOSITION**	**Prof Andy Gibson,** *Pro-Vice Chancellor (PVC - S&EF), Manchester Met University, UK*
10:30 - 11:00	**KEYNOTE SPEAKER 2**	**"5G/6G Satellite-Cellular Convergence Technologies, Use Cases and Applications"** – *Prof Ifiok Otung, Director of the Bradford-Renduchintala Centre for Space AI, University of Bradford, UK*
11:00 - 11:15	**VOTE OF THANKS**	**Dr Umar Raza | Short Courses Chair |** *Manchester Met University, UK*

ASSET 2.0 TECHNICAL PROGRAMME

The Second International Adaptive and Sustainable Science, Engineering and Technology (ASSET) Conference

Theme: Developing ASSETs for the Present and the Future Generations

Ikot Akpaden, Nigeria and Manchester, UK; Tuesday, 18 – Thursday, 20 July 2023; www.assetgrp.org; info@assetgrp.org

11:15 - 11:30	TEA & COFFEE BREAK / TECHNICAL SESSION	
11:30 – 12:00	Track Chair(s): **Dr Muhammad Ijaz**, Manchester Met University (MMU), UK & **Dr Ubong Ukommi**, *Akwa Ibom State University, Nigeria* AKSU Lecture Theatre / MMU Lecture Theatre, Business School (G33, G34 & G35) / UoM Lecture Theatre / MS Teams	
11:30 – 12:00	KEYNOTE SPEAKER 3	**"Semiconductor & Electronic Solutions for Satellite-Cellular 5G/6G Communication Convergence Use Cases and Applications"** – *Dr Chris Clifton, Chief Technology Officer & Technology Office Director, Sony Europe, UK*
12:00 – 12:30	Track Chair(s): **Dr Muhammad Ijaz**, *MMU, UK* & **Dr Ubong Ukommi**, *Akwa Ibom State University, Nigeria* [AKSU Lecture Theatre / MMU Lecture Theatre, Business School (G33, G34 & G35) / UoM Lecture Theatre / MS Teams]	
12:00 – 12:30	KEYNOTE SPEAKER 4	**"Reconfigurable Multiband Virtual Antenna for Low-Power Internet of Things Sensors Applications"** – *Prof Jaume Anguera, FIEEE, Associate Professor and CTO/Founder, Ignion, Barcelona, Spain*
12:30 – 12:55	Track Chair(s): **Dr Muhammad Ijaz**, *MMU, UK* and **Dr Ubong Ukommi**, AKSU, Nigeria	
12:30 – 12:30	SPEAKER 1	**"Designing Systems for Satellite-Cellular Communication Convergence Use Cases and Applications"** – *Dr Ayodeji Sunday and Glenn Howard, Keysight Technologies UK Ltd, UK*
13:05 – 13:35	LUNCH BREAK and POSTER SESSION	
	AFTERNOON SESSION	
13:35 – 14:20	Track Chair(s): **Dr Laurence Stamford**, *The University of Manchester, UK* & **Dr Itoro Esiet**, *University of Uyo, Nigeria* [AKSU Lecture Theatre / MMU Lecture Theatre, Business School (G33, G34 & G35) / UoM Lecture Theatre / MS Teams]	
13:35 – 14:20	KEYNOTE SPEAKER 5	**"Achieving the Net Zero Vision: The Role of Additively Manufactured Electronics Technology"** – *Dr Rafael Del Rey, Nano Dimension, UK*
14:25 – 15:25	Track Chair(s): **Dr Ubong Ukommi**, *AKSU and* **Dr Perpetual Idehen**, *UoM, UK* Venue: MMU Lecture Theatre, Business School (G33, G34 & G35) / MS Teams Room 1 / AKSU Lecture Theatre / MS Teams Room 2	
14:25 – 14:55	SPEAKER 2	**"A Laboratory Testbed for Unmanned and Autonomous Vehicles for University Teaching and Research Activities"** – *Dr Amirpasha Javid, Director of Academic Partnerships, Quanser Consulting Inc., Canada*

ASSET 2.0 TECHNICAL PROGRAMME

The Second International Adaptive and Sustainable Science, Engineering and Technology (ASSET) Conference

Theme: Developing ASSETs for the Present and the Future Generations

Ikot Akpaden, Nigeria and Manchester, UK; Tuesday, 18 – Thursday, 20 July 2023; www.assetgrp.org; info@assetgrp.org

14:55 – 15:25	SPEAKER 3	**"5G Starter Kits for 5G Private Mobile Network"** – *Stephen Muldowney, Athonet, UK*
15:25 -15:55	Track Chair: **Dr Umar Raza** *Manchester Met University, UK* **and Dr Ubong Ukommi**, *Akwa Ibom State University, Nigeria*	
15:25 – 15:30	INTRODUCTION	**"Hybrid Wireless Power Transfer for Passive Electronic Appliances"** – *Engr Stephen Alabi, SmOp Cleantech Ltd, UK*
15:30 – 16:00	SPEAKER 4	**"The Role of ASSETs in Building Capacities and Capabilities for Internationalised Education and Research"** – *Dr Eric Lou, Manchester Met University, UK*
16:00 – 16:55	EVENING SESSION	
16:00– 16:55	Track Chair: **Dr Olamide Jogunola**, *Manchester Met University, UK* Venue: UoM Lecture Theatre / MS Teams Room 1	Track Chair: **Dr Ubong Ukommi**, *Akwa Ibom State University, Nigeria* Venue: AKSU Lecture Theatre / MS Teams Room 2
15:55 – 16:10	**ASSET_2023_paper_11:** Novel Techniques for 5G/6G Power Amplifier Operation Constraints Mitigation for Satellite-Cellular Applications	**ASSET_2023_paper_3:** Development of Compact Design Formulae for Aperture Coupled MPA for Satellite Communication Applications
16:10 – 16:25	SPEAKER 5	**"A Sustainability Approach to the Design and Operation of Medical Devices" – Dr** *Hosam A. Aleem, The University of Manchester, UK*
16:25 – 16:55	Track Chair: **Dr Muhammad Ijaz** and **Engr Sunday Enahoro**, *MMU, UK*	
16:25 – 16:55	KEYNOTE SPEAKER 6	**"Digital Twins and Industry: A Case Study in Manufacturing"** – *Dr Aris Christos Alexoulis Chrysovergis, Manchester Met University / Siemens PLC, UK*
16:00 – 18:00	Short Courses and POSTER SESSION	
16:00 – 18:00	Short Courses Chair: **Dr Umar Raza**, *MMU, UK and* **Dr Felicia Umanah**, *AKSU, Nigeria*	
16:00 – 18:00	**Short Course Talk (SCT) 1:** **Artificial Intelligence, Big Data Analytics, Internet of Things Sensors, Digital Twins and Biomedical Systems**	**16:00 – 16:30** [1] SCT-1A: **"Age-related Changes of the Muscles and Human Body Systems"** – *Prof Hans Degens, Manchester Met University, UK* **16:30 – 17:00** [2] SCT-1B: **"The Adoption of Artificial Intelligence-enabled Internet of Things Sensors"** – *Dominic Heyhoe, Software AG, UK*

ASSET 2.0 TECHNICAL PROGRAMME

The Second International Adaptive and Sustainable Science, Engineering and Technology (ASSET) Conference

Theme: Developing ASSETs for the Present and the Future Generations

Ikot Akpaden, Nigeria and Manchester, UK; Tuesday, 18 – Thursday, 20 July 2023; www.assetgrp.org; info@assetgrp.org

		17:00 – 17:30 **[3] SCT-1C: "The Role of AI and Big Data in Sustainable Food Security for All** – *Prof Liangxiu Han, Co-Director of Centre for Advanced Computational Science and Deputy Director of ManMet Crime and Wellbeing / Big Data Centre, Manchester Met University, UK*
		17:30 – 18:00 **[4] SCT-1D: "Basics of Designing Self-driving Vehicles"** – *Prof Faisal Riaz, ORIC Director, Mirpur University of Science & Technology, Pakistan*
DAY 2: WEDNESDAY, 19 July 2023		
8:45 – 09:00	**Login via MS Teams Link sent and walk-in registration**	
09:00 – 10:00	Programme Director: **Dr Imeh Okop,** *Akwa Ibom State University, Nigeria* and **Dr Perpetual Eze-Idehen,** *The University of Manchester, UK*	
09:00 – 09:20	**INTRODUCTION AND WELCOME ADDRESS**	**Prof William Holderbaum**, *Manchester Met University, UK*
09:25 – 10:00	**KEYNOTE SPEAKER 7**	**Prof Charles E. A. Uko,** *University of Uyo, Nigeria*
10:00 – 10:30	**KEYNOTE SPEAKER 8**	**"Advanced Visible Light Communication Technology Systems for 5G/6G Satellite-Cellular Convergence Uses Cases and Applications"** – *Prof Ismael Soto, Research Director of Getic, Universidad de Santiago de Chile, Chile*
10:40 - 10:00	**TEA & COFFEE BREAK** **TECHNICAL SESSION**	
11:00 – 12:00	Track Chairs: **Dr Umar Raza** and **Arslan Altaf,** *MMU, UK*	Track Chairs: **Dr Itoro Eslet,** *University of Uyo, Nigeria;* **Dr Ubong Ukommi,** AKSU, Nigeria
	Venue: MMU Lecture Theatre, Business School (G33, G34 & G35) / MS Teams Room 1	**Venue: AKSU Lecture Theatre / MS Teams Room 2**
11:00 – 11:15	**ASSET_2023_paper_12:** Deep Learning-enabled Smart Wearables to Assist People with Autism Spectrum Disorder in Dynamic Environments	**ASSET_2023_paper_14:** Downlink Power Budget and Bit Error Analysis for Lora-Based Sensor Node–To–Satellite Link in the Industrial, Scientific, and Medical Frequency Bands
11:15 – 11:30	**ASSET_2023_poster_13:** Hybrid Deep Learning for Drones-IoTs Emotion Recognition for Mobile Hospital / Ambulance Application	**ASSET_2023_paper_15:** Evaluation of Energy Consumption and Battery Lifespan for Lora IoT

ASSET 2.0 TECHNICAL PROGRAMME

The Second International Adaptive and Sustainable Science, Engineering and Technology (ASSET) Conference

Theme: Developing ASSETs for the Present and the Future Generations

Ikot Akpaden, Nigeria and Manchester, UK; Tuesday, 18 – Thursday, 20 July 2023; www.assetgrp.org; info@assetgrp.org

11:30 – 11:45	**ASSET_2023_paper_23:** Improving the Perception of Moronic Systems whether Human or Self-driving Vehicles using Intelligent Agents: A Review	**ASSET_2023_paper_4:** A Comparative Analysis of Different Polygon-Shaped Slotted Anisotropic Unit-cell for Holographic Metasurface Application
11:45 – 12:00	**ASSET_2023_poster_26:** UCoD: Ensemble BERT for Hierarchical Classification of the Urdu Disinformation Corpus	**ASSET_2023_paper_19:** A Proficient Algorithm for Selecting Cluster Leaders in Wireless Sensor Networks
12:00 – 13:00	Track Chair(s): **Prof Isaac Oluwatayo,** *University of Venda, South Africa,* **Dr Robin Curtis** **[AKSU Lecture Theatre / MMU Lecture Theatre, Business School (G33, G34 & G35) / UoM Lecture Theatre / MS Teams]**	
12:00 – 12:30	**KEYNOTE SPEAKER 9**	**"The Role of Big Data Analytics, IoTs and Artificial Intelligence in ASSET-based Smart Infrastructure"** – *Engr Paul Sheedy, Unifi.id, UK*
12:30 – 13:00	**KEYNOTE SPEAKER 10**	**"Advances in optical wireless communications and applications for 6G networks"** – *Dr Muhammad Ijaz, The Manchester Metropolitan University, UK*
13:00 – 13:30	**LUNCH BREAK and POSTER SESSION**	
	AFTERNOON SESSION	
13:35 – 14:20	Track Chair(s): **Dr Perpetual Idehen,** *The UoM, UK* **and Dr Angham Sabagh,** *MMU, UK* **[AKSU Lecture Theatre / MMU Lecture Theatre, Business School (G33, G34 & G35) / UoM Lecture Theatre / MS Teams]**	
13:35 – 14:20	**KEYNOTE SPEAKER 11**	**"Tackling Engineering Challenges using a Life Cycle Approach: How can Sustainability be Integrated into Engineering Innovation?" – Dr Laurence Stamford,** *Head of Environmental Sustainability / Chair of the School of Engineering Board, The University of Manchester, UK*
14:25 – 15:25	Track Chairs: **Dr Perpetual Eze-Idehen,** *The University of Manchester, UK*	Track Chairs: **Dr Imeh Okop,** *Akwa Ibom State University, Nigeria*
	Venue: UoM Lecture Theatre / MS Teams Room 1	**Venue: AKSU Lecture Theatre / MS Teams Room 2**
14:25 – 14:40	**ASSET_2023_paper_20:** Metasurface-backed Slotted Polarization Diversified Microstrip Patch Antenna for Biomedical Multi-Communication Applications	**ASSET_2023_paper_17:** Preparation of Ecofriendly and Sustainable Adsorbent from Gmelina arborea pulping black liquor and Preliminary Study on its Potentials in the Treatment of Dyes Effluents
14:40 – 14:55	**ASSET_2023_paper_25:** A Reconfigurable Load-Modulated Balanced 9-GHz Power Amplifier Design for Radar Applications	**ASSET_2023_paper_18:** Scientific Analysis of Engineering Materials Derivable From Smart Electronically-Calibrated Non-Woody Agricultural Waste for Production of Pulp
14:55 – 15:25	Track Chair(s): **Dr Robin Curtis and Dr Felicia Umanah, Akwa Ibom State University, Nigeria** **[AKSU Lecture Theatre / MMU Lecture Theatre, Business School (G33, G34 & G35) / UoM Lecture Theatre / MS Teams]**	
14:55 – 15:25	**KEYNOTE SPEAKER 12**	**"Advances in optical wireless communications and applications for 6G networks"** – *Dr Muhammad Ijaz, The Manchester Metropolitan University, UK*
	EVENING SESSION	

ASSET 2.0 TECHNICAL PROGRAMME

The Second International Adaptive and Sustainable Science, Engineering and Technology (ASSET) Conference

Theme: Developing ASSETs for the Present and the Future Generations

Ikot Akpaden, Nigeria and Manchester, UK; Tuesday, 18 – Thursday, 20 July 2023; www.assetgrp.org; info@assetgrp.org

15:25 – 16:40	**Track Chairs: Dr Robin Curtis** and **Dr perpetual Eze-Idehen** *The UoM, UK* **Venue: UoM Lecture Theatre / MS Teams Room 1**	**Track Chairs: Dr Ubong Ukommi, AKSU, Nigeria and Mr Tshitso Mosolodi,** *SN, South Africa* **Venue: AKSU Lecture Theatre / MS Teams Room 2**
15:25 – 15:40	**ASSET_2023_paper_10:** Multiband Monopole Antenna Design for Sub-6 GHz 5G Internet-of-Things Applications	**ASSET_2023_paper_21:** Experimental Analysis of Useful Manufacturing Materials Embedded in Agricultural Wastes for Paper Industry Applications
15:55 – 16:10	**ASSET_2023_paper_1:** Path Loss Prediction of 5G in the 24.25-27.5 GHz Band based on Machine Learning	**ASSET_2023_paper_27:** Human Health Risk Assessment of Trace Metals Due to Dietary Intake of Edible Fishes Obtained from Qua Iboe River, Ibeno Local Government Area, Akwa Ibom State.
16:10 – 16:40	**Track Chair(s): Dr Olawole Kuti, MMU and Dr Angham Sabagh,** MMU **[AKSU Lecture Theatre / MMU Lecture Theatre, Business School (G33, G34 & G35) / UoM Lecture Theatre / MS Teams]**	
16:10 – 16:40	**KEYNOTE SPEAKER 13**	**"Optimising Machine Learning Methods for Inferencing Internet of Things and Hardware"** – *Dr Kelvin Anoh, University of Chester, UK*
	Short Courses AND **POSTER SESSION**	
16:00 – 18:00	**Short Courses Chair: Dr Mohammad Ijaz,** *Manchester Met Metropolitan University, UK and* **Dr Anyanime Akpan,** *AKSU, Nigeria,* **[AKSU Lecture Theatre / MMU Lecture Theatre, Business School / UoM Lecture Theatre / MS Teams]**	
16:00 – 18:00	**Short Course 2:** **5G/6G Radio Access Technologies / Satellite-Cellular Convergence / Metamaterials & Metasurfaces / Holographic Beamforming Technologies / Smart Manufacturing / Sustainable Energy Systems**	**16:00 – 16:30** **[1] SCT-2A: "Quantitative Visualization of Microwave Circuits Under Realistic Operating Conditions for 5G/6G Communications"** – *Prof Peter Aaen, Colorado School of Mines, USA* **16:30 – 17:00** **[2] SCT-2B: "The Role of Artificial Intelligence in Space Missions: Existing and Emerging Capabilities and Applications"** – *Prof Sang Choi, NASA Inventor/Scientist, NASA Langley Research Centre, Virginia, USA* **17:00 – 17:30** **[3] SCT-2C: "Strategy to Reduce Carbon Emission in Port Terminal"**– *Prof William Holderbaum, Manchester Met University, UK* **17:30 – 18:00** **[4] SCT-2D: "Holographic Beamformer Metasurfaces Design for Mobile Communication Applications"** – *Prof Samik Chakraborty, Regent Education and Research Foundation, Kolkata, India*
	ASSET DINNER AND AWARDS CEREMONY	
18:00 – 19:30	**Programme Director(s): Dr Umar Raza and Dr M Ijaz** *(MMU, UK);* **Dr Laurance Stamford** and **Dr Perpetual Idehen (UoM)) Mr Tshitso Mosolodi,** *Snake Nation, South Africa,* **[MMU Lecture Theatre, Business School (G33, G34 & G35) / AKSU Lecture Theatre / MS Teams Room]**	

ASSET 2.0 TECHNICAL PROGRAMME

The Second International Adaptive and Sustainable Science, Engineering and Technology (ASSET) Conference

Theme: Developing ASSETs for the Present and the Future Generations

Ikot Akpaden, Nigeria and Manchester, UK; Tuesday, 18 – Thursday, 20 July 2023; www.assetgrp.org; info@assetgrp.org

18:00 – 19:30	**Opening Remarks**	**Prof Nyaudoh Ukpabio Ndaeyo,** *Vice Chancellor, University of Uyo, Nigeria*
	Dinner Talk	**"The Need for ASSETs for Trans-Generational Global Co-existence"** *by Prof Isaac Oluwatayo, University of Venda, South Africa*
	Closing Remarks	**Mr Tshitso Mosolodi, Country Director – Snake Nation**
	DAY 3: THURSDAY, 20 July 2023	
8:45 – 09:00	**Login via MS Teams Link sent and walk-in registration**	
09:00 – 10:00	**Programme Director(s): Dr Itoro Esiet,** University of Uyo, Nigeria and **Dr Qiuyu Wang,** Manchester Met University, UK **[AKSU Lecture Theatre / MMU Lecture Theatre, Business School (G33, G34 & G35) / UoM Lecture Theatre / MS Teams]**	
09:00 – 09:20	**INTRODUCTION AND WELCOME ADDRESS**	**Prof Akinloye Fatai Lawal,** *Federal University of Technology, Akure Nigeria*
09:25 – 10:00	**KEYNOTE SPEAKER 14**	**"Sustainable Hybrid Energy Harvesting for Energy-Efficient Communication Networks"** – *Prof Bamidele Adebisi, Manchester Met University, UK*
10:00 - 11:40	**TEA & COFFEE BREAK** **TECHNICAL SESSION**	
10:00 – 11:40	**Track Chair(s): Dr Muhammad Ijaz,** *MMU, UK* and **Dr Imeh Okop,** *AKSU, Nigeria* **[AKSU Lecture Theatre / MMU Lecture Theatre, Business School (G33, G34 & G35) / UoM Lecture Theatre / MS Teams]**	
10:05 – 10:35	**KEYNOTE SPEAKER 15**	**"The Role of Internationalised Industry-linked University Innovation and Enterprise Education in Raising Up Job-Creating Graduates"** – *Prof Temitayo Shenkoya, Higher Education Programme Director, British Council, Nigeria*
10:40 – 11:10	**KEYNOTE SPEAKER 16**	**"The Role of Artificial Intelligence Ethics in Developing ASSETs"** – *Prof Keeley Crockett, Manchester Met University, UK*

ASSET 2.0 TECHNICAL PROGRAMME

The Second International Adaptive and Sustainable Science, Engineering and Technology (ASSET) Conference

Theme: Developing ASSETs for the Present and the Future Generations

Ikot Akpaden, Nigeria and Manchester, UK; Tuesday, 18 – Thursday, 20 July 2023; www.assetgrp.org; info@assetgrp.org

10:00 – 12:25	Track Chair(s): **Dr Robin Curtis** *UoM, UK* and **Dr Anyanime Akpan**, *AKSU, Nigeria*, **Business *School* Lecture Theatre (G33, G34 & G35)** / MS Teams	Track Chair(s): **Dr Felicia Umanah**, *AKSU, Nigeria and* **Eng Fanuel Elias**, *MMU, UK* **Lecture Theatre Lecture Theatre** / MS Teams
11:10 – 11:25	**ASSET_2023_paper_5:** Theoretical Analysis and Design Equation of Rotated Dual Split Elliptical Split Ring Resonator (RDS-ESRR)	**ASSET_2023_paper_2:** Establishment of the Anammox Wastewater Treatment Process in the Expanded Bed Biofilm Reactor with ABDite
11:25 – 11:40	**ASSET_2023_paper_6:** Rectifier and Reconfigurable Impedance Matching Network Analysis for Wireless Sub-6 GHz 5G/Wi-Fi 6/6E Energy Harvester	**ASSET_2023_paper_16:** Vibration Analysis for Predictive Maintenance and Improved Machine Reliability of Electric Motors in Centrifugal Pumps
11:40 – 11:55	**ASSET_2023_paper_8:** A DC-50GHz SPDT Switch for 5G Wireless Applications	**ASSET_2023_paper_22:** Evaluation of Standalone Solar Photovoltaic Power System with Load Scheduling for Automated Teller Machine Gallery
11:55 – 12:10	**ASSET_2023_paper_9:** Multi-Radio Frequency Antenna for Sub-6 GHz 5G Carrier and Data Links Margins Enhancement	**ASSET_2023_paper_24:** Design and Carbon Credits Evaluation Of A Grid Connected Solar Photovoltaic Power System for Campus Street Light: Case Study of University of Uyo Main Campus at Nsukara Offot in Uyo, Akwa Ibom State
12:10 – 12:30	**ASSET Conference Organising Committee Meeting & Networking Session** [AKSU Lecture Theatre / MMU Lecture Theatre, Business School **(G33, G34 & G35)** / UoM Lecture Theatre / MS Teams]	
12:30 – 13:00	**CLOSING SESSION AND LUNCH**	
12:30 – 13:00	Session Chair: **Dr Gloria Iyawa**, *University of Salford, UK* and **Dr Angham Sabagh**, *MMU, UK* **[AKSU Lecture Theatre / MMU Lecture Theatre, Business School (G33, G34 & G35) / UoM Lecture Theatre /** MS Teams]	
12:30 – 13:00	**Closing Remarks**	**Dr Perpetual Eze-Idehen** – *Programme Director, The University of Manchester, UK*
	Vote of Thanks	**Dr Imeh Joseph Okop** – *Local Organising Committee Chair, Akwa Ibom State University, Nigeria*
	Announcements	**Dr Sunday Cookey Ekpo** – *General Chair and Conference Proceedings Editor-in-Chief, MMU, UK*
12:30-14:00	**NETWORKING AND LUNCH**	

ASSET 3.0 !!!

[Tuesday, 16 – Thursday, 18 July 2024; Uyo, Nigeria and Manchester, UK]

ASSET 2.0 TECHNICAL PROGRAMME

The Second International Adaptive and Sustainable Science, Engineering and Technology (ASSET) Conference

Theme: Developing ASSETs for the Present and the Future Generations

Ikot Akpaden, Nigeria and Manchester, UK; Tuesday, 18 – Thursday, 20 July 2023; www.assetgrp.org; info@assetgrp.org

Speakers' Introduction

Dr. Sang H. Choi
Senior Lead Scientist
NASA Langley Research Center, Hampton, Virginia 23681

Dr. Choi is a senior lead scientist who leads a research team in advanced electronic and energetic materials at NASA Langley Research Center. At NASA, he has developed satellite ERBE sensors, solar-pumped lasers, plasma switches, iodine lasers, microwave reflectometers & rectennas, solar thermal rocket, bionano battery, smart optics, nano-energetic propellants, micro-spectrometer, neural probe, rhombohedral hybrid bandgap engineering, single crystal SiGe growth, and nuclear thermionic avalanche cells. Also Dr. Choi has served for NASA's Quantum Technology Committee, NASA's Risk-Taking Success Strategy Committee, NASA LaRC Ideation-Innovation Blue Sky Team, and NASA LaRC Honor Awards panel. Dr. Choi received 72 various awards from NASA. Dr. Choi also received 3 of Nano50 Awards for his innovative bionano technology in 2006 and 2007, respectively, and a Nano50 Award cited as an "Innovator of the Year" in 2008. He won R&D100 Award in 2009, the 2010 SOLAR Award in 2010, and 1st prize in the 2017 Create the Future Design Contest in Electronics. Dr. Choi published over 220 technical papers and reports. He received 59 U.S. patents and 6 patents pending out of his 169 inventions. He has total 34 invited, featured, plenary, or keynote talks on various technical areas. Dr. Choi has given 74 invited talks on general subject areas and has 43 news media captures. Dr. Choi served as the Editor-in-Chief of Recent Progress in Space Technology from 2009 to 2016. He is an editorial board member for 5 different journals. Dr. Choi has served as the conference chairs and session chairs for numerous professional events.

His honors are:

- ***a Fellow of National Academy of Inventors (NAI)***
- ***a Fellow of SPIE***
- ***an Associate Fellow of AIAA***
- ***an Inductee of NASA Inventors Hall of Fame, 2020***
- ***NASA Exceptional Service Medal, 2021***

ASSET 2.0 TECHNICAL PROGRAMME

The Second International Adaptive and Sustainable Science, Engineering and Technology (ASSET) Conference

Theme: Developing ASSETs for the Present and the Future Generations

Ikot Akpaden, Nigeria and Manchester, UK; Tuesday, 18 – Thursday, 20 July 2023; www.assetgrp.org; info@assetgrp.org

Keeley Crockett SMIEEE SFHEA is a Professor in Computational Intelligence at Manchester Metropolitan University and Chair of the IEEE Technical Committee SHIELD (Ethical, Legal, Social, Environmental and Human Dimensions of AI/CI). She has over 27 years' experience of research and development in Ethical and responsible AI, computational intelligence algorithms and applications, including adaptive psychological profiling, fuzzy systems, semantic similarity, and dialogue systems. Keeley has led work on Place based practical Artificial Intelligence, facilitating a parliamentary inquiry with Policy Connect and the All-Party Parliamentary Group on Data Analytics (APGDA), leading to the inquiry report "Our Place Our Data: Involving Local People in Data and AI-Based Recovery". She obtained STRENGTH IN PLACES POLICY funded engagement work with Greater Manchester businesses on "SME Readiness for Adoption of Ethical Approaches to AI Development and Deployment" and has contributed to the recent APGDA: AI and Ethics Repot (launched 19-06-23). She is currently the PI on the EPSRC "PEAs in Pods: Co-production of community based public engagement for data and AI research." Grant, Co-I on The Alan Turing Institute "People-powered AI: responsible research and innovation through community ideation and involvement" Grant, and PI on the Innovate UK Knowledge Transfers Partnership with My First Five Years. Keeley is an International Collaborator on COMPROMISE 2021 – 2024: Enhancing COMmunication PROtocols with Machine Learning while Protecting SEnsitive Data (Ministerio De Ciencia e Innovación, Spain). She was Co-academic lead of the GM AI Foundry and is Co-Lead for AI/Cybersecurity on the new Centre for Digital Innovation (CDI). She was also elected to the IEEE Computational intelligence member ADCOM (2023-25), Chair of the IEEE Computational Intelligence Diversity and Inclusion subcommittee and the Co-Chair of the IEEE Women in Engineering Educational Outreach, and she is a U.K. STEM Ambassador. She has 28 PhD completions and will be technical Co-Chair at IEEE FUZZ 2024 (IEEE WCCI 2024) Japan.

Prof. Keeley A Crockett SFHEA SMIEEE
Professor in Computational Intelligence
Lead Data and AI Ethics Theme / Lead Machine Intelligence Theme
Chair IEEE Technical Committee on Ethical, Legal, Social, Environmental and Human Dimensions of AI/CI (SHIELD)
Chair IEEE Computational Intelligence Society Diversity and Inclusion subcommitee
Department of Computing and Mathematics
Manchester Metropolitan University
Chester Street, Manchester, M1 5GD
Email: K.Crockett@mmu.ac.uk
Keeley Crockett's Virtual Office Hours on Teams please email K.Crockett@mmu.ac.uk for an appointment

ASSET 2.0 TECHNICAL PROGRAMME

The Second International Adaptive and Sustainable Science, Engineering and Technology (ASSET) Conference

Theme: Developing ASSETs for the Present and the Future Generations

Ikot Akpaden, Nigeria and Manchester, UK; Tuesday, 18 – Thursday, 20 July 2023; www.assetgrp.org; info@assetgrp.org

PEAs in Pods
Community | Engagement | Research

EPSRC PEAs in Pods: Co-production Of Community Based Public Engagement For Data And AI Research

Abstract – The ethical, social, and legal landscape of Artificial Intelligence (AI) driven systems is rapidly changing. Since the General Data Protection Regulation (2018), stakeholders developing AI systems have faced numerous challenges in the interpretation and implementation of Article 22, specifically concerning an individual's rights in the context of automated decision making, the ability of AI to explain decisions, explanation of the logic involved, and to develop models using only "correct" data. This has and is causing major challenges due to the lack of legal guidance, case law and ethical principles about the use of AI in different contexts. In 2023, the passing of the EU AI Act and other Global "AI and Digital" legalisation promises to also be a big disrupter to all stakeholders working in the field as adopting a risk-based approach to AI becomes the norm. In this talk I will first briefly overview why ethics matters in developing ASSETS and outline my transformational journey. I will provide an overview of the policy and impact landscape and how adopting consequence scanning can help organisations start to adopt a risk-based approach to AI. Responsible innovation and ethical tech are essential to build public trust, yet how can businesses and researchers in academia harness People Power through community ideation and involvement? In this talk I will share two projects that involve public engagement and co-production as mechanisms to build citizen confidence and trust.

ASSET 2.0 TECHNICAL PROGRAMME

The Second International Adaptive and Sustainable Science, Engineering and Technology (ASSET) Conference

Theme: Developing ASSETs for the Present and the Future Generations

Ikot Akpaden, Nigeria and Manchester, UK; Tuesday, 18 – Thursday, 20 July 2023; www.assetgrp.org; info@assetgrp.org

Speaker: Prof. Liangxiu Han, Manchester Metropolitan University

Short Bio: Prof. Liangxiu Han

website: http://www2.docm.mmu.ac.uk/STAFF/L.Han/

Prof. Liangxiu Han has a PhD in Computer Science from Fudan University, Shanghai, P.R. China (2002). Prof. Han is currently a Professor of Computer Science at the Department of Computing and Mathematics, Manchester Metropolitan University. She is a co-Director of Centre for Advanced Computational Science and Deputy Director of ManMet Crime and Well-Being Big Data Centre. Han's research areas mainly lie in the development of novel big data analytics/Machine Learning/AI, and development of novel intelligent architectures that facilitates big data analytics (e.g., parallel and distributed computing, Cloud/Service-oriented computing/data intensive computing) as well as applications in different domains (e.g. Precision Agriculture, Health, Smart Cities, Cyber Security, Energy, etc.) using various large scale datasets such as images, sensor data, network traffic, web/texts and geo-spatial data. As a Principal Investigator (PI) or Co-PI, Prof. Han has been conducting research in relation to big data/Machine Learning/AI, cloud computing/parallel and distributed computing (funded by EPSRC, BBSRC, Innovate UK, Horizon 2020, British Council, Royal Society, Industry, Charity, respectively, etc.).

Prof. Han has served as an associate editor/a guest editor for a number of reputable international journals and a chair (or Co-Chair) for organisation of a number of international conferences/workshops in the field. She has been invited to give a number of keynotes and talks on different occasions (including international conferences, national and international institutions/organisations). Prof. Han is a member of EPSRC Peer Review College, an independent expert of European Commission for proposal evaluation/mid-term project review, and British Council Peer Review Panel.

Talk title: Precision Agriculture: A Big Data Driven, AI enabled Approach

Abstract:

The convergence of big data and artificial intelligence (AI) in agriculture offers tremendous potential in tackling food security challenges. Through the utilization of these technologies, agricultural practices can be optimized to enhance productivity and ensure a sustainable food supply. By leveraging big data and AI, farmers can make data-driven decisions regarding resource allocation, crop management, and yield optimization. This enables efficient use of resources, minimizes waste, and maximizes agricultural output to meet the growing global food demand. This talk will focus on our latest developments in scalable deep learning/AI from big data in the context of precision agriculture to address food security and sustainability.

ASSET 2.0 TECHNICAL PROGRAMME

The Second International Adaptive and Sustainable Science, Engineering and Technology (ASSET) Conference

Theme: Developing ASSETs for the Present and the Future Generations

Ikot Akpaden, Nigeria and Manchester, UK; Tuesday, 18 – Thursday, 20 July 2023; www.assetgrp.org; info@assetgrp.org

Dr. Jaume Anguera, IEEE Fellow, founder and CTO at the technology company Ignion (Barcelona, Spain). Associate Professor at Ramon LLull University and a member of the Smart Society research group. He is an inventor of more than 170 granted patents, most of them licensed to telecommunication companies. Among his most outstanding contributions is that of the inventor of Antenna Booster Technology, a technology that fostered the creation of Ignion. The wireless industry has adopted many of these products worldwide to allow wireless connectivity to IoT devices through a miniature component called an antenna booster that is ten times smaller than conventional antennas. Author of more than 280 widely cited scientific papers and international conferences (h-index 53). Author of 7 books. He has participated in more than 22 competitive research projects financed by the Spanish Ministry, CDTI, CIDEM (Generalitat de Catalunya), and the European Commission for an amount exceeding $13M as a principal researcher in most of them. He has taught over 40 antenna courses worldwide (USA, China, Korea, India, UK, France, Poland, Czech Republic, Tunisia, Perú, Brazil, Canada, Spain). With over 23 years of R&D experience, he has developed part of his professional experience with Fractus in South Korea in designing miniature antennas for large Korean companies such as Samsung and LG. Since 2017 he has been with Ignion in the role of CTO. He leads the company's R&D activity to create new products, envisage new technologies, and provide technology strategy to scale the company's business. He has received several national and international awards (ex. 2004 Best Ph. D Thesis -two prizes, one given by Telefónica Mobile, 2004 IEEE New Faces of Engineering, 2014 Finalist European Patent Award). He has directed the master/doctorate thesis to more than 160 students, many of them have received awards for their thesis (COIT, COITT, Ministry of Education). His biography appears in Who'sWho in the World and Who'sWho in Science and Engineering. He is an associate editor of the IEEE Open Journal on Antennas and Propagation, Electronics Letters, and a reviewer in several IEEE and other scientific journals. He is an IEEE Antennas and Propagation Distinguished Lecturer and vice-chair of the working group "Software and Modeling" at EurAAP. More info at http://users.salleurl.edu/~jaume.anguera/

ASSET 2.0 TECHNICAL PROGRAMME
The Second International Adaptive and Sustainable Science, Engineering and Technology (ASSET) Conference
Theme: Developing ASSETs for the Present and the Future Generations

Ikot Akpaden, Nigeria and Manchester, UK; Tuesday, 18 – Thursday, 20 July 2023; www.assetgrp.org; info@assetgrp.org

Prof Uko obtained a BSc (**First Class Honours**) in Civil Engineering in 1978 at the University of Nigeria, Nsukka and was the **best graduating student in the department for that year.** In 1980, **Prof Uko** won a **Commonwealth Scholarship** to pursue an MSc programme in Construction Engineering at the University of Leeds (UK) and completed the course with **Distinction** in 1981. He was awarded a University Scholarship to do a doctoral degree in Structural Engineering, which he completed in 1984. **Prof. Uko** worked as a Lecturer in the Department of Civil engineering at the University of Port Harcourt until July 1988 when he was appointed **Commissioner for Lands and Housing** in Uyo, Akwa Ibom State. In 1995, Prof. Uko joined the University of Uyo as the pioneer lecturer in the Faculty of Engineering. In1998, he joined the **Petroleum (Special) Trust Fund [PTF]** and worked there till August 1999 when he joined the **Special Projects Unit of the Presidency**, from where he moved to serve as the **Technical Assistant to the Executive Director [PROJECTS]** in the Niger Delta Development Commission, [NDDC], a position he held until May 2002. **Prof. Uko** returned to the University of Uyo in June 2002 and served as **the Acting Head of Department of Civil Engineering** until September 2009. In 2003, he was promoted **Associate Professor** and in 2007 he became a **Professor of Structural Engineering. He was appointed Dean of the Faculty of Engineering** in 2010. He has to his credit two books and a number of publications in local and international journals. **Prof Uko** is a registered engineer with the **Council for the Regulation of Engineering in Nigeria [COREN]**. He is also a member of the **Nigerian Society of Engineers** and the **American Concrete Institute**. He is married to **Mrs Patricia Charles Uko** and they both have five (5) children. He served as the **Deputy Vice Chancellor [ADMIN]** of the **University of Uyo** between 2011 and 2012. He has served as consultant on many road, building and bridge projects and has also developed a computer software for Concrete Mix Design **[COMBA]** using the Nigerian Concrete Mix Design Manual published by COREN. Professor Uko was the Vice President of the **Council for the Regulation of Engineering in Nigeria [COREN]** from 2013 to 2016.

Talk Title: Smart Foundations for Engineering Structures

Foundations or substructures of engineering structures are very important elements in the structure. The integrity of the structure largely depends on the performance of the foundation. When Jesus said in the Bible that a wise man builds his house on the rock, he was very correct. But where rocks cannot be found, the geotechnical and structural engineers have to find ways of overcoming the challenges posed by substandard foundation materials. In this paper, a review of the foundation system used for the Burg Khalifa building in Dubai is presented with emphasis on the smart system that must be kept working for the integrity of the edifice. The paper also presents an insight into current efforts at the University of Uyo, Uyo to develop MICPU, an inverted pyramidal frustum isolated concrete foundation system that has prospects of cutting the cost of building foundations by up to 40%. The paper ends by recommending increased use of artificial intelligence [AI] by engineers in solving challenges encountered in the design and construction of sustainable foundations for engineering structures.

ASSET 2.0 TECHNICAL PROGRAMME
The Second International Adaptive and Sustainable Science, Engineering and Technology (ASSET) Conference
Theme: Developing ASSETs for the Present and the Future Generations
Ikot Akpaden, Nigeria and Manchester, UK; Tuesday, 18 – Thursday, 20 July 2023; www.assetgrp.org; info@assetgrp.org

Age-related changes of the muscles and human body systems

Prof Hans Degens

Department of Life Sciences, Research Centre for Musculoskeletal Science and Sports Medicine, Manchester Metropolitan University, Manchester, UK; Institute of Sport Science and Innovations, Lithuanian Sports University, Kaunas, Lithuania

We rarely consider that the birth of a person is something miraculous! Even though we can be in awe of new life, it seldom crosses our mind that this new life represents an extreme case of rejuvenation. Generation upon generation has displayed this phenomenon where offspring have a similar total life expectancy as their parents, and perhaps even more startling is the fact that we all developed from a single cell (the zygote) -that is barely visible with the naked eye- into the person we are now. This incredible feat of increasing complexity and specialisation during embryological development and maturation is baffling, yet we are unable to thwart the almost imperceptible but progressive deterioration after reaching adulthood, ultimately resulting in our physical demise.

The muscular system is no exception and with increasing age, we become progressively weaker. At some stage, the loss of muscle power is such that it drops below a certain threshold and causes mobility limitations and a reduced quality of life, the 'disability threshold'. Part of this is due to a loss of muscle mass, but also the maximal power generating capacity per unit of muscle mass is reduced. But, are muscle weakness and death inevitable? Is it all maybe because we become less active in old age? Would maintaining our activity levels help maintain muscle function and even open the chance to live eternally? While there is no doubt that resistance exercise increases the power generating capacity even in the oldest and frail old, the gains appear to be less in old age and in some cases, exercise even leads to further muscle weakness rather than gains in strength!

While exercise appears to improve our physical abilities and quality of life, there is also evidence that sedentary behaviour during the day has a negative impact on many health parameters. It is, however, a challenge to adequately monitor sedentary behaviour and any changes in the patterns of sedentary behaviour. Also, how to analyse muscle structure and function during ageing is a challenge and these may be greatly facilitated by the use of computational analyses and computer models.

Overall, it appears there is a progressive, and apparently irreversible, ageing process. The good news is, however, that regular exercise and perhaps other future interventions or lifestyle changes, are able to capitalise on the adaptability of physiological systems and delay the decrement of our functional capacity below a 'disability threshold'.

ASSET 2.0 TECHNICAL PROGRAMME

The Second International Adaptive and Sustainable Science, Engineering and Technology (ASSET) Conference

Theme: Developing ASSETs for the Present and the Future Generations

Ikot Akpaden, Nigeria and Manchester, UK; Tuesday, 18 – Thursday, 20 July 2023; www.assetgrp.org; info@assetgrp.org

Prof. Dr. Faisal Riaz (PhD in Cyber Physical Systems)
Professor in the Department of Computer Science and Information Technology (CS & IT),
Mirpur University of Science and Technology (MUST),
Email: faisal.riaz@must.edu.pk

Dr. Faisal Riaz is a distinguished professor at Mirpur University of Science and Technology, where he leads the Control, Automotive, and Robotics (CAR) lab. He holds a Ph.D. in the field of affective computing-inspired autonomous vehicles from Iqra University in Islamabad, Pakistan. Dr. Riaz has a strong background in conducting quality research and possesses expertise in focused groups, interviewing, and teaching. His doctoral research focused on proposing an agent-based modelling paradigm that simulates the decision-making process of an artificial human driver during collision avoidance. One of Dr. Riaz's notable achievements is the invention of the world's first human factor-inspired autonomous vehicle, called EMO. Furthermore, Dr. Faisal Riaz has also played a pivotal role in the development of the first autonomous industrial logistics vehicle in Pakistan, known as Automaton. These groundbreaking achievements showcase his expertise and innovative mindset in applying autonomous technologies to real-world industrial settings. Dr. Riaz's research team in the CAR lab focuses on leveraging machine-learning techniques for various aspects of self-driving car navigation, including perception, localization, and motion planning. Their goal is to pioneer cutting-edge solutions that integrate artificial intelligence and human-inspired approaches to enhance the capabilities and safety of autonomous vehicles.

Short Course Title: **"Basics of Designing Self-Driving Vehicles"**

Abstract
Certainly! Designing self-driving cars involves a comprehensive and multidisciplinary approach. It typically begins with understanding the principles of artificial intelligence, computer vision, and machine learning. These concepts are crucial for developing algorithms that enable the car to perceive and interpret its surroundings accurately. Additionally, knowledge of robotics and control systems is essential for designing the vehicle's hardware and ensuring precise and safe maneuvering. The course would also cover topics like perception module design, motion planning, decision-making and actuator control. By combining these skills, students can gain a solid foundation in designing self-driving cars and contribute to the exciting field of autonomous transportation.

ASSET 2.0 TECHNICAL PROGRAMME
The Second International Adaptive and Sustainable Science, Engineering and Technology (ASSET) Conference
Theme: Developing ASSETs for the Present and the Future Generations
Ikot Akpaden, Nigeria and Manchester, UK; Tuesday, 18 – Thursday, 20 July 2023; www.assetgrp.org; info@assetgrp.org

Prof Isaac Busayo Oluwatayo

Prof IB Oluwatayo is a full Professor of Agricultural Economics in the Department of Agricultural Economics and Agribusiness, University of Venda, South Africa. Prior to joining University of Venda, he has taught at the University of Limpopo, South Africa, University Ibadan, Nigeria and University of Ado-Ekiti (Now Ekiti State University), Nigeria. Prof IB Oluwatayo has supervised more than 60 undergraduate projects and 70 Dissertations and Theses (63 MSc and 7 PhDs). He is a well-travelled person and has over 100 articles comprising of over 80 journal articles in reputable local and international outlets, conference papers and book chapters. Prof IB Oluwatayo's research interests span through poverty analysis and food security issues, economics of climate change and smallholder agriculture, gender and financial inclusion issues. He is a consultant to both national and international organisations, member of editorial advisory board of Journals and has won several academic and research awards.

Talk Title: The Need for Adaptable and Sustainable Science, Engineering and Technologies for Trans-Generational Global Co-existence

Isaac Oluwatayo
Department of Agricultural Economics and Agribusiness, University of Venda, South Africa.
Email: isaac.oluwatayo@univen.ac.za

There is no gainsaying the fact that science has redefined the course of human existence (Raja and Nagasubramani, 2018). While the world today celebrates the significant achievements/ transformation and ease that science has brought to mankind, the world still faces increasing challenges and adaptable and sustainable science, engineering and technologies (ASSET) are critical for achieving transgenerational prosperity and global co-existence. The need to leverage technology in advancing economic development, social interactions, data analytics and management and climate resilience is fast gaining relevance in literature (Bernai, 2013; Morrar et al. 2017). Adaptive and sustainable science, engineering and technology has provided a platform for nations to champion their development in a sustainable manner. The concept of adaptive and sustainable science, engineering and climate resilience is multidisciplinary, multidimensional and interconnected.

ASSET 2.0 TECHNICAL PROGRAMME

The Second International Adaptive and Sustainable Science, Engineering and Technology (ASSET) Conference

Theme: Developing ASSETs for the Present and the Future Generations

Ikot Akpaden, Nigeria and Manchester, UK; Tuesday, 18 – Thursday, 20 July 2023; www.assetgrp.org; info@assetgrp.org

Dr Laurence Stamford
The University of Manchester, UK

Dr Laurence Stamford (LS; Department of Chemical Engineering) is a Senior Lecturer in Sustainable Chemical Engineering at The University of Manchester. He has specialised in sustainability assessment for more than a decade, focusing on a variety of technologies and sectors spanning energy (nuclear, gas, renewables, storage), chemicals, materials and food systems. He uses a variety of techniques including environmental, economic and social life cycle assessment (LCA, LCC and SLCA, respectively), scenario analysis and decision analysis, among others. His research has been funded by EPSRC, NERC, BBSRC, EC FP7, EC H2020 and Euratom, and has led to >50 journal publications (H index = 25) and consultation by government bodies (e.g. Scottish EPA, Parliamentary Office of Science and Technology). His publications and consultation were key contributors to the JRC review of nuclear energy that led to its inclusion in the EU Taxonomy in 2022. Internationally, he has worked directly with >70 companies and institutions across Europe, the USA and New Zealand. His recent grant awards address nuclear waste, circular economy, and energy generation (EU H2020 PRE-Disposal management of radioactive waste, £200k of £24m; EPSRC IAA Life cycle Innovation in variable Frequency drives for elevators (LIFT), £70k; NERC The social construction of unconventional gas extraction: Towards a greater understanding of Socio-economic impact of unconventional gas development, £160k of £460k). LS has a track record in the use of life cycle sustainability assessment to aid the development of new technological solutions, including recent publications in materials (Nature Sci Rep, 2019, 9, 19033; Sci Total Environ, 2020, 742, 140510) and the energy sector (J Ind Ecol, 2022, 26, 964-979; Energy Convers Manage, 2021, 230, 113838).

Talk Title: Tackling engineering challenges using a life cycle approach: how can sustainability be integrated into engineering innovation?

Sustainability is a frequently cited concept but is less frequently quantified and acted upon with any rigour. The field of life cycle sustainability assessment incorporates some relatively standardised tools, such as environmental life cycle assessment (LCA), and some less formal approaches, such as social impact assessment. Elements of this – particularly LCA – have been used in virtually all sectors and are increasingly mandated via regulation (e.g. the EU Product Environmental Footprint initiative). However, sustainability assessment is typically applied after-the-fact to existing commercial systems. This talk aims to demonstrate that greater sustainability gains can be realised by using life cycle thinking continuously as a way to direct R&D. To this end, we will consider two cases: first, the use of LCA to develop circular economy business models in the power electronics sector, including evaluation of 18 different environmental impact categories in the search for more sustainable approaches for elevator variable frequency drive units; second, the design of energy, water and food provision systems in economically developing countries via life cycle assessment, social impact assessment, techno-economic analysis, and scenario analysis. Through these cases we will investigate methods, sustainability impacts, and the benefits of an integrated approach.

Talk Title: Optimisation of Embedded Machine Learning for Edge Inferencing

Dr Kelvin Anoh, PhD, MSc, BSc (Hons), FHEA, MIET, MIEEE

ASSET 2.0 TECHNICAL PROGRAMME
The Second International Adaptive and Sustainable Science, Engineering and Technology (ASSET) Conference
Theme: Developing ASSETs for the Present and the Future Generations
Ikot Akpaden, Nigeria and Manchester, UK; Tuesday, 18 – Thursday, 20 July 2023; www.assetgrp.org; info@assetgrp.org

Senior Lecturer in Electrical and Electronic Engineering
Google Scholar: https://tinyurl.com/2vzkhyhp; **Researchgate**: https://tinyurl.com/vskux653

Abstract: Edge computing, a scheme that requires data collection, processing and decision-making at the points closer to where data are generated, is continuously evolving. The most recent evolution requires introducing artificial intelligence on edge devices (or the so-called edge AI) so that data can be collected, processed, and decisions made at the edges of communication networks. Implicitly, machine learning algorithms that were traditionally deployed on high-powered computing devices and the cloud, such as servers, are now being required on edge devices such as IoT, sensors, mobile phones, and robots, including satellites. Some of the significant problems with such evolution are that the edge devices can be power-constrained, computationally incapable and might not provide the required results in real time. Deep Neural Networks are complex machine learning algorithms that are quite power-hungry and significantly challenging to deploy on edge devices. Therefore, the machine learning algorithms and the hardware architectures must be independently optimised to qualify the edge AI devices to operate optimally at the edges of communication networks. This talk will present the state-of-the-art machine learning schemes that are being used for different applications and how they can be optimised for applications on edge devices to enable real-time inferencing with tolerable latency. The talk will also present specific examples of optimisation models that can be used to enhance embedded machine learning techniques for energy-efficient and low-latency inferencing of edge devices. Beyond this level, enabling machine learning algorithms for immediate deployment on physical hardware (e.g., CPU, GTU/TPU, FPGA and ASIC) is significantly challenging. The talk will discuss the optimisation of hardware architectures that can enhance edge inferencing, edge AI and decision-making and the speed at which they are made at the edges. When combined, the optimisation of machine learning algorithms and the optimisation of hardware architectures for edge AI will reduce telecommunication network costs, minimise data traffic over networks, enhance data security and integrity, reduce the power consumption of data networks and increase battery lives.

ASSET 2.0 TECHNICAL PROGRAMME

The Second International Adaptive and Sustainable Science, Engineering and Technology (ASSET) Conference

Theme: Developing ASSETs for the Present and the Future Generations

Ikot Akpaden, Nigeria and Manchester, UK; Tuesday, 18 – Thursday, 20 July 2023; www.assetgrp.org; info@assetgrp.org

Peter H. Aaen received the B.A.Sc. degree in engineering science and the M.A.Sc. degree in electrical engineering from the University of Toronto, Toronto, ON, Canada, in 1995 and 1997, respectively, and the Ph.D. degree in electrical engineering from Arizona State University, Tempe, AZ, USA, in 2005. He was the Manager of the RF Division, RF Modeling and Measurement Technology Team, Freescale Semiconductor, Inc., Tempe, AZ, USA, a company which he joined in 1997, then the Semiconductor Product Sector, Motorola, Inc. In 2013, he joined the Faculty of Engineering and Physical Sciences, University of Surrey, Guildford, U.K., where he was a Reader of microwave semiconductor device modeling. He was also the Director of the Nonlinear Microwave Measurement and Modeling Laboratory, a joint University of Surrey/National Physical Laboratory, and the Director of National Physical Laboratory – South of England, Guildford U.K. In 2019, he joined the Colorado School of Mines as a Professor and Head of the Electrical Engineering Department.

He has co-authored Modeling and Characterization of RF and Microwave Power FETs (Cambridge University Press, 2007). Dr. Aaen is a member of the Microwave Theory and Techniques and Electron Device Societies, served as an Executive Committee Member and Vice-President of the Automatic RF Techniques Group, and was the Chair of the IEEE Technical Committee (MTT-1) on Computer-Aided Design.

Talk Title: Quantitative Visualization of Microwave Circuits Under Realistic Operating Conditions

Abstract: Microwave circuits, especially power transistors, are essential components of mobile communication as they amplify signals to be transmitted wirelessly from the base-station to subscriber terminals. Their compact design is increasingly difficult, as mobile network operators require operation at higher output power and frequency while simultaneously demanding a reduction in the circuit size. Shrinking the device and increasing operational frequency results in significant internal electromagnetic coupling and increased power results in higher temperatures, both of which are detrimental to performance. These requirements conflict with one another, and the device physics, electromagnetic fields and distributed temperatures co-couple to limit the overall efficiency. In this presentation, I'll discuss our efforts to develop computational multiphysics simulations techniques and the application of electromagnetic field and temperature measurements to quantitively visualize the internal operations of these microwave devices under realistic operating conditions. These new techniques provide vast new amounts of data and insights that will enable future higher efficiency semiconductor devices.

ASSET 2.0 TECHNICAL PROGRAMME

The Second International Adaptive and Sustainable Science, Engineering and Technology (ASSET) Conference

Theme: Developing ASSETs for the Present and the Future Generations

Ikot Akpaden, Nigeria and Manchester, UK; Tuesday, 18 – Thursday, 20 July 2023; www.assetgrp.org; info@assetgrp.org

Prof Temitayo Shenkoya

Director of Higher Education at the British Council, Nigeria

Temitayo is a professor of science, technology and innovation, and the Director of Higher Education at the British Council, Nigeria. He has a bachelor's degree in Mathematics and Computer Science, a master's degree in Industrial Mathematics and a doctoral degree in Public Policy (S&T policy). He is a registered Projects in Controlled Environments (PRINCE2 - 2017) practitioner and has over 16 years of experience in programme / project management at the United Nations, international NGOs, public organisations, and grassroots organisations (including over 6 years of international experience in South Korea). Temitayo's expertise is in delivering high risk sustainable development / research projects in education, science and technology, energy, climate, water, and green growth. His research interests are in the areas of national innovation systems, regional innovation systems, innovation clusters, metaverse, IoT, Blockchain, AI, big data, and knowledge management. He has several published research articles and books in high impact peer-reviewed journals.

Talk Title: The Role of Internationalised Industry-linked University Innovation and Enterprise Education in Raising Up Job-Creating Graduates

Abstract: Within the National Innovation System (NIS), the three main key players are the government, university, and industry players (Shenkoya, 2021). For any NIS to function properly effectively and efficiently, the communication and interaction between these key players must be seamless. The role of the university within the innovation ecosystem is well known and critical to driving innovation within the NIS. The university plays a fundamental role in knowledge diffusion, human capacity development, and building research excellence - useful in solving societal issues as well as policy development and implementation (Hwang et al., 2023; Shenkoya & Kim, 2021). Furthermore, the university serves as the powerhouse in the relationship amongst the dynamic three.

ASSET 2.0 TECHNICAL PROGRAMME

The Second International Adaptive and Sustainable Science, Engineering and Technology (ASSET) Conference

Theme: Developing ASSETs for the Present and the Future Generations

Ikot Akpaden, Nigeria and Manchester, UK; Tuesday, 18 – Thursday, 20 July 2023; www.assetgrp.org; info@assetgrp.org

Professor William Holderbaum

Professor in Control Engineering

Manchester Metropolitan University, Science & Engineering
John Dalton Building, Chester Street, Manchester, M1 5GD
Email: W.Holderbaum@mmu.ac.uk
Phone: +44 (0) 161 247 1555

URL: https://www2.mmu.ac.uk/soe/staff/profile/index.php?id=2548
http://www.reading.ac.uk/sse/about/staff/w-holderbaum.aspx
IMA Conference on Mathematics of Robotics:http://ima.org.uk/conferences/conferences_calendar/mathematics_of_robotics.html
Mathematics of Robotics: http://oxfordjournals.org/our_journals/imamci/mathematics-of-robotics-special-issue.html
Control of Energy Storage: http://www.mdpi.com/journal/energies/special_issues/control-energy-storage
Rehabilitation Engineering FES: http://fesreading.uk/

Talk Title: Strategy to Reduce Carbon Emission in Port Terminal

Abstract: In low voltage networks, Energy Storage Systems (ESSs) play a significant role in increasing energy cost savings, peak reduction, and energy efficiency whilst reinforcing the electrical network infrastructure. This talk will give an outline of an optimal management system based on different control strategies (Optimal Control and Model Predictive Control) for the control of an ESS equipped with a network of electrified Rubber Tyre Gantry (RTG) cranes. The stochastic management system aims to improve the reliability and economic performance, for given ESS parameters, of a single crane or a network of cranes by considering the uncertainty in the RTGs electrical demand. A specific case study is presented using real operational data of the RTGs network in the Port of Felixstowe, UK, and the results of the stochastic control system are compared to a standard set-point controller. The results of the proposed control strategies indicate that the stochastic management system successfully increases the electric energy cost savings, the peak demand reductions and successfully outperforms a comparable set-point controller.

ASSET 2.0 TECHNICAL PROGRAMME

The Second International Adaptive and Sustainable Science, Engineering and Technology (ASSET) Conference

Theme: Developing ASSETs for the Present and the Future Generations

Ikot Akpaden, Nigeria and Manchester, UK; Tuesday, 18 – Thursday, 20 July 2023; www.assetgrp.org; info@assetgrp.org

Stephen Muldowney is a Senior Sales Manager with Athonet, the global leader in private cellular network technology. Stephen has more than 25 years' experience at the forefront of the global telecommunications sector working with some of the largest vendors and with the latest advances in mobile and wireless technologies with a particular focus on LTE/5G private mobile networks.

Abstract:

- Introduction to Athonet
- Why did we launch our 5G Starter Kits?
- Who are our 5G Starter Kits aimed at and for?
- What is a 5G Starter Kit?
- What are the 5G-SA Starter Kits options?
- What does a 5G Starter Kit network look like?

Twitter: @Athonet

athonet

ASSET 2.0 TECHNICAL PROGRAMME

The Second International Adaptive and Sustainable Science, Engineering and Technology (ASSET) Conference

Theme: Developing ASSETs for the Present and the Future Generations

Ikot Akpaden, Nigeria and Manchester, UK; Tuesday, 18 – Thursday, 20 July 2023; www.assetgrp.org; info@assetgrp.org

Paul Sheedy
Founder/ CEO, Unifi.id

A serial entrepreneur in innovative technology. His is experienced in loyalty data solutions globally, digital comms, design and detection systems. Previously, Paul was the co-founder of EYC.

Paper ID:	ASSET_2023_paper_1
Paper Title:	Path Loss Prediction of 5G in the 24.25-27.5 GHz Band based on Machine Learning
Author(s):	Zaid Abedeen, Sunday Ekpo, Muhammad Ijaz, Umar Raza, Liangxiu Han and Stephen Alabi

ASSET 2.0 TECHNICAL PROGRAMME

The Second International Adaptive and Sustainable Science, Engineering and Technology (ASSET) Conference

Theme: Developing ASSETs for the Present and the Future Generations

Ikot Akpaden, Nigeria and Manchester, UK; Tuesday, 18 – Thursday, 20 July 2023; www.assetgrp.org; info@assetgrp.org

Paper ID:	ASSET_2023_paper_2
Paper Title:	Establishment of the Anammox Wastewater Treatment Process in the Expanded Bed Biofilm Reactor with ABDite
Author(s):	Comfort Anyanwu, Devine Akhidime, Mike Dempsey, Louise Melling and Andrew Dean

Paper ID:	ASSET_2023_paper_3
Paper Title:	Development of Compact Design Formulae for Aperture Coupled MPA for Satellite Communication Applications
Author(s):	Amit Ghosh, Swarnadipto Ghosh, Dipankar Saha, Samik Chakraborty, Sunday Cookey Ekpo and Stephen Alabi

Paper ID:	ASSET_2023_paper_4
Paper Title:	A Comparative Analysis of Different Polygon-Shaped Slotted Anisotropic Unit-cell for Holographic Metasurface Application

ASSET 2.0 TECHNICAL PROGRAMME

The Second International Adaptive and Sustainable Science, Engineering and Technology (ASSET) Conference

Theme: Developing ASSETs for the Present and the Future Generations

Ikot Akpaden, Nigeria and Manchester, UK; Tuesday, 18 – Thursday, 20 July 2023; www.assetgrp.org; info@assetgrp.org

Author(s):	Swarnadipto Ghosh, Dipankar Saha, Ayona Chakraborty, Samik Chakraborty, Sunday Cookey Ekpo and Fanuel Elias

Paper ID:	ASSET_2023_paper_5
Paper Title:	Theoretical Analysis and Design Equation of Rotated Dual Split Elliptical Split Ring Resonator (RDS-ESRR)
Author(s):	Dipankar Saha, Swarnadipto Ghosh, Ayona Chakraborty, Samik Chakraborty, Sunday Cookey Ekpo and Stephen Alabi

Paper ID:	ASSET_2023_paper_6
Paper Title:	Rectifier and Reconfigurable Impedance Matching Network Analysis for Wireless Sub-6 GHz 5G/Wi-Fi 6/6E Energy Harvester
Author(s):	Fanuel Elias, Sunday C. Ekpo, Stephen Alabi, Dipankar Saha, Samik Chakraborty, Swarnadipto Ghosh, Mfonobong Charles Uko, Muhammad Ijaz and Umar Raza

Paper ID:	ASSET_2023_paper_8

ASSET 2.0 TECHNICAL PROGRAMME

The Second International Adaptive and Sustainable Science, Engineering and Technology (ASSET) Conference

Theme: Developing ASSETs for the Present and the Future Generations

Ikot Akpaden, Nigeria and Manchester, UK; Tuesday, 18 – Thursday, 20 July 2023; www.assetgrp.org; info@assetgrp.org

Paper Title:	A DC-50GHz SPDT Switch for 5G Wireless Applications
Author(s):	Mfonobong Uko, Sunday Ekpo, Fanuel Elias, Dipankar Saha, Swarnadipto Ghosh, Muhammad Ijaz, Samik Chakraborty and Andrew Gibson

Paper ID:	ASSET_2023_paper_9
Paper Title:	Multi-Radio Frequency Antenna for Sub-6 GHz 5G Carrier and Data Links Margins Enhancement
Author(s):	Sunday Enahoro, Sunday Ekpo, Andy Gibson, Kin Kee Chow, Helen Ji and Khaled Rabie

Paper ID:	ASSET_2023_paper_10
Paper Title:	Multiband Monopole Antenna Design for Sub-6 GHz 5G Internet-of-Things Applications
Author(s):	Sunday Enahoro, Sunday Ekpo, Andy Gibson, Kin Kee Chow, Helen Ji and Khaled Rabie

ASSET 2.0 TECHNICAL PROGRAMME
The Second International Adaptive and Sustainable Science, Engineering and Technology (ASSET) Conference

Theme: Developing ASSETs for the Present and the Future Generations

Ikot Akpaden, Nigeria and Manchester, UK; Tuesday, 18 – Thursday, 20 July 2023; www.assetgrp.org; info@assetgrp.org

Paper ID:	ASSET_2023_paper_11
Paper Title:	Novel Techniques for 5G/6G Power Amplifier Operation Constraints Mitigation for Satellite-Cellular Applications
Author(s):	Arslan Altaf, Sunday C. Ekpo, Muhammad Ijaz, Andy Gibson and Peter H. Aaen

Paper ID:	ASSET_2023_paper_12
Paper Title:	Deep Learning-enabled Smart Wearables to Assist People with Autism Spectrum Disorder in Dynamic Environments
Author(s):	Przemyslaw Danowski, Sunday Ekpo, Muhammad Ijaz, Umar Raza and Raul Carbrero

Paper ID:	ASSET_2023_paper_13
Paper Title:	Hybrid Deep Learning for Drones-IoTs Emotion Recognition for Mobile Hospital / Ambulance Application
Author(s):	Przemyslaw Danowski, Sunday Ekpo and Umar Raza

ASSET 2.0 TECHNICAL PROGRAMME
The Second International Adaptive and Sustainable Science, Engineering and Technology (ASSET) Conference

Theme: Developing ASSETs for the Present and the Future Generations

Ikot Akpaden, Nigeria and Manchester, UK; Tuesday, 18 – Thursday, 20 July 2023; www.assetgrp.org; info@assetgrp.org

Paper ID:	ASSET_2023_paper_14
Paper Title:	Downlink Power Budget and Bit Error Analysis For Lora-Based Sensor Node–to–Satellite Link in the Industrial, Scientific, and Medical Frequency Bands
Author(s):	Anietie Essien, Ubong Ukommi and Emmanuel Ubom

Paper ID:	ASSET_2023_paper_15
Paper Title:	Evaluation of Energy Consumption and Battery Lifespan for Lora IoT Multisensor Node for Precision Farming Application
Author(s):	Israel Oduoye and Ubong Ukommi

Paper ID:	ASSET_2023_paper_16
Paper Title:	Vibration Analysis for Predictive Maintenance and Improved Machine Reliability of Electric Motors in Centrifugal Pumps
Author(s):	Fortune Jameson, Emmanuel Ubom and Ubong Ukommi

ASSET 2.0 TECHNICAL PROGRAMME

The Second International Adaptive and Sustainable Science, Engineering and Technology (ASSET) Conference

Theme: Developing ASSETs for the Present and the Future Generations

Ikot Akpaden, Nigeria and Manchester, UK; Tuesday, 18 – Thursday, 20 July 2023; www.assetgrp.org; info@assetgrp.org

Paper ID:	ASSET_2023_paper_17
Paper Title:	Preparation of Eco-friendly and Sustainable Adsorbent from Gmelina arborea pulping black liquor and Preliminary Study on its Potentials in the Treatment of Dyes Effluents
Author(s):	Aniekan Akpakpan, Itoro Udo, Nyeneime Akpanudo and Okon E. Okon

Paper ID:	ASSET_2023_paper_18
Paper Title:	Scientific Analysis of Engineering Materials Derivable from Smart Electronically-Calibrated Non-Woody Agricultural Waste for Production of Pulp
Author(s):	Itoro Udo, Enefiok Usungurua, Sunday Ekpo, Perpetual Eze-Idehen and Joseph Sam

Paper ID:	ASSET_2023_paper_19
Paper Title:	VERD: A Proficient Algorithm for Selecting Cluster Leaders in Wireless Sensor Networks
Author(s):	Emmanuel Ogungbemi, Kufre Udofia, Philip Asuquo and Kingsley Udofia

ASSET 2.0 TECHNICAL PROGRAMME

The Second International Adaptive and Sustainable Science, Engineering and Technology (ASSET) Conference

Theme: Developing ASSETs for the Present and the Future Generations

Ikot Akpaden, Nigeria and Manchester, UK; Tuesday, 18 – Thursday, 20 July 2023; www.assetgrp.org; info@assetgrp.org

Paper ID:	ASSET_2023_paper_20
Paper Title:	Metasurface-backed Slotted Polarization Diversified Microstrip Patch Antenna for Biomedical Multi-Communication Applications
Author(s):	Swarnadipto Ghosh, Dipankar Saha, Ayona Chakraborty, Samik Chakraborty, Fanuel Elias and Sunday Cookey Ekpo

Paper ID:	ASSET_2023_paper_21
Paper Title:	Experimental Analysis of Useful Manufacturing Materials Embedded in Agricultural Wastes for Paper Industry Applications
Author(s):	Enefiok Usungurua, Itoro Udo, Daniel Essang, Sunday Ekpo and Perpetual Eze-Idehen

Paper ID:	ASSET_2023_paper_22
Paper Title:	Evaluation of Standalone Solar Photovoltaic Power System with Load Scheduling for Automated Teller Machine Gallery

ASSET 2.0 TECHNICAL PROGRAMME

The Second International Adaptive and Sustainable Science, Engineering and Technology (ASSET) Conference

Theme: Developing ASSETs for the Present and the Future Generations

Ikot Akpaden, Nigeria and Manchester, UK; Tuesday, 18 – Thursday, 20 July 2023; www.assetgrp.org; info@assetgrp.org

Author(s):	Mfonobong Umoren, Eduediuyai Ekerette and Okpura Nseobong

Paper ID:	ASSET_2023_paper_23
Paper Title:	Improving the Perception of Moronic Systems whether Human or Self-driving Vehicles using Intelligent Agents: A Review
Author(s):	Fakhra Riaz, Faisal Riaz, Asma Jabeen, Sidra Kouser, Saeid Jamali, William Holderbaum and Umar Raza

Paper ID:	ASSET_2023_paper_24
Paper Title:	Design and Carbon Credits Evaluation of a Grid Connected Solar Photovoltaic Power System for Campus Street Light: Case Study of University of Uyo Main Campus at Nsukara Offot in Uyo, Akwa Ibom State
Author(s):	Iniobong Abasi-Obot, Eduediuyai Dan and Mfonobong Umoren

Paper ID:	ASSET_2023_paper_25

ASSET 2.0 TECHNICAL PROGRAMME

The Second International Adaptive and Sustainable Science, Engineering and Technology (ASSET) Conference

Theme: Developing ASSETs for the Present and the Future Generations

Ikot Akpaden, Nigeria and Manchester, UK; Tuesday, 18 – Thursday, 20 July 2023; www.assetgrp.org; info@assetgrp.org

Paper Title:	A Reconfigurable Load-Modulated Balanced 9-GHz Power Amplifier Design for Radar Applications
Author(s):	Joseph Skippings, Sunday Ekpo and Andy Gibson

Paper ID:	ASSET_2023_paper_26
Paper Title:	UCoD: Ensemble BERT for Hierarchical Classification of the Urdu Disinformation Corpus
Author(s):	Umar Farooq, Omer Beg, Faisal Riaz, Saeid Jamali, William Holderbaum and Umar Raza

Paper ID:	ASSET_2023_paper_27
Paper Title:	Human Health Risk Assessment of Trace Metals Due to Dietary Intake of Edible Fishes Obtained from Qua Iboe River, Ibeno Local Government Area, Akwa Ibom State
Author(s):	Uwem Ubong, Akanimo Akpan and Ifiok Ekwere

ASSET 2.0 TECHNICAL PROGRAMME
The Second International Adaptive and Sustainable Science, Engineering and Technology (ASSET) Conference
Theme: Developing ASSETs for the Present and the Future Generations
Ikot Akpaden, Nigeria and Manchester, UK; Tuesday, 18 – Thursday, 20 July 2023; www.assetgrp.org; info@assetgrp.org

ASSET 2.0 TECHNICAL PROGRAMME
The Second International Adaptive and Sustainable Science, Engineering and Technology (ASSET) Conference
Theme: Developing ASSETs for the Present and the Future Generations
Ikot Akpaden, Nigeria and Manchester, UK; Tuesday, 18 – Thursday, 20 July 2023; www.assetgrp.org; info@assetgrp.org

About Akwa Ibom State and Greater Manchester

Akwa Ibom State falls within the tropical zone with a dominant vegetation of green foliage of trees and shrubs. As the State lies within the tropical rain forest zone, vegetables, seafood and domestic animals dominate her traditional diets. Akwa Ibom attracts both local and international tourists and holiday makers. Akwa Ibom State is the highest oil-producing state in Nigeria and has large deposits of crude oil and natural gas.

Greater Manchester's contributions to the UK's innovation and education span over the last 200 years and the Manchester city-region is a globally recognised centre of cultural, industrial and technological excellence. Manchester is a student-focused city with enriching sporting and buzzing social scenes. It is one of the world's top 3 student cities (QS Best Student Cities 2023) that showcase dynamic real international cuisine and cultural events.

Keynote Speeches

The Role of Artificial Intelligence Ethics in Developing ASSETs

Keeley Crockett
Department of Computing and Mathematics
Manchester Metropolitan University,
Manchester, UK

Abstract The ethical, social and legal landscape of artificial intelligence (AI)-driven systems is rapidly changing. Since the General Data Protection Regulation (2018), stakeholders developing AI systems have faced numerous challenges in the interpretation and implementation of Article 22, specifically concerning an individual's rights in the context of automated decision-making, the ability of AI to explain decisions, explanation of the logic involved and to develop models using only "correct" data. This has and is causing major challenges due to the lack of legal guidance, case law and ethical principles about the use of AI in different contexts. In 2023, the passing of the EU AI Act and other Global "AI and Digital" legalisation promises to also be a big disrupter to all stakeholders working in the field as adopting a risk-based approach to AI becomes the norm. In this talk I will first briefly overview why ethics matters in developing ASSETS and outline my transformational journey. I will provide an overview of the policy and impact landscape and how adopting consequence scanning can help organisations start to adopt a risk-based approach to AI. Responsible innovation and ethical tech are essential to build public trust, yet how can businesses and researchers in academia harness People Power through community ideation and involvement? In this talk I will share two projects that involve public engagement and co-production as mechanisms to build citizen confidence and trust.

Precision Agriculture: A Big Data-Driven, AI-Enabled Approach

Liangxiu Han
Department of Computing and Mathematics
Manchester Metropolitan University,
Manchester, UK

Abstract The convergence of big data and artificial intelligence (AI) in agriculture holds significant promise for addressing global food security challenges. These transformative technologies offer the means to optimise agricultural practices, boost productivity and ensure a sustainable food supply. By harnessing the power of big data and AI, farmers gain the capability to make data-driven decisions pertaining to resource allocation, crop management and yield optimisation. This data-driven, AI-enabled approach fosters efficient resource utilisation, minimises waste, and maximises agricultural output, thereby meeting the increasing global demand for food.

Our latest advancements in scalable deep learning and AI within the realm of precision agriculture, aimed at addressing food security and sustainability. Leveraging our profound expertise in Big Data and Artificial Intelligence analytics, we tackle pressing societal challenges within the agricultural domain. Our approach involves the integration of extensive datasets sourced from satellites, weather monitoring and agronomic measurements, complemented by precise data collected directly from farmers' fields – right down to individual plant precision. To enable this precision data collection, we employ cutting-edge technologies such as autonomous robotics, drones, edge computing devices, IoT sensors and mobile devices deployed within the fields.

Our innovative Edge-to-Cloud systems offer continuous real-time data collection and analysis, furnishing farmers with immediate insights into their crop status. This empowers them with actionable advice to optimise crop production, reducing the reliance on costly chemical and biological inputs. Moreover, it saves time and curtails emissions, contributing to climate change mitigation. When seamlessly synchronised with our optimised cloud data processing pipelines, our analytics ensure that aggregated precision data is consistently updated and examined for deeper insights. This data-driven approach informs policy decisions, creating a virtuous feedback loop that perpetuates benefits to farmers and aligns with evolving societal needs. The integration of autonomous AI precision agriculture represents a pivotal step toward a more secure and sustainable future for global food production.

Keywords Big Data Analytics/Machine Learning/AI, Precision Agriculture, Smart Farming.

References

1. Shi, Y., Han, L., González-Moreno, P., et al.: A fast Fourier convolutional deep neural network for accurate and explainable discrimination of wheat yellow rust and nitrogen deficiency from Sentinel-2 time-series data[J]. Front. Plant Sci. **14**, (2023). https://doi.org/10.3389/fpls.2023.1250844
2. Zhang, X., Han, L., et al.: The self-supervised spectral–spatial vision transformer network for accurate prediction of wheat nitrogen status from UAV imagery. Remote Sens. **14**(6), 1400 (2022)
3. Shi, Y., Han, L., Han, L., Chang, S., Hu, T., Dancey, D.: A Latent Encoder Coupled Generative Adversarial Network (LE-GAN) for efficient hyperspectral image super-resolution. IEEE Trans. Geosci. Remote Sens., 1–1 (2022). https://doi.org/10.1109/tgrs.2022.3193441
4. A novel CropdocNet for automated potato late blight disease detection from the unmanned aerial vehicle-based hyperspectral imagery. Remote Sens. **14**(2), 396 (2022). https://doi.org/10.3390/rs14020396
5. Shi, Y. Han, L., Huang, W., et al.: A Biologically Interpretable Two-stage Deep Neural Network (BIT-DNN) for vegetation recognition from hyperspectral imagery. IEEE Trans. Geosci. Remote Sens. **60**, 4401320. ISSN 0196-2892 (2021). https://doi.org/10.1109/TGRS.2021.3058782
6. Zhang, X., Han, L., et al.: How well do deep learning-based methods for land cover classification and object detection perform on a high resolution remote sensing imagery. Remote Sens. **12**(3), 417 (2020)
7. Zhang, X., Han, L., Dong, Y., Shi, Y., Huang, W., Han, L.H., et al.: A deep learning-based approach for automated yellow rust disease detection from high-resolution hyperspectral UAV images. Remote Sens. **11**(13), 1554 (2019)
8. Ma, H., Huang, W., Jing, Y., Yang, C., Han, L. Dong, Y., Ye, H., Shi, Y., Zheng, Q., Liu, L., Ruan, C.: Integrating growth and environmental parameters to discriminate powdery mildew and aphid of winter wheat using bi-temporal Landsat-8 imagery. Remote Sens. **11**(7), 846 (2019)
9. Robot enabled machine vision for nitrogen diagnosis, https://n2visionai.wordpress.com, last accessed 23 Sept 2023
10. Agrione: IoT enabled, AI-driven Smart Agriculture, http://agrione.farm, last accessed 23 Sept 2023
11. Agrilyse: https://play.google.com/store/apps/details?id=uk.co.mmu.tams.agrilyse, last accessed 23 Sept 2023
12. EPIC: An automated diagnostic tool for Potato Late Blight disease detection from images https://epic337255300.wordpress.com, last accessed 23 Sept 2023
13. Han, L., Haleem, S.M., Taylor, M.: Automatic detection and severity assessment of crop diseases using image pattern recognition. In: Studies in Computational Intelligence, p. 647. Springer, ISBN: 978-3-319-33351-9 (2016)

Reconfigurable Multiband Virtual Antenna for Low-Power Internet of Things Sensors Applications

Jaume Anguera
Associate Professor and CTO/Founder, Ignion, Barcelona, Spain

Abstract Calling all Antenna, Microwave, RF, Wireless, and Electronics engineers! Discover the cutting-edge world of Antenna Booster Technology for designing wireless devices with embedded antenna boosters. Antenna boosters are off-the-shelf electrically small components that can seamlessly integrate into any wireless device, enabling operation across a wide frequency range (0.4–10.6 GHz) through a well-designed matching network. Unlike traditional antenna designs that often require geometry modifications for different frequency bands, antenna booster frequency bands can be easily adjusted solely through the appropriate matching-network design. This approach is especially beneficial for RF/microwave and wireless engineers familiar with matching network designs used in various telecommunication systems, such as filters and amplifiers, making the adoption of antenna boosters a more straightforward and faster process. In this session, attendees will gain valuable insights into the physical principles behind antenna boosters and how they revolutionise the design of wireless devices, including IoT applications. Learn how passive matching networks and reconfigurable architectures enable wireless devices to incorporate antenna boosters covering single-band or multi-band applications. The presentation offers a comprehensive overview of Antenna Booster Technology, starting from its fundamental principles and extending to its latest applications in IoT devices. Join us to explore the exciting possibilities that antenna boosters bring to the world of wireless communication. Discover how this technology is shaping the future of IoT devices and beyond.

Smart Foundations for Engineering Structures

Charles Etim Akpan Uko
Department of Civil Engineering
University of Uyo, Uyo, Nigeria

Abstract Foundations or substructures of engineering structures are very important elements in the structure. The integrity of the structure largely depends on the performance of the foundation. When Jesus said in the Bible that a wise man builds his house on the rock, he was very correct. But where rocks cannot be found, the geotechnical and structural engineers have to find ways of overcoming the challenges posed by substandard foundation materials. In this paper, a review of the foundation system used for the Burj Khalifa building in Dubai is presented with emphasis on the smart system that must be kept working for the integrity of the edifice. The paper also presents an insight into current efforts at the University of Uyo, Uyo to develop Mfonobong, Imaobong, Charles, Pat-Bassey and Utibe (MICPU), an inverted pyramidal frustum isolated concrete foundation system that has prospects of cutting the cost of building foundations by up to 40%. The paper ends by recommending increased use of artificial intelligence (AI) by engineers in solving challenges encountered in the design and construction of sustainable foundations for engineering structures.

Basics of Designing Self-driving Vehicles

Faisal Riaz
Department of Computer Science and Information Technology (CS & IT)
Mirpur University of Science and
Technology (MUST), New Mirpur,
Pakistan

Abstract Certainly! Designing self-driving cars involves a comprehensive and multidisciplinary approach. It typically begins with understanding the principles of artificial intelligence, computer vision and machine learning. These concepts are crucial for developing algorithms that enable the car to perceive and interpret its surroundings accurately. Additionally, knowledge of robotics and control systems is essential for designing the vehicle's hardware and ensuring precise and safe manoeuvring. The course would also cover topics like perception module design, motion planning, decision-making and actuator control. By combining these skills, students can gain a solid foundation in designing self-driving cars and contribute to the exciting field of autonomous transportation.

Tackling Engineering Challenges Using a Life Cycle Approach: How Can Sustainability Be Integrated into Engineering Innovation

Laurence Stamford
Head of Environmental Sustainability / Chair of the School of Engineering Board
The University of Manchester,
Manchester, UK

Abstract Sustainability is a frequently cited concept but is less frequently quantified and acted upon with any rigour. The field of life cycle sustainability assessment incorporates some relatively standardised tools, such as environmental life cycle assessment (LCA), and some less formal approaches, such as social impact assessment. Elements of this – particularly LCA – have been used in virtually all sectors and are increasingly mandated via regulation (e.g., the EU Product Environmental Footprint initiative). However, sustainability assessment is typically applied after-the-fact to existing commercial systems. This talk aims to demonstrate that greater sustainability gains can be realised by using life cycle thinking continuously as a way to direct R&D. To this end, we will consider two cases: first, the use of LCA to develop circular economy business models in the power electronics sector, including evaluation of 18 different environmental impact categories in the search for more sustainable approaches for elevator variable frequency drive units; second, the design of energy, water and food provision systems in economically developing countries via life cycle assessment, social impact assessment, techno-economic analysis and scenario analysis. Through these cases we will investigate methods, sustainability impacts and the benefits of an integrated approach.

Optimisation of Embedded Machine Learning for Edge Inferencing

Kelvin Anoh
Senior Lecturer in Electrical and Electronic Engineering
University of Chichester, Chichester, West
Sussex, UK

Abstract Edge computing, a scheme that requires data collection, processing and decision-making at the points closer to where data are generated, is continuously evolving. The most recent evolution requires introducing artificial intelligence on edge devices (or the so-called edge AI) so that data can be collected, processed and decisions made at the edges of communication networks. Implicitly, machine learning algorithms that were traditionally deployed on high-powered computing devices and the cloud, such as servers, are now being required on edge devices such as IoT, sensors, mobile phones and robots, including satellites. Some of the significant problems with such evolution are that the edge devices can be power-constrained, computationally incapable and might not provide the required results in real time. Deep neural networks are complex machine learning algorithms that are quite power-hungry and significantly challenging to deploy on edge devices. Therefore, the machine learning algorithms and the hardware architectures must be independently optimised to qualify the edge AI devices to operate optimally at the edges of communication networks. This talk will present the state-of-the-art machine learning schemes that are being used for different applications and how they can be optimised for applications on edge devices to enable real-time inferencing with tolerable latency. The talk will also present specific examples of optimisation models that can be used to enhance embedded machine learning techniques for energy-efficient and low-latency inferencing of edge devices. Beyond this level, enabling machine learning algorithms for immediate deployment on physical hardware (e.g., central processing unit (CPU), graphical process unit (GPU) / tensor processing unit (TPU), field programmable gate Array (FPGA) and application-specific integrated circuit (ASIC)) is significantly challenging. The talk will discuss the optimisation of hardware architectures that can enhance edge inferencing, edge AI and decision-making and the speed at which they are made at the edges. When combined, the optimisation of machine learning algorithms and the optimisation of hardware architectures for edge AI will reduce telecommunication network costs, minimise data traffic over networks, enhance data security and integrity, reduce the power consumption of data networks and increase battery lives.

Quantitative Visualisation of Microwave Circuits Under Realistic Operating Conditions

Peter H. Aaen
Electrical Enngineering Department Head
C Colorado School of Mines, Illinois, CO,
USA

Abstract Microwave circuits, especially power transistors, are essential components of mobile communication as they amplify signals to be transmitted wirelessly from the base-station to subscriber terminals. Their compact design is increasingly difficult, as mobile network operators require operation at higher output power and frequency while simultaneously demanding a reduction in the circuit size. Shrinking the device and increasing operational frequency results in significant internal electromagnetic coupling and increased power results in higher temperatures, both of which are detrimental to performance. These requirements conflict with one another, and the device physics, electromagnetic fields and distributed temperatures co-couple to limit the overall efficiency. In this presentation, I'll discuss our efforts to develop computational multiphysics simulation techniques and the application of electromagnetic field and temperature measurements to quantitively visualise the internal operations of these microwave devices under realistic operating conditions. These new techniques provide vast new amounts of data and insights that will enable future higher efficiency semiconductor devices.

The Role of Internationalised Industry-Linked University Innovation and Enterprise Education in Raising Up Job-Creating Graduates

Temitayo Shenkoya
Higher Education Programme Director, British Council, Lagos, Nigeria

Abstract Within the National Innovation System (NIS), the three main key players are the government, university and industry players [15]. For any NIS to function properly, effectively and efficiently, the communication and interaction between these key players must be seamless. The role of the university within the innovation ecosystem is well known and critical to driving innovation within the NIS. The university plays a fundamental role in knowledge diffusion, human capacity development and building research excellence – useful in solving societal issues as well as policy development and implementation [8, 16]. Furthermore, the university serves as the powerhouse in the relationship amongst the dynamic three.

Internationalised Industry-Linked Education

The players from the industry are key funders of research within the university and are key stakeholders in absorbing innovation and technology – developed within the university via technology transfer [1, 5, 15]. The industry is also an employer of labour trained by the university. In some case, players from the industry – also known as firms – have their own research outfit that manages their research endeavours, but this is reserved for big firms and family-owned corporations [7].

Within this lecture, the role of the university in knowledge transfer and capacity building of human resources – through internationalised industry linked innovation and enterprise education will be focused on. The internationalised approach is particularly interesting because the world is now connected than ever and can be viewed as a global village. Internationalisation focuses on enhancing the education experience of student and knowledge diffusion – through exchange and collaboration on an international scale [6]. Likewise, fundamental aspects of innovation and enterprise education helps to build the spirit of entrepreneurship in students. Several studies have shown that in most developed countries, the government and policy makers have prioritised entrepreneurial studies within their universities in a bid to improve entrepreneurship and startup development within their NIS [4, 10, 12].

The impact of the integration of innovation and enterprise education in the curriculum in higher education institutions will be examined next. A study that focused on the impact of entrepreneurship education (EE) on entrepreneurial mindset (EM) of higher education students in China concluded that EE was important in building EM which had a positive impact of the development of new enterprises [2]. In the United Kingdom (UK), the approach of the government is not just to ensure that entrepreneurship is a core of the curriculum, within the higher education system, but the UK government has already set up entrepreneurial universities in various regions [14]. This style of entrepreneurial studies is leading innovation, economic growth and regional growth in the UK [14]. Another study on the impact of entrepreneurship universities argued that these universities have contributed to knowledge transfer from the university to the industry, while the university learns from the industry [9].

Considering internationalised industry linked innovation education, one of the approaches of universities in the UK is the development of living labs within universities. Living labs across the UK is leading sustainable development within the Sustainable Development Goals (SDG) framework in Bulgaria, United States of America (USA) and the UK [13]. Living labs create an open knowledge environment where innovation is developed in a real-life situation that fosters collaboration amongst individual, cooperations and the civil society [11]. They do this by combining research with real life experience.

Based on the experience and lesson learnt by the UK government, through the British Council, it has been supporting the Nigerian government to achieve a systemic and institutional change to improve the education ecosystem in Nigeria. While the British Council has embarked on various programmes to achieve this, only the impact of a select few will be discussed herein. The first programme that will be discussed is the Innovation for African Universities (IAU) programme. The main aim of the IAU is to strengthen the capacity of universities across Sub-Saharan Africa – through capacity building and collaboration between UK universities and their counterpart in sub-Saharan Africa (SSA), while focusing on building the entrepreneurship ecosystems in these countries. Over the years through this programme, some of the success stories in Nigeria have been a change in the entrepreneurial mindset of higher education students in Nigeria and a growth in the number of student startups.

The British Council is also supporting the efforts of the Nigerian government to rejuvenate its curriculum to imbibe the concept of entrepreneurship for higher education institutions in the country. The British Council has delivered and funded numerous conferences, workshops and training programmes for key players within the innovation ecosystem in Nigeria and continues to do more [3]. Experts from the UK have been actively engaged to support and transfer knowledge from the UK to Nigeria. While the new curriculum developed by the National Universities Commission (NUC) is being finalised, the British Council is actively supporting the NUC to fast track the process and ensure that the mission and vision of implementing the curriculum are achieved.

Conclusion

Conclusively, the concept of embedding innovation and enterprise education with an internationalised industry linked focus is a proven way of support sustainable development within the society. This is mainly achieved through building entrepreneurial skills in students in universities to be able develop and run their own startups, SMEs and firms. These startups do not just contribute to sustainable development – through paying taxes and solving a societal need – they are also key employers of labour. While developed countries have largely adopted this model within their higher education systems, countries like Nigeria are just beginning the process and the British Council posits itself as a development partner that will support the government in achieving this goal.

References

1. Belitski, M., Aginskaja, A., Marozau, R.: Commercializing university research in transition economies: technology transfer offices or direct industrial funding? Res. Policy **48**(3), 601–615 (2019)
2. Boldureanu, G., Ionescu, A.M., Bercu, A.M., Bedrule-Grigoruță, M.V., Boldureanu, D.: Entrepreneurship education through successful entrepreneurial models in higher education institutions. Sustainability **12**(3), 1267 (2020)
3. British Council: Going Global Partnerships – Innovation for African Universities. https://www.britishcouncil.org.ng/programmes/education/higher-and-further-education/going-global-partnerships-innovation-african (2023)
4. Colombo, M.G., Piva, E.: Start-ups launched by recent STEM university graduates: the impact of university education on entrepreneurial entry. Res. Policy **49**(6), 103993 (2020)
5. Good, M., Knockaert, M., Soppe, B., Wright, M.: The technology transfer ecosystem in academia. An organizational design perspective. Technovation **82**, 35–50 (2019)
6. De Wit, H.: Internationalization in higher education, a critical review. SFU Educ. Rev. **12**(3), 9–17 (2019)
7. Hu, Q., Hughes, M.: Radical innovation in family firms: a systematic analysis and research agenda. Int. J. Entrepreneurial Behav. Res. **26**(6), 1199–1234 (2020)
8. Hwang, K.Y., Cho, D.W., Shenkoya, T.: Industry-Academic cooperation in Korea: an analysis of the effect of human resources. Technol. Anal. Strateg. Manag. **35**(3), 339–351 (2023)
9. Klofsten, M., Fayolle, A., Guerrero, M., Mian, S., Urbano, D., Wright, M.: The entrepreneurial university as driver for economic growth and social change-key strategic challenges. Technol. Forecast. Soc. Change **141**, 149–158 (2019)
10. Li, L., Kang, K., Sohaib, O.: Investigating factors affecting Chinese tertiary students' online-startup motivation based on the COM-B behaviour changing theory. J. Entrep. Emerg. Econ. **15**(3), 566–588 (2023)
11. Ruijer, E., Meijer, A.: Open government data as an innovation process: lessons from a living lab experiment. Public Perform. Manag. Rev. **43**(3), 613–635 (2020)
12. Metcalf, L.E., Katona, T.M., York, J.L.: University startup accelerators: startup launchpads or vehicles for entrepreneurial learning? Entrep. Educ. Pedagogy **4**(4), 666–701 (2021)

13. Purcell, W.M., Henriksen, H., Spengler, J.D.: Universities as the engine of transformational sustainability toward delivering the sustainable development goals: "Living labs" for sustainability. Int. J. Sustain. High. Educ. **20**(8), 1343–1357 (2019)
14. Sánchez-Barrioluengo, M., Benneworth, P.: Is the entrepreneurial university also regionally engaged? Analysing the influence of university's structural configuration on third mission performance. Technol. Forecast. Soc. Change **141**, 206–218 (2019)
15. Shenkoya, T.: A study of startup accelerators in Silicon Valley and some implications for Nigeria. Afr. J. Sci. Technol. Innov. Dev. **13**(3), 303–314 (2021)
16. Shenkoya, T., Kim, E.: The impact of technology transfer/policies on the economic catch-up of the Korean National Innovation System and its implications for Nigeria. Afr. J. Sci. Technol. Innov. Dev. **13**(6), 685–699 (2021)

Strategy to Reduce Carbon Emission in Port Terminal

William Holderbaum
Manchester Metropolitan University, Science & Engineering
John Dalton Building, Manchester, UK

Abstract In low voltage networks, energy storage systems (ESSs) play a significant role in increasing energy cost savings, peak reduction and energy efficiency whilst reinforcing the electrical network infrastructure. This talk will give an outline of an optimal management system based on different control strategies (Optimal Control and Model Predictive Control) for the control of an ESS equipped with a network of electrified rubber tyre gantry (RTG) cranes. The stochastic management system aims to improve the reliability and economic performance, for given ESS parameters, of a single crane or a network of cranes by considering the uncertainty in the RTGs electrical demand. A specific case study is presented using real operational data of the RTGs network in the Port of Felixstowe, UK, and the results of the stochastic control system are compared to a standard set-point controller. The results of the proposed control strategies indicate that the stochastic management system successfully increases the electric energy cost savings, the peak demand reductions and successfully outperforms a comparable set-point controller.

Introduction to ASSET Conference 2023

Adaptive This involves coming up with ideas, science, engineering, technology and innovations that are flexible and can fit within the rapidly changing world that, in many ways, is changing faster than it has ever done before.

Sustainable Contextualising this in terms of the United Nations' Sustainable Development Goals (SDGs), we consider how the ASSET 2023 research outputs presentations fit within the SDGs. We can identify at least one SDG that each presentation contributes to. The listed ASSET 2.0 conference themes make direct contributions to at least 9 of the SDGs and indirect contributions to all the 17 SDGs.

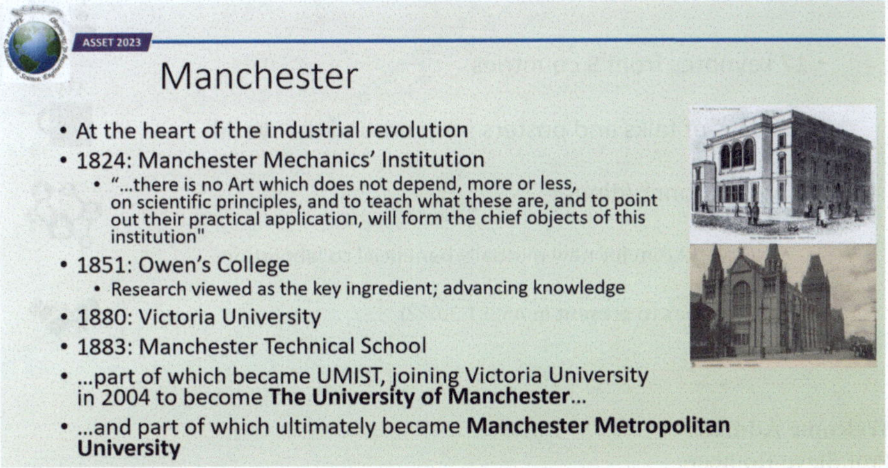

Expertise ASSET 2.0 draws experts from five (5) continents of the world (Africa, Asia, Europe, North America and South America) being represented.

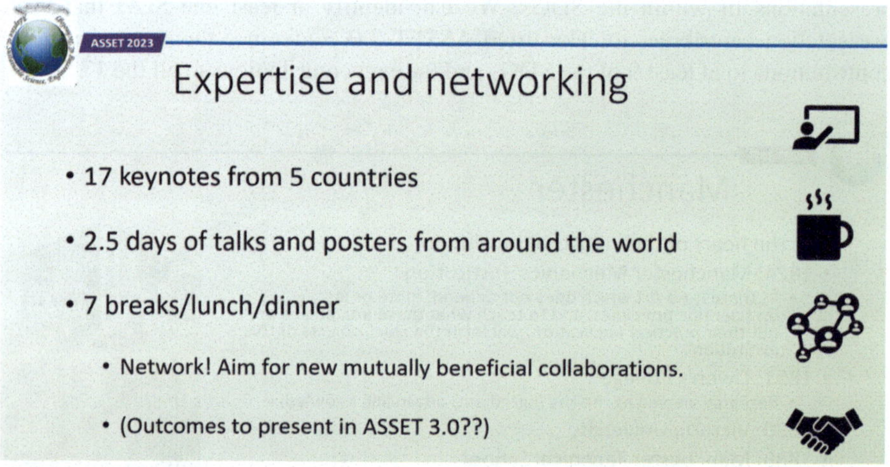

Welcome Address

Prof Steve Rothberg
Provost / Deputy Vice Chancellor, Manchester Met University, UK
S.Rothberg@mmu.ac.uk

Conference Theme Exposition

Prof Andy Gibson
Pro-Vice Chancellor (PVC – Sci. & Eng. Faculty), Manchester Met University, UK
A.Gibson@mmu.ac.uk

***Opening Keynote Speech*: Scientific Innovation and Challenges for Space Exploration**

Prof Sang Choi
NASA Inventor/Scientist, NASA Langley Research Centre, Virginia, 23681-2199, USA
Sang.h.Choi@nasa.gov

Head of Environmental Sustainability / Chair of the School of Engineering Board, The University of Manchester (UoM), Manchester, UK

Laurence Stamfordlaurence.stamford@manchester.ac.uk

Contents

Age-Related Changes of the Muscles and Human Body Systems

Hans Degens

1 Introduction

We rarely consider that the birth of a person is something miraculous! Even though we can be in awe of new life, it seldom crosses our mind that this new life represents an extreme case of rejuvenation. Generation upon generation has displayed this phenomenon where offspring have a similar total life expectancy as their parents, and perhaps even more startling is the fact that we all developed from a single cell (the zygote) – that is barely visible with the naked eye – into the person we are now. This incredible feat of increasing complexity and specialisation during embryological development and maturation is baffling, yet we are unable to thwart the almost imperceptible but progressive deterioration after reaching adulthood, ultimately resulting in our physical demise. Ageing indeed is rather an enigma, and how it came about is an unsolved puzzle [1], particularly as it seems organisms are programmed for survival, not death [2]. The deterioration in bodily functions, such as muscle strength, will over time result in a diminished ability to perform daily life activities and consequently a reduction in the quality of life. The good news is that over the years the average life expectancy has increased [3], but this is not always accompanied by a similar increase in healthy life expectancy, and in some cases, there is even a reduction in healthy life expectancy [4]. A reduced quality of life particularly occurs when one is not anymore able, or almost unable, to perform activities of daily life, or in other words crosses a 'disability threshold'. First the rate

H. Degens (✉)
Department of Life Sciences, Manchester Metropolitan University, Manchester, UK

Institute of Sport Science and Innovations, Lithuanian Sports University, Kaunas, Lithuania
e-mail: h.degens@mmu.ac.uk

S. C. Ekpo (ed.), *The Second International Adaptive and Sustainable Science, Engineering and Technology Conference*, Signals and Communication Technology,
https://doi.org/10.1007/978-3-031-53935-0_1

of ageing will be discussed, followed by a short discussion of the possibility to modulate the rate of ageing, concluding with how to capitalise on the plasticity of many physiological systems to extend the healthy life expectancy.

2 The Rate of Ageing

2.1 Are There Differences in the Rate of Ageing?

As average life expectancy in the world is rising [3], it may come as a surprise that even in ancient Greece (e.g. Alexis, a poet) and Rome (e.g. Terentia, the wife of Cicero) there were centenarians. Recently, there has even been a global reduction in the maximum age at death [5], and combined these observations indicate that while the average life expectancy has been rising over the centuries, the maximum life expectancy has remained stubbornly constant [3]. Another observation is that while genetics [6] and certain behaviours, such as smoking and an unhealthy diet, can cause variations in the rate of ageing between people [7], it seems that in a given population the rate of ageing, expressed, e.g. as the annual % decline in muscle strength, is usually rather uniform [8]. The rate of ageing seems even rather similar between different physiological systems, such as respiratory and muscle function [9], and for the rate of decline in aerobic and anaerobic power [10] suggesting that there is a common factor underlying the rate of ageing in all those systems. Such a common denominator may be an exponential random accumulation of unrepaired (micro)damage in the cells [1,11]. In line with this, it has been observed that the life expectancy of an organism is positively related to their cellular DNA repair capacity [12].

2.2 Modulating the Rate of Ageing

As master athletes have a better exercise capacity than age-matched non-athletes and have astonishing athletic achievements rivalling those seen in the first Olympic games in 1896 [13], it has often been claimed that regular exercise can reverse or attenuate the rate of ageing. However, the rates of decline in maximal oxygen consumption [13], muscle power [14] and muscle fibre size [15] are in absolute terms even faster in athletes than in non-athletes! This then may suggest that the rate of ageing is even faster in athletes than non-athletes, arguing against the idea that exercise attenuates the rate of ageing and extends lifespan and healthy lifespan. In fact, in flies, it has been shown that less active flies lived longer [16], but of course a bed-bound life is not very appealing.

Perhaps there is some other message to be detected in the master athletes. While their absolute rate of ageing is higher, they are in general in a much better condition than age-matched non-athletes [17]. To reconcile the apparent faster absolute rate of

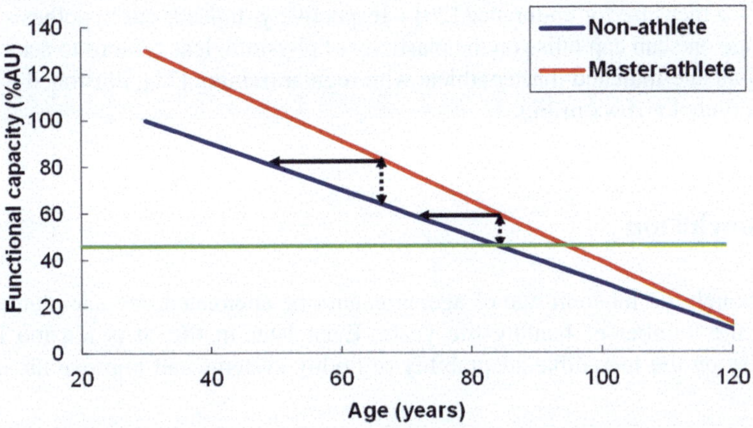

Fig. 1 Illustration of the age-related decline in a physiological function, such as muscle power, maximal oxygen uptake or muscle mass. The physiological function in non-athletes (blue line) at age 30 years is set at 100%. In the illustration, master athletes (red line) have a 30% higher value than age-matched non-athletes at the age of 30 years. The annual rate of decline is assumed to be 1% of the value at the age of 30 years for both athletes and non-athletes [10, 18]. The dotted two-headed arrow indicates the gain or loss of function with training or cessation of training, respectively. The solid black arrows indicate the degree of 'rejuvenation' that can be achieved with master athletics or training. This arrow gets shorter with increasing age, indicating that the 'rejuvenation' realised by master athletics or regular training decreases with increasing age. The green line indicates the 'disability threshold' below which function will impair the ability to perform daily life activities. Master athletes (91 years for the red line) cross this disability threshold later in life than non-athletes (80 years). However, non-athletes, even if they have crossed the disability threshold, can by training come above the disability threshold at least till the age of 91 years in this example (red line)

ageing with the health benefits seen in the athletes, we need to look at the rate of ageing as the annual percentage rate of decline, and then it appears that athletes and non-athletes show a similar rate of age-related decline, as reflected by the converging of the lines in Fig. 1. It can further be seen that athletes cross the disability threshold (green line) later than non-athletes, clearly providing an extended healthy lifespan, but not so much an extended total lifespan, as the inherent rate of ageing apparently cannot be attenuated by regular exercise.

2.3 *Capitalising on the Adaptability of the Human Body*

Maybe you have never exercised regularly or gave up on it a long time ago, and reading the above may make you think that there is no hope for you to recover what you lost over the years due to ageing. There is, however, good news, as the cardiovascular and muscular system have an enormous ability to adapt to altered functional demands. Indeed, even in old age, resistance and endurance exercise can induce a significant improvement in muscle strength [19] and maximal oxygen uptake – a measure for endurance [20] – respectively. Indeed, our hypothesis is that at any age one can capitalise on the plasticity of physiological systems to achieve the level of an age-matched master athlete with regular training [21], illustrated with the upward dotted arrows in Fig. 1.

3 Conclusion

Even though the inherent rate of ageing cannot be attenuated, regular exercise can extend the number of healthy life years. Even later in life, it is not too late to capitalise on the incredible adaptability of bodily systems and improve the quality of life.

References

1. Degens, H.: The origin of senescence and death: an evolutionary enigma. Creation Res. Soc. Q. **44**, 280–286 (2008)
2. Kirkwood, T.B.: Evolution of ageing. Mech. Ageing Dev. **123**(7), 737–745 (2002)
3. Fries, J.F.: Aging, natural death, and the compression of morbidity. N. Engl. J. Med. **303**(3), 130–135 (1980)
4. Taylor, J.A., Greenhaff, P.L., Bartlett, D.B., Jackson, T.A., Duggal, N.A., Lord, J.M.: Multisystem physiological perspective of human frailty and its modulation by physical activity. Physiol. Rev. **103**(2), 1137–1191 (2023)
5. Dong, X., Milholland, B., Vijg, J.: Evidence for a limit to human lifespan. Nature. **538**(7624), 257–259 (2016)
6. Khanal, P., Williams, A.G., He, L., Stebbings, G.K., Onambele-Pearson, G.L., Thomis, M., Degens, H., Morse, C.I.: Sarcopenia, obesity, and sarcopenic obesity: relationship with skeletal muscle phenotypes and single nucleotide polymorphisms. J. Clin. Med. **10**(21), 4933 (2021)
7. Degens, H., Korhonen, M.T.: Factors contributing to the variability in muscle ageing. Maturitas. **73**(3), 197–201 (2012)
8. Degens, H., Onambélé-Pearson, G.L.: A simple framework to distinguish 'individualistic' from a 'uniform rate' of ageing within or between study populations. GeroScience. **46**, 2765–2769 (2023)
9. Ocana, P.D., Darabseh, M.Z., Ishihara, K., Aburub, A., Zambolin, F., Montgomery, G., Mills, R., Scorcelletti, M., Cameron, J., Ganse, B., Degens, H., Bagley, L.: Age-related declines in muscle and respiratory function are proportionate to declines in performance in master track cyclists. Eur. J. Appl. Physiol. **121**(12), 3447–3457 (2021)
10. Bagley, L., McPhee, J.S., Ganse, B., Muller, K., Korhonen, M.T., Rittweger, J., Degens, H.: Similar relative decline in aerobic and anaerobic power with age in endurance and power master athletes of both sexes. Scand. J. Med. Sci. Sports. **29**(6), 791–799 (2019)
11. Degens, H.: Determinants of skeletal muscle hypertrophy and the attenuated hypertrophic response at old age. Sport. Med. Doping Stud. **1**, 003 (2012)
12. Cortopassi, G.A., Wang, E.: There is substantial agreement among interspecies estimates of DNA repair activity. Mech. Ageing Dev. **91**(3), 211–218 (1996)
13. Tanaka, H., Seals, D.R.: Endurance exercise performance in Masters athletes: age-associated changes and underlying physiological mechanisms. J. Physiol. **586**(1), 55–63 (2008)

14. Pearson, S.J., Young, A., Macaluso, A., Devito, G., Nimmo, M.A., Cobbold, M., Harridge, S. D.: Muscle function in elite master weightlifters. Med. Sci. Sports Exerc. **34**(7), 1199–1206 (2002)
15. Messa, G.A.M., Piasecki, M., Rittweger, J., McPhee, J.S., Koltai, E., Radak, Z., Simunic, B., Heinonen, A., Suominen, H., Korhonen, M.T., Degens, H.: Absence of an aging-related increase in fiber type grouping in athletes and non-athletes. Scand. J. Med. Sci. Sports. **30**(11), 2057–2069 (2020)
16. Sohal, R.S., Buchan, P.B.: Relationship between physical activity and life span in the adult housefly, *Musca domestica*. Exp. Gerontol. **16**(2), 157–162 (1981)
17. Kettunen, J.A., Kujala, U.M., Kaprio, J., Sarna, S.: Health of master track and field athletes: a 16-year follow-up study. Clin. J. Sport Med. **16**(2), 142–148 (2006)
18. Charlier, R., Knaeps, S., Mertens, E., Van Roie, E., Delecluse, C., Lefevre, J., Thomis, M.: Age-related decline in muscle mass and muscle function in Flemish Caucasians: a 10-year follow-up. Age. **38**(2), 36 (2016)
19. Harridge, S.D., Kryger, A., Stensgaard, A.: Knee extensor strength, activation, and size in very elderly people following strength training. Muscle Nerve. **22**(7), 831–839 (1999)
20. Hepple, R.T., Mackinnon, S.L., Goodman, J.M., Thomas, S.G., Plyley, M.J.: Resistance and aerobic training in older men: effects on VO2peak and the capillary supply to skeletal muscle. J. Appl. Physiol. **82**(4), 1305–1310 (1997)
21. Ganse, B., Degens, H.: Current insights in the age-related decline in sports performance of the older athlete. Int. J. Sports Med. **42**(10), 879–888 (2021)

The Need for Adaptable and Sustainable Science, Engineering, and Technologies for Trans-generational Global Coexistence

Isaac Busayo Oluwatayo

1 Introduction

There is no gainsaying the fact that science has redefined the course of human existence [14]. While the world today celebrates the significant achievements/transformation and ease that science has brought to mankind, the world still faces increasing challenges, and adaptable and sustainable science, engineering, and technologies (ASSET) are critical for achieving trans-generational prosperity and global coexistence. The need to leverage technology in advancing economic development, social interactions, data analytics and management, and climate resilience is fast gaining relevance in literature [3, 12]. Adaptive and sustainable science, engineering, and technology have provided a platform for nations to champion their development in a sustainable manner. The concept of adaptive and sustainable science, engineering, and climate resilience is multidisciplinary, multidimensional, and interconnected.

1.1 Understanding Adaptive Science, Engineering, and Technology: What Is Adaptive Science?

Adaptive science is the study of how living systems evolve and change over time to survive in their environment [9]. Adaptive science represents a dynamic, flexible, and context-specific that acknowledges the ever-evolving nature of challenges. It

I. B. Oluwatayo (✉)
Department of Agricultural Economics and Agribusiness, University of Venda, Thohoyandou, South Africa
e-mail: isaac.oluwatayo@univen.ac.za

S. C. Ekpo (ed.), *The Second International Adaptive and Sustainable Science, Engineering and Technology Conference*, Signals and Communication Technology,
https://doi.org/10.1007/978-3-031-53935-0_2

diverges from the traditional linear approach to scientific research, which assumes that problems can be resolved through hypothesis formulation, experimental designs to and result analysis [5]. While the traditional approach has been successful in addressing past societal issues, it may be deficient in dealing with problems that are complex, multidimensional, and volatile. Conversely, adaptive science embraces the uncertainties and complexities inherent in the problems of today. The contemporary problems include climate change, the COVID-19 pandemic, ecosystem conservation, food and nutrition insecurity, and poverty among others [6, 13]. It enables constant revision of research plans and strategies in response to emerging data and the prevailing realities in a particular period. For instance, in addressing productivity and disease management issues, adaptive science has been applied through artificial insemination, selective breeding, and embryonic transfer [8]. In addition, the adoption of drones and satellite imagery aids in assessing soil health, thus providing valuable information for irrigation, crop or livestock selection, and pest management decisions [7]. These approaches allow farmers to raise animals or cultivate crops tailored to specific needs, while sustainably maximizing productivity.

1.2 What Is Adaptive Engineering?

Adaptive engineering encompasses the design and advancement of products and systems that are flexible and customizable and can adapt to evolving conditions. As a result, it revolves around the principles of resilience, ongoing innovation, scalable designs, and timely incorporation of feedback from end users or clients. Adaptive engineering has been applied in various areas including addressing mobility issues through the deployment of electric vehicles, revitalizing urban spaces to align with changing traffic patterns and population growth, facilitating laser surgeries, developing prosthetic and industry 4.0/Internet of Things that is fast-changing marketing, and creating sustainable livelihood opportunities [1]. In addition, adaptive engineering tackles energy poverty through innovative clean energy solutions and adaptable building designs. Therefore, adaptive engineering places the end user at the center, fosters innovation, prioritizes resilience, and leverages forward-thinking approaches and responsiveness in tackling contemporary societal challenges.

1.3 What Is Adaptive Technology?

Adaptive technology involves the use of technology to create products and systems that can adapt to the needs of people with disabilities and special needs. It leverages devices, software, and models to support the unique requirements and realities of individuals with special needs. Adaptive technology provides a unique platform for people with disabilities to lead a decent life and foster inclusion. Adaptive technology has been applied to communication, mobility, education, and daily living of people with special needs. This includes text-to-speech or eye-tracking technology, hearing aids, screen readers and magnifiers, prosthetic limbs, and learning and educational aids among others.

The theme of this conference could not have been more timely as the world tackles wicked and mega problems. It is instructive to note that words such as innovative approaches, responsiveness to emerging needs, and sustainability are central to the concept of adaptive science, technology, and engineering.

2 Sustainability and Adaptive Science, Engineering, and Technology

Sustainability entails assuring the needs of current generations without compromising the needs of future generations, while ensuring a balance between economic growth, environmental care, and social well-being [15]. Sustainability improves the quality of our lives, protects the ecosystem, and preserves natural resources for future generations. In the corporate world, for instance, sustainability is associated with an organization's holistic approach, taking cognizance of everything, from manufacturing to logistics to customer service [2, 11].

Science and technology for a sustainable future on the other hand assist us to understand our environment and the impact of our activities on it, which is essential for making smart decisions about how to use our resources. This can also help us proffer solutions to emerging and persistent problems, such as climate change and pollution, and improve the quality of life of our people.

Sustainable engineering refers to the integration of social, environmental, and economic considerations into product, process, and energy system design methods. It encourages the consideration of the complete product and process life cycle during the design effort. In short, sustainable engineering is very pivotal to minimizing energy and materials consumption, which increases operational efficiency of organizations by reducing costs.

3 The Importance of Trans-generational Global Coexistence

The world as a global village is interconnected, interrelated, and interdependent. Therefore, it follows that global problems exist in specific geographies in varying forms and severity. We need to prioritize trans-generational global coexistence for the following reasons:

- *Global challenges:* The global community is faced with several challenges that are currently undermining shared prosperity and sustainable development. Climate change, water and food security, poverty and inequality, conflict, migration, growth of slums, and global health crises are just a few of the challenges we face globally [4, 10]. While some of these problems are self-inflicted, others are natural disasters that are difficult to predict and manage. Trans-generational solutions are therefore required in tackling these challenges head-on.

- *Intergenerational existence:* Our future depends on finding innovative solutions that can help us coexist with one another and with the environment across generations.
- *The benefits of cooperation:* A shared understanding of the challenges and opportunities can help communities to come together and create practical and effective solutions for coping with change.
- *Relevance to societal needs:* Trans-generational global coexistence is relevant to the societal needs of these times including mutual respect, empathy, collaborations, and peaceful coexistence among people of diverse cultures and different generations across the globe. Global coexistence fosters transfer of technology and shared learning, inclusion, shared prosperity, empathy, collaboration, and sustainability.

Therefore, this paper highlights the challenges the world face today; the place of sustainable science, engineering, and technologies; and their impact on trans-generational global coexistence and documents examples of some successful implementation/initiatives.

4 Challenges Facing Global Coexistence

The world faces social, economic, and environmental challenges that threaten global coexistence. Climate change, food insecurity, inequality, and economic instability are just a few examples. To address these challenges, we need adaptable and sustainable science, engineering, and technologies.

4.1 Manifestations of These Global Challenges

- *Climate change:* environmental degradation, drought, desertification, flooding, infrastructure damage, excessive heat, too much cold, disease multiplications leading to loss of lives and properties.
- *Food insecurity:* emanating from poor yield, inadequate food/scarcity manifesting through rising food prices, malnutrition, nutrient deficiencies, obesity, ill health, etc.
- *Economic instability:* poor government revenue, slow growth rate (declining GDP), high mortality rate, rising unemployment, service delivery protests, and civil unrest.

5 Importance of Adaptable Technologies

Adaptable technologies allow for flexibility, which is essential for meeting changing needs. They can be modified as situations evolve and can be tailored to specific contexts and users. Examples include open-source software, modular construction, and reconfigurable equipment.

6 Importance of Sustainable Technologies

- *Environmental benefits* – sustainable technologies help to minimize environmental impact and promote conservation of resources. Examples include renewable energy, efficient buildings, and circular economy practices.
- *Social benefits* – sustainable technologies can improve social well-being by creating jobs and reducing poverty. This includes sustainable agriculture, water management, and ecotourism.
- *Economic benefits* – sustainable technologies can also generate economic growth and reduce the cost of doing business. Examples include green manufacturing, waste reduction, and smart transportation.

7 The Challenges Faced by Adaptation in Science, Engineering, and Technology

- *Knowledge gaps and poor awareness:* Generally, most global problems are difficult to address because of varying perspectives and different levels of understanding of the immediate and remote causes of the problem, misconceptions, and lack of awareness. For instance, climate change issues would have been better managed if there were no misconceptions or information asymmetry.
- *Technological obsolescence:* With technology advancing at a rapid pace, it can be challenging to design products and systems that will remain relevant over time.
- *Multidisciplinary collaborations:* In addressing contemporary problems, it is difficult to harness collaborations across disciplines due to differences in language, perspectives, and methods.
- *Financial constraints:* Implementing adaptive solutions can often be expensive, requiring substantial investment of time and resources.
- *Cultural barriers:* The acceptance of adaptive technologies may be limited by cultural attitudes toward change, particularly in more traditional societies.

7.1 *Impact of Science and Engineering on Trans-generational Global Coexistence*

Science and engineering are critical enablers of trans-generational global coexistence through shared learning, collaborations, and innovations that are solving the most important problems confronting our world. Science and engineering have transformed the world, but they can also contribute to global challenges if not used responsibly. We need to prioritize research and development that addresses societal needs and promotes ethical and equitable practices. Only then can science and engineering contribute to a sustainable and peaceful global community.

8 Examples of Successful Adaptable and Sustainable Technologies

Renewable energy – wind turbines, solar panels, and hydroelectric power are just a few examples of clean and renewable energy sources. Their use is increasing around the world, reducing carbon emissions and reliance on fossil fuels (Fig. 1).

Modular construction – modular buildings are constructed from prefabricated components, improving efficiency and cost-effectiveness. They can be used for a variety of purposes, from housing to industrial buildings (Fig. 2).

Sustainable agriculture – sustainable agriculture practices, such as precision farming, crop rotation, and natural pest control, improve yield and reduce environmental impact. These practices can help increase profitability and support local communities (Fig. 3).

Fig. 1 Wind turbines

Fig. 2 Modular building

Fig. 3 Sustainable agriculture

8.1 Call to Action

Improving accessibility to adaptable and sustainable technologies can be done through the following:

- *Education* – education is crucial for raising awareness about adaptable and sustainable technologies. From primary school to university, students should learn about these technologies and their potential benefits. According to Dr. Ellen R Auster (professor of strategy and entrepreneurship at the Schulich School of Business), education and research play a vital role in the creation of adaptive science, engineering, and technology. Thus, investing in these areas will aid in the development of sustainable solutions that can address global challenges.
- *Policy* – policy makers should prioritize funding and support for adaptable and sustainable technologies. This includes subsidies, tax incentives, and regulations that encourage responsible innovation and implementation.
- *Collaboration* – collaboration between industry, government, and civil society is essential for creating and implementing adaptable and sustainable technologies. This includes networking, knowledge sharing, and joint investment.

9 Examples of Successful Adaptive Science, Engineering, and Technology Solutions (Figs. 4, 5 and 6)

10 Conclusion and Future Considerations

In conclusion, adaptable and sustainable science, engineering, and technologies for trans-generational global coexistence are critical as the global community navigates through social, public health, and economic challenges in a changing world. The world as a global village underscores interconnectedness and interdependence of societies regardless of the location. The twenty-first-century challenges such as climate change, the COVID-19 pandemic, insecurity, and food and nutrition insecurity are evolving hence the need to develop and deploy solutions that are dynamic, inclusive, and far-reaching.

Therefore, *ASSET* is an essential tool for trans-generational global coexistence in our ever-changing world. Trans-generational global coexistence is hinged on multidisciplinary collaborations and shared learning. Through adaptable and sustainable approaches, our world can solve critical problems, create sustainable

Fig. 4 Biodegradable plastic – biodegradable plastic can help reduce the amount of plastic waste in the environment, making it an effective adaptive solution to the problem of plastic pollution

Fig. 5 Drip irrigation systems – drip irrigation systems provide an efficient way of watering crops, reducing water usage and promoting crop growth in areas with water scarcity

Fig. 6 Solar panel farms – solar panel farms provide a large, clean source of energy, which can help countries transition away from fossil fuels to promote renewable and sustainable energy sources

opportunities, and provide a framework for future generations to tackle the problems of their time.

As we strive to build a prosperous, equitable, and resilient world, it is crucial to embrace adaptable and sustainable science, engineering, and technologies. Adopting adaptable and sustainable science, engineering, and technologies is essential as we create a world that is prosperous, equitable, and resilient. It is up to us to embrace its principles and continuously adapt to meet the challenges of the present and future. This can be done through the following:

- *Impactful research:* Adaptable and sustainable science, engineering, and technologies hold promise for addressing global challenges and promoting transgenerational global coexistence. Investing in scientific research and development can help us find more adaptive solutions to global challenges.
- *Innovation and responsibility:* Innovative approaches must be coupled with responsible practices to ensure ethical and equitable outcomes for all. Using innovative solutions can improve existing products and services, while creating new ones that better meet our needs.
- *Local and international collaboration:* Collaboration across sectors and continents is required to achieve these objectives. Collaboration between multiple stakeholders, including individuals, corporations, and governments is essential for the success of adaptive science, engineering, and technology.

References

1. Aceto, G., Persico, V., Pescapé, A.: The role of information and communication technologies in healthcare: taxonomies, perspectives, and challenges. J. Netw. Comput. Appl. **107**, 125–154 (2018)
2. Alshehhi, A., Nobanee, H., Khare, N.: The impact of sustainability practices on corporate financial performance: literature trends and future research potential. Sustain. For. **10**(2), 494 (2018)
3. Bernai, R.R.: Managing the risks of extreme events and disasters to advance climate change adaptation. Econ. Energy Environ. Policy. **2**(1), 101113 (2013)
4. Giles-Corti, B., Vernez-Moudon, A., Reis, R., Turrell, G., Dannenberg, A.L., Badland, H., et al.: City planning and population health: a global challenge. Lancet. **388**(10062), 2912–2924 (2016)
5. Hekler, E.B., Klasnja, P., Riley, W.T., Buman, M.P., Huberty, J., Rivera, D.E., Martin, C.A.: Agile science: creating useful products for behavior change in the real world. Transl. Behav. Med. **6**(2), 317–328 (2016)
6. Ingutia, R.: The impacts of COVID-19 and climate change on smallholders through the lens of SDGs; and ways to keep smallholders on 2030 agenda. Int. J. Sustain. Dev. World Ecol. **28**(8), 693–708 (2021)
7. Javaid, M., Haleem, A., Khan, I.H., Suman, R.: Understanding the potential applications of artificial intelligence in agriculture sector. Adv. Agrochem. **2**(1), 15–30 (2023)
8. Kumar Patel, G., Haque, N., Madhavatar, M., Kumar Chaudhari, A., Kumar Patel, D., Bhalakiya, N., et al.: Artificial insemination: a tool to improve livestock productivity. J. Pharmacogn. Phytochem. **6**(6S), 307313 (2017)
9. Levin, S., Xepapadeas, T., Crépin, A.S., Norberg, J., De Zeeuw, A., Folke, C., et al.: Social-ecological systems as complex adaptive systems: modeling and policy implications. Environ. Dev. Econ. **18**(2), 111–132 (2013)
10. Marx, B., Stoker, T., Suri, T.: The economics of slums in the developing world. J. Econ. Perspect. **27**(4), 187–210 (2013)
11. Montiel, I., Delgado-Ceballos, J.: Defining and measuring corporate sustainability: are we there yet? Organ. Environ. **27**(2), 113–139 (2014)
12. Morrar, R., Arman, H., Mousa, S.: The fourth industrial revolution (Industry 4.0): a social innovation perspective. Technol. Innov. Manag. Rev. **7**(11), 12–20 (2017)
13. Odularu, G., Aluko, O.A., Odularu, A., Akokuwebe, M., Adedugbe, A.: Conclusion: fostering nutrition security, climate adaptation and sustainable agriculture strategies amid COVID-19 pandemic. In: Nutrition, Sustainable Agriculture and Climate Change in Africa: Issues and Innovative Strategies, pp. 175–182. Palgrave Macmillan (2020)
14. Raja, R., Nagasubramani, P.C.: Impact of modern technology in education. J. Appl. Adv. Res. **3**(1), 33–35 (2018)
15. Tomislav, K.: The concept of sustainable development: from its beginning to the contemporary issues. Zagreb. Int. Rev. Econ. Bus. **21**(1), 67–94 (2018)

Path Loss Prediction of 5G in the 24.25–27.5 GHz Band Based on Machine Learning

Zaid Abedeen, Sunday Cookey Ekpo, Fanuel Elias, Muhammad Ijaz, Umar Raza, Stephen Alabi, and Liangxiu Han

1 Introduction

Data traffic has increased significantly over time due to the increased demand for mobile data, leading to congestion within the sub-6 GHz frequencies [1]. Due to the considerably greater bandwidth available, interests have been focused on frequencies above the 6 GHz band, also known as millimetre waves [2]. A crucial frequency in the 5G communication spectrum is the 26 GHz 3GPP band n258 whereby gigabit speeds are achievable. This spectrum falls between 24.5 and 27.5 GHz [3]. However, the millimetre-wave spectrum has, up until recently, remained unoccupied for mobile communications due to its poor propagation properties which include low penetration, diffraction around obstacles, high path loss, and air and rain absorption [4, 5]. For mmWave frequencies, an accurate path loss model must be created due to the low spectral efficiency to select the best locations for 5G base stations. Path loss (PL) can be defined as the loss or attenuation a propagating electromagnetic signal encounters from the transmitter to the receiver [6]. Machine learning implementations in wireless communication networks are growing in popularity due to advancements in technology, data availability, and computational power. These methods have appeared as promising tools for radio propagation prediction that could increase accuracy [7].

Z. Abedeen (✉) · S. C. Ekpo · F. Elias · M. Ijaz · U. Raza · L. Han
Manchester Metropolitan University, Manchester, UK

S. Alabi
Wilsons Park, Manchester, UK

S. C. Ekpo (ed.), *The Second International Adaptive and Sustainable Science, Engineering and Technology Conference*, Signals and Communication Technology,
https://doi.org/10.1007/978-3-031-53935-0_3

The proposed hybrid machine learning model has numerous applications across various domains. For wireless network planning, accurate path loss estimation helps optimise the placement of base stations, antenna configurations, and transmission power levels for an efficient and reliable connectivity. Respecting the autonomous vehicles, the machine learning model can be implemented into a microcontroller in the vehicle; the predetermined path of the vehicle when driving can be used to predict when the vehicle should smoothly transition between base stations and select the suitable frequency to use to ensure zero connection loss in the journey. Satellites and airborne assets can benefit from this implementation as well. The model is also able to be implemented into mobile phones which can benefit users in regions of poor cellular signal coverage, identifying channels and areas with better signal strength and connectivity.

2 Conventional 5G/6G Signal Propagation Modelling

Empirical models work through curve-fitting methods using data collected from measurements in a path loss environment. The key parameters include frequency, distance between transmitter (Tx) and receiver (Rx), and path loss factor. However, these models are often not ideal for the planning of base stations due to their limited accuracy [8]. An example of an empirical path loss model used for millimetre frequencies is the 3GPP TR 38.901 channel model (used for designing, improving, and assessing fifth-generation (5G) systems for the 0.5–100 GHz frequency band) [9]. If the scenario is in an urban-macro environment with a line-of-sight condition, the path loss would be calculated using (1):

$$\mathrm{PL} = 28.0 + 22\log_{10}(d_{3\mathrm{D}}) + 20\log_{10}(fc) \tag{1}$$

where:

$d_{3\mathrm{D}}$ = three-dimensional separation distance (m)
fc = carrier frequency (Hz)

Deterministic methods are based on models such as raytracing and finite-distance time-domain. It can provide a path loss value with high accuracy at any location. However, the method is not computationally efficient and requires knowledge of the site geometry and material dielectric properties, making it a time-consuming computational process [10]. In machine learning techniques for path loss modelling, attributes such as the frequency, environment, and the separation between the transmitter and the receiver are used as the input features with the determined path loss value being the intended output. It is seen as a regression problem. The learned relationship between input and output allows it to outperform empirical models in some cases [8].

3 Methodology

3.1 Path Loss Model

The proposed path loss model is a hybrid system consisting of two machine learning models: an environment classifier to identify the environment of the scenario and the path loss prediction model. The path loss prediction model is also able to work independently without the input of environment classifier. TensorFlow will be used to classify the environment taking in images as its input. It is a deep learning framework that is well-suited for image classification tasks due to having extensive support for building and training convolutional neural networks (CNNs), which are highly effective in processing and extracting features from images [11]. XGBoost will be used to predict path loss; it is renowned for its outstanding accomplishment in a variety of machine learning tasks, including regression issues. XGBoost leverages gradient boosting algorithms, which iteratively improve the model's performance by minimising the loss function, resulting in better predictive power compared with other models [12]. Figure 1 shows the proposed flowchart of the model development process.

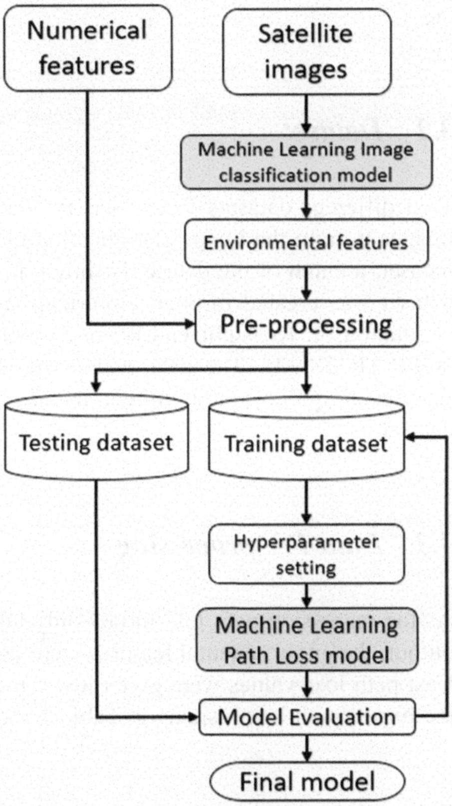

Fig. 1 Depicts the development process

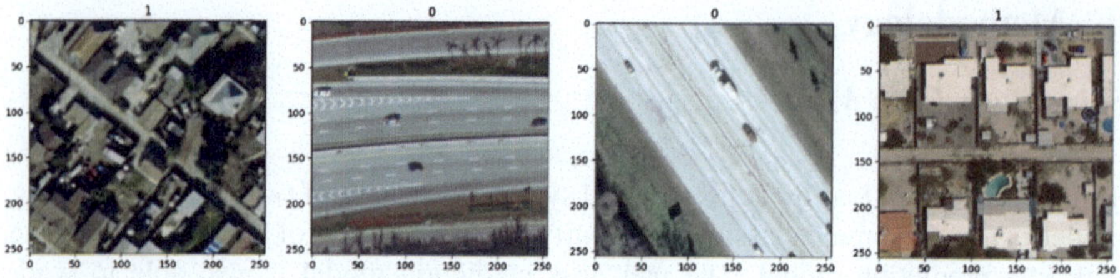

Fig. 2 Batch from the UC Merced land use dataset

Table 1 Dataset summary

Frequency (GHz)	Scenario	Indoor/outdoor	Line of sight	Distance (m)	Path loss (dB)
24.25	Uma	Outdoor	NLOS	10	−101.392
26	Uma	Indoor	NLOS	50	−162.625
25.25	Uma	Outdoor	LOS	88	−111.957

Table 2 Pre-processed dataset summary

Frequency (GHz)	Scenario	Distance (m)	Path loss (dB)
24.25	2	10	−101.392
26	3	50	−162.625
25.25	1	88	−111.957

3.2 Dataset

Two different datasets were used for the models. The first consisted of satellite images to train the image classification model taken from the UC Merced land use dataset; a batch of the dataset is shown in Fig. 2 [13]. For the path loss model, the dataset was created through empirical-based simulations. This was done through simulation on Keysight PathWave SystemVue using the empirical channel model 3GPP TR 38.901. The dataset consists of path loss outputs from different input features: frequency, environment, and distances. Table 1 presents a summary of the dataset.

3.3 Data Pre-processing

As the regression path loss model only takes numerical inputs to give a numerical output, the environmental features were assigned a number. The scenarios with the least path loss values were given lower numbers compared with the scenarios with the higher path loss responses. Table 2 shows the pre-processed dataset.

3.4 *Model Training*

The XGBoost regressor model was then trained with the dataset; three key parameters are employed: the number of estimators is set to 100, indicating an ensemble of 100 decision trees; the learning rate is at 0.3, suggesting each tree contributes 30% of its prediction toward the final model; and a maximum tree depth of 6, constraining each individual tree's complexity. This configuration enables the model to learn intricate patterns relatively quickly, while maintaining moderate complexity.

3.5 *Performance Metrics*

To evaluate the effectiveness of the regression machine learning models, the root mean square error of the measurements (RMSE) is frequently employed. It provides a straightforward interpretation of the average prediction error in the same unit as the target variable and measures the square root of the average squared difference between the predicted and actual values. Using the same units enables an easy and judicious comparison and understanding of the magnitude of the prediction error [14].

4 Results and Discussion

4.1 *Path Loss Model Evaluation*

The results of the machine learning path loss approach are analysed and presented in this section. An evaluation was conducted by comparing the predicted output values with the values calculated using the SystemVue (this was not utilised during the model training process to test the path loss machine learning model). By employing the empirical SystemVue data as an independent benchmark, any deviations or disparities between the predicted and the test path loss values can be identified, allowing for a comprehensive evaluation of the machine learning model's performance. Besides, an investigation was carried out to analyse the impact of reducing the dataset size on the model's accuracy. This analysis aims to explore the relationship between the training dataset size and the model's predictive capability. It provides insights into the model's robustness and ability to adapt to varying training data sizes. The results demonstrate that the machine learning path loss model achieved an exceptional prediction accuracy compared with the SystemVue simulations (Fig. 3). Even when trained on only 8% of the dataset, the model exhibited a negligible root mean square error (RMSE) of less than 1 dB when tested on the remaining dataset. This validates the model's ability to accurately predict path loss values.

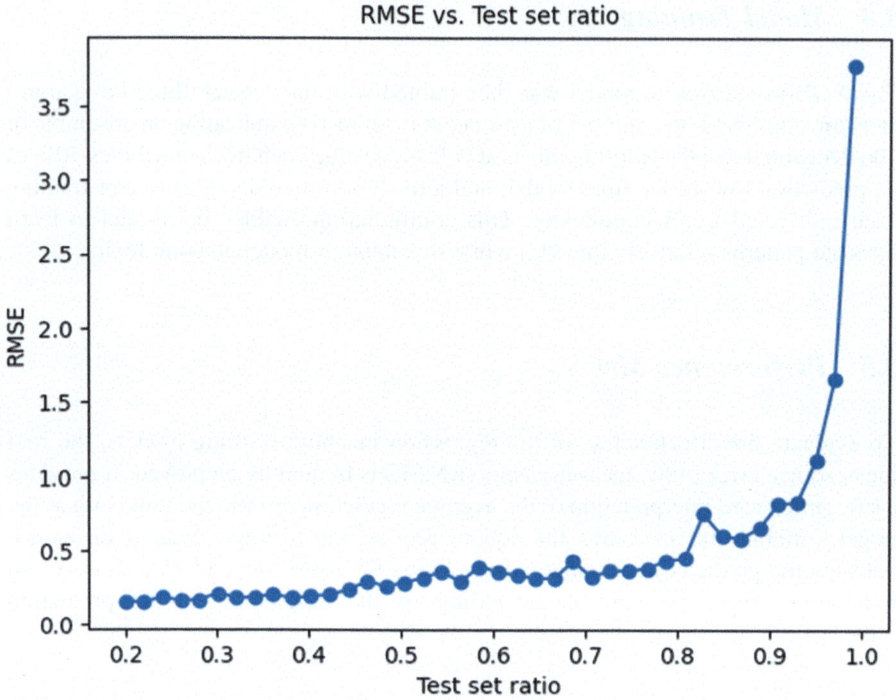

Fig. 3 RMSE vs train-test dataset ratio

A parameter sweep calculating the path loss when propagation distance changes from 10 to 100 m with 50 number of points was performed on both models. Figure 4 shows how the predicted output changes in response to the input distance when the frequency is 26 GHz in an urban-macro, indoor UT non-line-of-sight scenario. Figure 5 shows the simulation process time which was logged using the windows performance monitor tool.

The machine learning-based simulation exhibited superior performance compared to SystemVue in terms of execution time, whereby the machine learning model took 0.3 s and the SystemVue model took 86 s. Although CPU utilisation was similar, averaging 20%, the machine learning model demonstrated a 99.65% decrease in execution time compared to the SystemVue simulations. The reduced execution time could potentially result in significant time and resource savings when performing large-scale or complex simulations.

A dual parameter sweep (to change the distance and the frequency) was carried out (Fig. 6). The distance parameter sweep was from 10 to 100 m with 40 samples. The frequency parameter sweep was from 24.25 to 26 GHz with 40 samples. The model calculated 1600 different predictions in 0.67 s and generated a 3D contour plot after about 0.23 s.

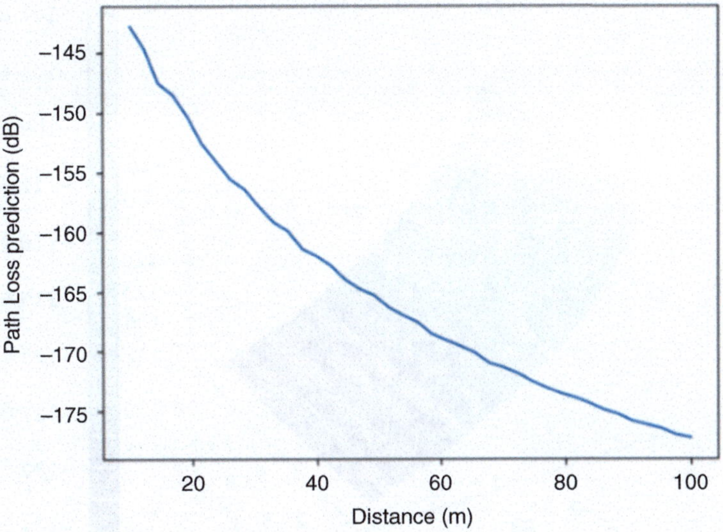

Fig. 4 Path loss vs distance

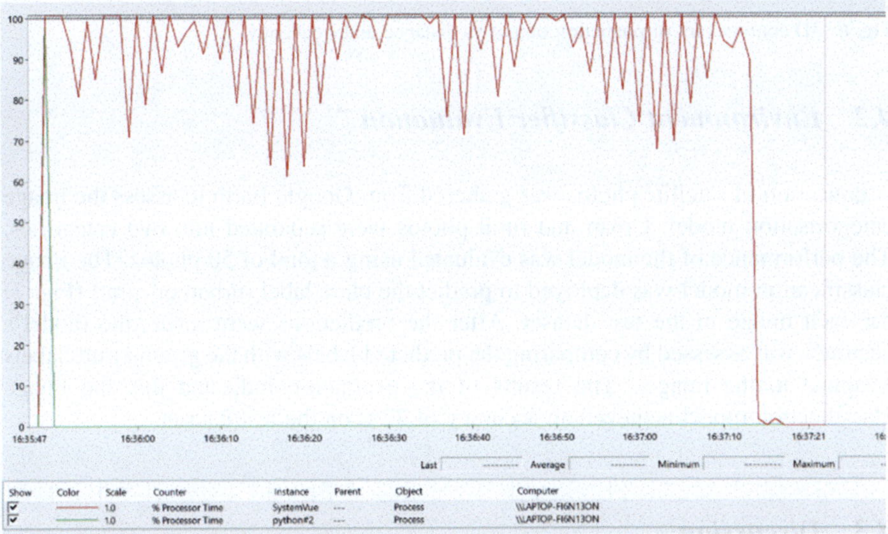

Fig. 5 Simulation times of SystemVue model vs machine learning model

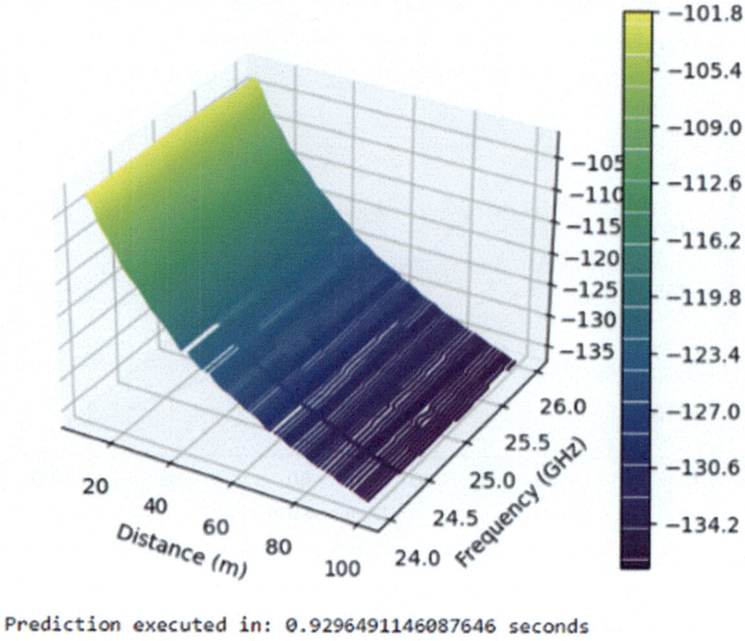

Fig. 6 3D contour plot of prediction output vs distance and frequency

4.2 Environment Classifier Evaluation

A collection of satellite photos was gathered from Google Earth to assess the image categorisation model. Urban and rural photos were separated into two categories. The performance of the model was evaluated using a total of 50 photos. The image classification model was deployed to predict the class label (urban or rural) (Fig. 7) for each image in the test dataset. After the predictions were made, the model's accuracy was assessed by comparing the predicted labels with the ground truth labels assigned to the images. The results of the evaluation indicated that the image classification model achieved an accuracy of 72% on the test dataset.

4.3 Discussion

The proposed machine learning-based PL determination would benefit hybrid RF/mmWave-Li-Fi systems [14–17] with hybrid reconfigurable high bandwidth channels, multiple concurrent multimedia variable data rates connections [14–16], ultra-low latency support occasioned by on-demand packet sizes, and operational adaptive beamforming modes for short- and long-range 5G/6G vehicular [14, 16] and satellite-cellular [18–23] communications and wireless energy harvesting [17] use cases/applications.

Fig. 7 TensorFlow model predicting the environment class

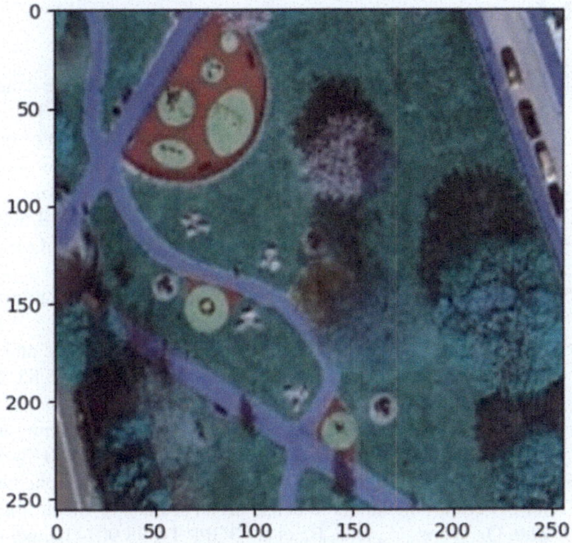


```
yhat = model.predict(np.expand_dims(resize/255, 0))
```

```
1/1 [==============================] - 0s 79ms/step
```

```
yhat
```

```
array([[0.28445885]], dtype=float32)
```

```
if yhat > 0.5:
    print(f'Predicted class is urban')
else:
    print(f'Predicted class is rural')
```

```
Predicted class is rural
```

5 Conclusion

This paper presents a hybrid machine learning model proposed to predict the path loss for millimetre-wave frequencies in the 26 GHz band. The model demonstrated its ability to provide accurate predictions. The model's computational latency of 0.23 s also offers significant time savings. However, certain limitations related to the specific frequency set, limited scenarios, and classification capabilities need to be addressed for further improvements. The dataset was created through empirical simulations rather than deterministic hardware. The future work will involve real-time measured path loss data rather than the empirical simulations. It will incorporate additional input features and scenario classifications to improve the model's versatility and enhance its overall performance. The addition of the environment classifier (s) to determine the line-of-sight and the non-line-of-sight conditions would also be beneficial.

References

1. Aijaz, A., Aghvami, H., Amani, M.: A survey on mobile data offloading: technical and business perspectives. IEEE Wirel. Commun. **20**, 104–112 (2013)
2. Sun, S., Rappaport, T., Thomas, T., et al.: Investigation of prediction accuracy, sensitivity, and parameter stability of large-scale propagation path loss models for 5G wireless communications. IEEE Trans. Veh. Technol. **65**, 2843–2860 (2016)
3. GSMA: http://www.springer.com/lncs. Last accessed 14 Sept 2023
4. Andrews, J.G., Buzzi, S., Choi, W., et al.: What will 5G be? IEEE J Sel Areas Commun. **32**, 1065–1082 (2014)
5. Ofcom: https://www.ofcom.org.uk/spectrum/information/innovation-licensing/enabling-5g-uk. Last accessed 15 Sept 2023
6. Durgin, G., Rappaport, T.S., Xu, H.: Measurements and models for radio path loss and penetration loss in and around homes and trees at 5.85 GHz. IEEE Trans. Commun. **46**, 1484–1496 (1998)
7. Seretis, A., Sarris, C.D.: An overview of machine learning techniques for radiowave propagation modeling. IEEE Trans. Antennas Propag. **70**, 3970–3985 (2022)
8. Jin, W., Kim, H., Lee, H.: A novel machine learning scheme for mmWave path loss modeling for 5G communications in dense urban scenarios. Electronics. **11**, 1809 (2022)
9. Zhu, Q., Wang, C., Hua, B., et al.: 3GPP TR 38.901 channel model. In: Wiley 5G Ref, pp. 1–35 (2021). https://doi.org/10.1002/9781119471509.w5gref048
10. Zhang, Y., Wen, J., Yang, G., et al.: Path loss prediction based on machine learning: principle, method, and data expansion. Appl. Sci. **9**, 1908 (2019)
11. Abadi, M.: TensorFlow: learning functions at scale. In: Proceedings of the 21st ACM SIGPLAN International Conference on Functional Programming, p. 1 (2016). https://doi.org/10.1145/2951913.2976746
12. nocode.ai: https://www.nocode.ai/why-xgboost-is-so-popular-among-data-scientists/. Last accessed 17 Sept 2023
13. Yang, Y., Newsam, S.: Bag-of-visual-words and spatial extensions for land-use classification. In: Proceedings of the 18th SIGSPATIAL International Conference on Advances in Geographic Information Systems, pp. 270–279 (2010)
14. Lau, I., Ekpo, S., Zafar, M., et al.: Hybrid mmWave-Li-Fi 5G architecture for reconfigurable variable latency and data rate communications. IEEE Access. **11**, 42850–42861 (2023)
15. Ekpo, S., George, D.: Impact of noise figure on a satellite link performance. IEEE Commun. Lett. **15**, 977–979 (2011)
16. Ekpo, S., Adebisi, B., Wells, A.: Regulated-element frost beamformer for vehicular multimedia sound enhancement and noise reduction applications. IEEE Access. **5**, 27254–27262 (2017)
17. Zafar, M., Ekpo, S., George, J., et al.: Hybrid power divider and combiner for passive RFID tag wireless energy harvesting. IEEE Access. **10**, 502–515 (2022)
18. Ekpo, S., George, D.: 4–8 GHz LNA design for a highly adaptive small satellite transponder using InGaAs pHEMT technology. In: 2010 IEEE 11th Annual Wireless and Microwave Technology Conference (WAMICON). Melbourne, FL (2010)
19. Ekpo, S.: Parametric system engineering analysis of capability-based small satellite missions. IEEE Syst. J. **13**, 3546–3555 (2019)
20. Ekpo, S.: Thermal subsystem operational times analysis for ubiquitous small satellites relay in LEO. Int. Rev. Aerosp. Eng. **11**, 48 (2018)
21. Uko, M., Ekpo, S.: 8–12 GHz pHEMT MMIC low-noise amplifier for 5G and fiber-integrated satellite applications. Int. Rev. Aerosp. Eng. **13**, 99 (2020)
22. Sowande, O., Idachaba, F., Ekpo, S., et al.: Sub-6 GHz 5G spectrum for satellite-cellular convergence broadband internet access in Nigeria. Int. Rev. Aerosp. Eng. **15**, 85 (2022)
23. Ekpo, S., George, D.: A system engineering analysis of highly adaptive small satellites. IEEE Syst. J. **7**, 642–648 (2013)

Establishment of the Anammox Wastewater Treatment Process in a Lab-Scale Expanded Bed Biofilm Reactor with ABDite®

Comfort Anyanwu, Devine Akhidime, Mike Dempsey, and Louise Melling

1 Introduction

Globally, about 80% of wastewater is discharged into natural water bodies without prior treatment, thereby increasing deterioration in water quality, with detrimental effects on the ecosystem and human health [1, 2]. The discharge of ammonia-rich wastewaters from human urea, nitrate fertilizer, metal finishing, processing crude oil, and pharmaceutical products into lakes, rivers, and streams upsets the nitrogen balance in these water bodies, causing several problems associated with the water pollution [3, 4].

1.1 Biological Wastewater Treatment

The reactions of the biological nitrogen cycle (Fig. 1) that removes nitrogen pollutants (ammonia, nitrite, and nitrate) from wastewater include nitrification, denitrification [5, 6], and the anaerobic ammonia oxidation (Anammox), which converts the nitrite from partial nitrification (PN) and ammonia to nitrogen gas [7].

C. Anyanwu (✉) · L. Melling
Department of Natural Sciences, Manchester Metropolitan University, Manchester, UK
e-mail: c.anyanwu@mmu.ac.uk; l.melling@mmu.ac.uk

D. Akhidime
Department of Life Science, Manchester Metropolitan University, Manchester, UK
e-mail: d.akhidime@mmu.ac.uk

M. Dempsey
Advanced Bioprocesses Development Ltd, Manchester, UK

S. C. Ekpo (ed.), *The Second International Adaptive and Sustainable Science, Engineering and Technology Conference*, Signals and Communication Technology,
https://doi.org/10.1007/978-3-031-53935-0_4

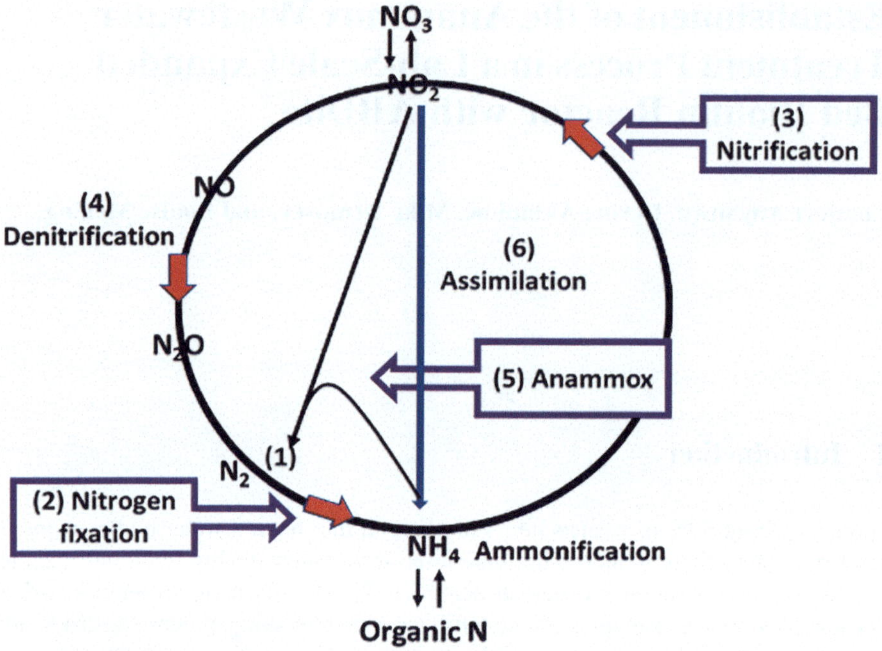

Fig. 1 Biological nitrogen cycle. (1) Nitrogen gas (N_2) fixation, (2) aerobic ammonium (NH_{4^+}) oxidation by ammonia-oxidizing bacteria (AOB) and archaea, (3) aerobic nitrite (NO_{2^-}) oxidation, (4) denitrification, (5) anaerobic ammonia oxidation (Anammox), and (6) dissimilatory nitrate and nitrite reduction [8, 9]

However, while partial nitrification (PN) conducted by AOB such as *Nitrosomonas* oxidizes ammonia to nitrite, in complete nitrification (CN), nitrite-oxidizing bacteria (NOB), e. g., *Nitrobacter*, converts nitrite to nitrate. The products of both processes represented in Eqs. 1, 2, 3 and 4 are toxic to aquatic lives when the treated wastewater is discharged [10, 11]. Equally, traditional nitrification-denitrification displayed in Eq. 5 that reduces nitrate to nitrogen gas requires an external source of carbon (e.g., methanol), which increases operational costs [12]:

$$NH_3 + O_2 + 2H + 2e^- \rightarrow NH_2OH + H_2O \quad (1)$$

$$NH_2OH + H_2O^- \rightarrow NO_{2^-} + 4e^- + 5H^+ \quad (2)$$

$$\tfrac{1}{2} O_2 + 2H^+ + 2e^{--} \rightarrow H_2O \quad (3)$$

$$NO_{2^-} + H_2O^- \rightarrow NO_{3^-} + 2H^+ + 2e^- \quad (4)$$

$$2NO_{3^-} + 10e^- + 12H + N_2 + 6H_2O \quad (5)$$

1.2 Anammox

The drawbacks of these WWT methods described have created the need for a WWT process that can remove the nitrogenous pollutants from wastewater and is cost-effective. Anammox is a biological WWT process, during which, nitrite and ammonia are converted directly into dinitrogen gas (N_2) under anoxic conditions [5]. The Anammox process produces up to 50% of the gaseous N_2 in the Earth's atmosphere [13] and accounts for nearly 50% of the nitrogen recycled in aquatic environments [14]. The Anammox wastewater process relies on partial nitrification (PN) of ammonia to nitrite by oxidizing bacteria (AOB); then, the remaining ammonia and nitrite are converted to molecular nitrogen gas [7, 15]:

$$NH_3 + NO_2^- + H^+ \rightarrow N_2 + 2H_2O \tag{6}$$

The Anammox process, which simultaneously removes both ammonia nitrogen (NH_3-N) and nitrite nitrogen (NO_2-N) from wastewater, is conducted by Anammox bacteria which are autotrophs that utilize inorganic carbon such as carbon dioxide and carbonates as energy sources for metabolic activity. They belong to a group of bacteria called Planctomycetes with five genera (*Anammoxoglobus*, *Brocadia*, *Kuenenia*, *Jettenia*, and *Scalindua*) identified. Anammox bacteria are obligate anaerobes which produce N_2 from ammonia and nitrite, with hydroxylamine and hydrazine as intermediates [16, 17].

The Anammox process is important in reducing the detrimental effects of discharging raw or partially treated wastewater into natural water bodies. Anammox removes up to 90% of nitrogen from wastewater at a lower cost, with reductions in energy (63%) and oxygen (60%) consumption and 70–80% sludge production. The Anammox process neither produces carbon dioxide nor requires an external organic carbon source, making it an eco-friendly, energy efficient, cost-effective, and sustainable wastewater treatment process [18].

Since the discovery of the Anammox process in the 1990s [19], over 100 full-scale WWT plants have been built around the world [20, 21]. While Germany, with about 25% of total installations, has benefited greatly from this WWT process, countries including the United States (7%) and the United Kingdom (<4%) are yet to fully maximize the potential of the Anammox wastewater treatment technology [18]. Furthermore, it is not known if developing countries, like Nigeria and others, have explored the Anammox WWT technology.

Evaluation of the top ten world Anammox research publications have revealed that although the Anammox process was discovered in the Netherlands and the first industrial application developed there, most research publications are from China (27.71%), then the Netherlands (23.8%) and followed by the United States (20.03%) [11, 22]. There is the need for an increase in Anammox wastewater treatment research both in the United Kingdom and developing countries.

1.3 Expanded Bed Biofilm Reactor and ABDite®

The Anammox process has contributed to the reduction of the adverse effects caused by the discharge of nitrogenous wastewater into aquatic environments and has successfully treated wastewater from agricultural, domestic, and industrial sources from laboratory to industrial scales around the world using different bioreactors and biomass growth support medium [23].

The EBBR that uses approximately 1.0 mm ABDite® growth medium to develop biofilms has been successful at laboratory and pilot scales for high-rate tertiary nitrification processes to convert wastewater ammonia to nitrate. The EBBR has met most of the criteria for sustainable WWT technology; it is simple, low-cost, and energy-efficient [24], yet despite these advantages, it is yet to be deployed for the Anammox wastewater treatment.

2 Experimental Design

Figure 2 shows the 0.7 L lab-scale EBBR designed for the Anammox wastewater treatment research. It was constructed using borosilicate glass components (QVF Corning Process Systems, UK), comprised of two vertical columns for bed expansion and bulk wastewater recirculation. The 8 × 4 cm expanded bed column and the 45 × 2 cm recirculation column were connected using silicon tubing (Silex, UK) to ensure effective mixing of influent and effluent synthetic wastewater (SWW).

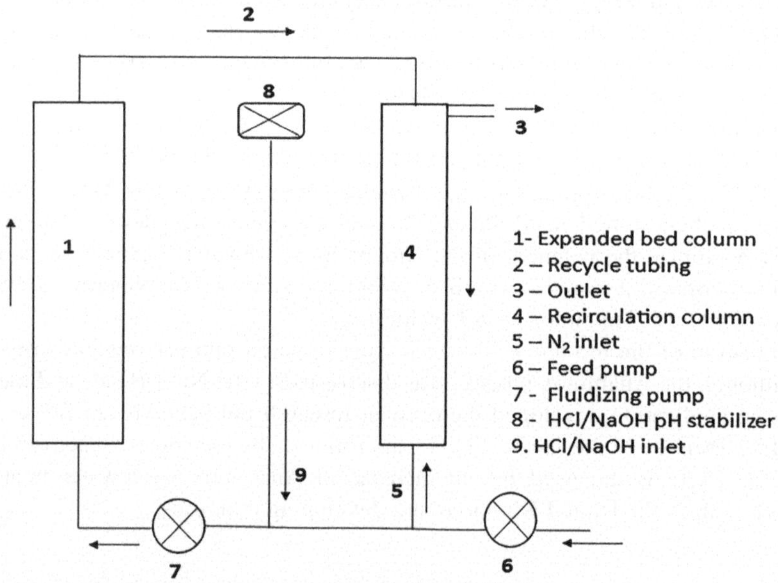

Fig. 2 Schematic diagram of the lab 0.7 L lab-scale EBBR

2.1 Setup of the EBBR-Anammox Process

The EBBR expanded column was filled to one third of bed height (12 cm) with 0.7–1 mm of porous ABDite® biomass support particles (ABD Ltd., Manchester, UK), inoculated with Anammox bacteria granules (Severn Trent, UK) and continuously supplied with high strength ammonia synthetic wastewater (SWW) and trace element solution [25].

The EBBR was maintained at 50% bed expansion from the baseline height using a peristaltic pump (Watson Marlow 120S, UK) by the upward flowing SWW at a velocity of 1 cm s^{-1} [24]. Temperature was kept at 30 °C by a water jacket connected to a thermo-circulator (Thermo Fisher, UK) and pH 8.3, using a pH controller (Electrolab 260, UK).

2.2 Sampling and Analysis

The transparent EBBR column was examined every 2–3 days for physical evidence of the Anammox process, including formation of brick red granules or brown biofilm/bioparticles on the ABDite® biomass and gas production. Physical parameters (bed height increase, pH, and temperature measurements) were recorded manually. Liquid samples (5 mL) for chemical analysis were collected after point 6, before the N_2 gas inlet at point 5 (Fig. 2). Concentrations of NH_3-N, NO_2-N, and NO_3-N were determined by standard spectrophotometric methods and the results used to calculate the NRR and NRE. The NRR was calculated as the difference between the influent nitrogen (ammonia + nitrite) load into and effluent nitrogen (ammonia + nitrite) from the Anammox bioreactor, which is based on the stoichiometry reaction represented by (8) [6]:

$$\mathrm{NRR} = (\mathrm{Load\ Nin} - \mathrm{Load\ Nout})/V\ \left(\mathrm{mgN\ L^{-1} d^{-1}}\right) \tag{7}$$

where:

- Load Nin (sum of inlet nitrogen fed into the reactor (ammonia + nitrite) (mgN d^{-1}))
- Load Nout (sum of nitrogen leaving the reactor (ammonia and nitrite) (mgN d^{-1}))
- *V*: volume of the Anammox bioreactor (L) [8]

In the EBBR, the volumetric Anammox rate was calculated and expressed as KgN m^{-3} EBBR d^{-1}. The total reaction volume was calculated based on the bioparticle bed height upon expansion by the fluidizing pump in the expanded column and is expressed as V (m^3):

$$\mathrm{EBBR\ reaction\ volume\ (V)} = \mathrm{EBH} \times \pi \times r^2\ \left(\mathrm{m^3}\right) \tag{8}$$

The dilution rate D (d^{-1}) of the EBBR system was calculated as dilution rate:

$$D = F/V \ (d^{-1}). \quad (9)$$

where:

F = influent flow rate (m^{-3} d^{-1})

Finally, the dilution rate (D) was multiplied by the concentration of nitrogen consumed by the Anammox biomass (KgN m^{-3} EBBR)

Therefore, the reactor volumetric Anammox rate = concentration of nitrogen [NH_3-N] and [NO_2-N] removed by the Anammox biomass (KgN m^{-3} EBBR) × D (d^{-1}):

$$\mathrm{NRR}\ (\mathrm{KgN\ m^{-3}\ d^{-1}}) = D \times ([\text{Total inlet } N] - [\text{Total Outlet } N]) \quad (10)$$

ABDite® biomass support, suspected to be rich in Anammox bacteria biofilms (bioparticles), was harvested for scanning electron microscope (SEM) and genomic DNA extraction for further molecular analysis from the top of the top of the expanded bed column (Fig. 2) via suction, by inserting a silicon rubber tube into the sampling depth (17–20 cm). The suction point was sealed with metal clips to maintain anaerobic conditions.

3 Results and Discussion

The EBBR-Anammox wastewater treatment research lasted approximately 560 days in continuous culture from the start-up until the bioreactor became unstable due to damage from prolonged operation. Within the period, evidence suggesting the establishment and effectiveness of the Anammox process using the EBBR technology was collected. The results were evaluated against data from previous Anammox studies that used different bioreactors and biomass growth medium [25, 26].

3.1 *Physical Evidence*

Physical evidences to establish that the Anammox process was occurring in the lab-scale EBBR were increase in static bed height (SBH) and expanded bed heights (EBH), gas production, and brick red Anammox bacteria granules and biofilm development on the ABDite® growth support medium.

Figure 3a shows the presence of Anammox granules and biofilms developed on the ABDite® biofilm support particles and production of gas suspected, but unconfirmed to be nitrogen gas in the EBBR. These established characteristics of

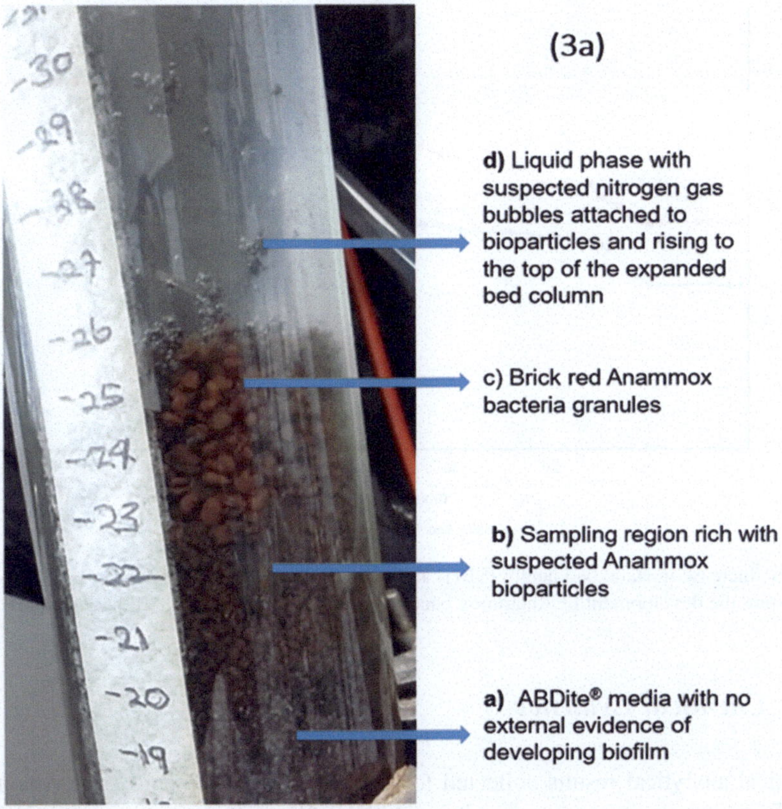

Fig. 3a Brick red Anammox granules, bioparticles, and suspected N_2 gas production in the EBBR, indicating the establishment of the Anammox process in the EBBR

the Anammox process within a system suggest that the operating conditions in the EBBR supported the development of active Anammox bacteria biomass needed to remove the nitrogen pollutants from the synthetic wastewater (SWW).

The growth of biofilm on the ABDite® particles resulted in a 63% increase in the expanded bed height (EBH) and 54% increase in the static bed height (SBH) shown in Fig. 3b. These results further indicated that the Anammox process was being established in the EBBR column. Using an Anammox enrichment system and a sequencing batch reactor (SBR), respectively, nitrogen gas production was reported as the main evidence for the occurrence of the Anammox process in those systems as observed in the EBBR [25, 26].

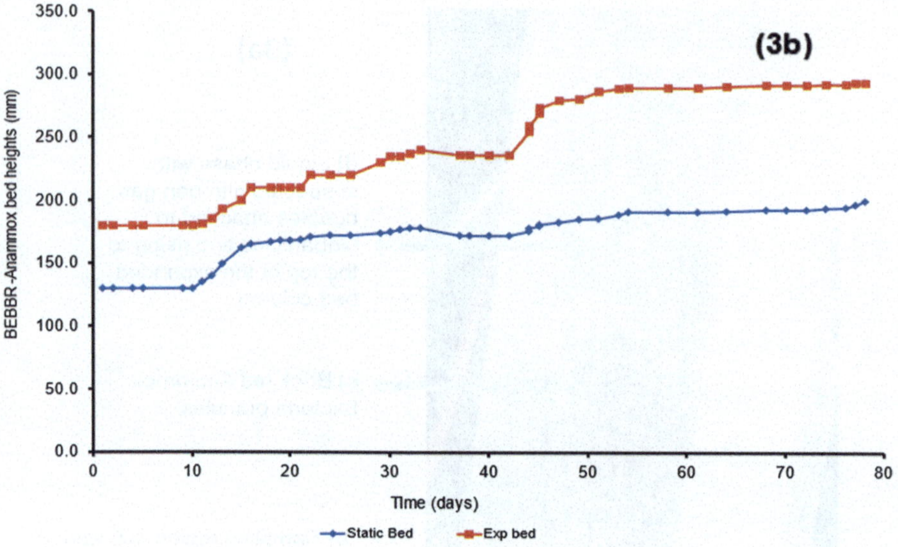

Fig. 3b Increase in static bed height (SBH) and expanded bed height (EBH) over time (80 days), suggesting the development of Anammox bacteria biomass on the EBBR ABDite® support

3.2 *Chemical Evidence*

Chemical analytical results collected to verify the establishment of the Anammox WWT process were ammonia nitrogen (NH_3-N) and nitrite nitrogen (NO_2-N) concentrations removed from the EBBR effluent, nitrogen removal rate (NRR), and % nitrogen removal efficiency (NRE).

Figures 4a and 4b show that from the start-up to day 78 (a) and day 139–317 (b), 82–362 mg L^{-1} (NH_3-N) and 156–449 mg L^{-1} (NO_2-N) were removed from the EBBR SWW, and 14–26 mg L^{-1} NO_3-N was produced.

The nitrogen removal rate (NRR) analysis indicated that the EBBR-Anammox process successfully achieved 6.1 KgN m^{-3} d^{-1} as the highest nitrogen removal rate (NRR) from a loading rate (NLR) of 9.0 KgN m^{-3} d^{-1} (Fig. 4c) and nitrogen removal efficiency (NRE) between 90% and 100% during steady operation day 139–317 (Fig. 4d).

The results suggested that the adapted EBBR system was effective at removing the nitrogen pollutants from the supplied wastewater. The EBBR-Anammox with ABDite NRR and NRE were within the range and compared well with results from other Anammox technologies, for example, the following NRR were reported for GLR (1.78 KgN m^{-3} d^{-1}), AnMBR (5.0 KgN m^{-3} d^{-1}), and SBR (0.98 KgN m^{-3} d^{-1}) [27–29].

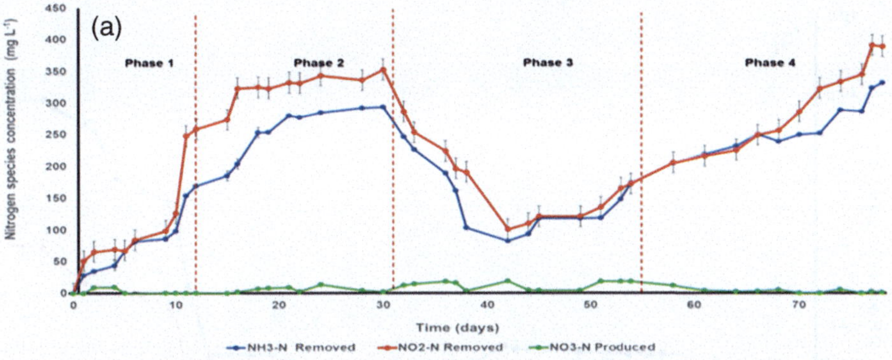

Fig. 4a Ammonia and nitrite removed from EBBR effluent and nitrate nitrogen produced between days 0 and 78. (Each data point represents a mean of 2 replicate samples ($n = 2$); error bars represent +/− 1SD of the mean)

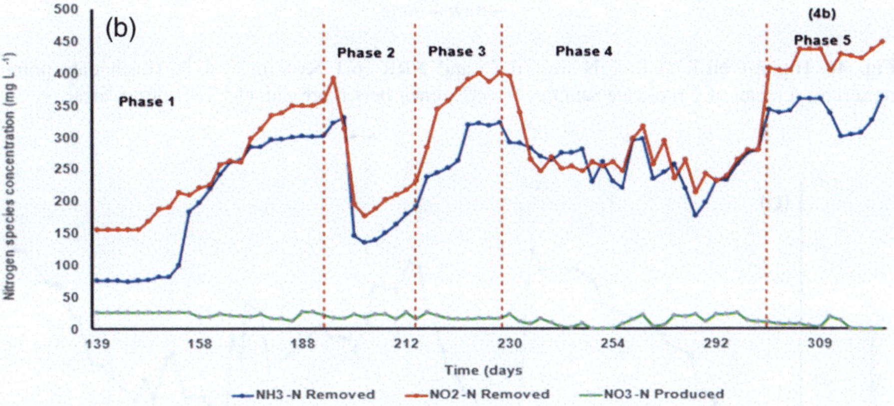

Fig. 4b Ammonia and nitrite removed from EBBR effluent and nitrate nitrogen produced between days 139 and 317. Removal of ammonia and nitrite and nitrate production simultaneously characterize the establishment of the Anammox process

The pioneers of the Anammox process from Delft University reported that simultaneous conversion of NH_3-N and NO_2-N to nitrogen gas and production of about 10% NO_3-N as observed in the EBBR study are indicators of the Anammox reaction within a bioreactor that was initially designed for an aerobic WWT process to convert ammonia to nitrate [30] but adapted for the Anammox process in the current study.

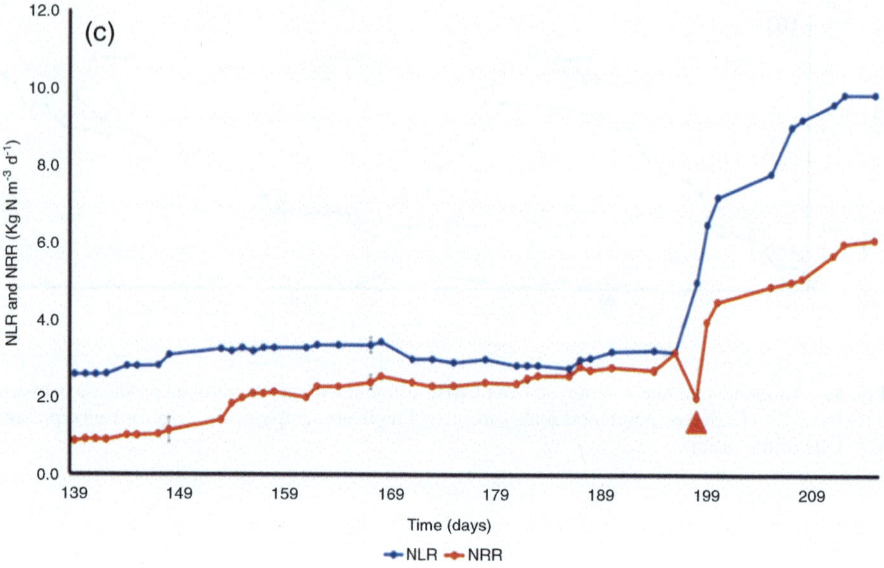

Fig. 4c Highest NLR (9.8 KgN m^{-3} d^{-1}) and NRR (6.1 KgN m^{-3} d^{-1}). (Each data point represents a mean of 2 replicate samples ($n = 2$); error bars represent +/− 1SD of the mean)

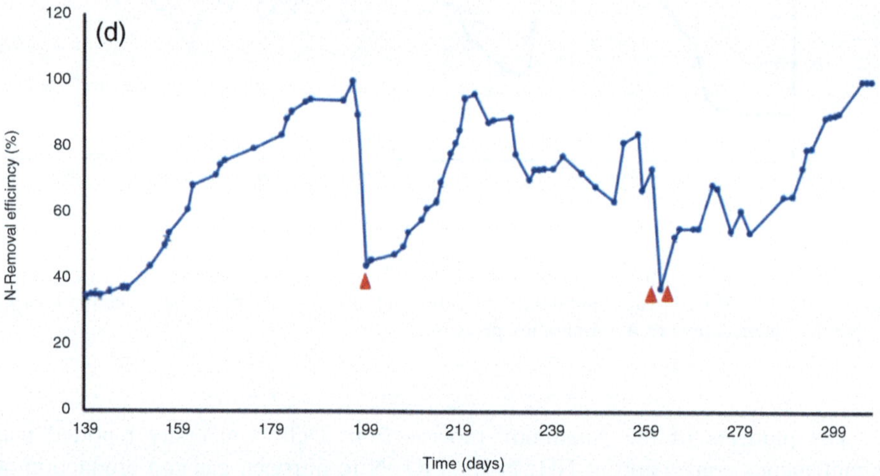

Fig. 4d Maximum NRE (>90–100%) between days 139 and 317. (Each data point represents a mean of 2 replicate samples ($n = 2$); error bars represent +/− 1SD of the mean). Indicate *period of instability*

3.3 Scanning Electron Microscopic Evidence

Figure 5a–g shows SEM images of uncolonized ABDite® (a) and the development of Anammox bacteria biofilm on the ABDite® medium to form bioparticles (b), then the cauliflower-like structure characteristic of Anammox bacteria biofilms (c–d) and distinct bacteria-like cells suspected to be Anammox bacteria (e–g). The SEM micrographs further strengthened the evidence that the EBBR ABDite® was suitable for the growth of Anammox bacteria like other biomass support medium such as foam, coconut husk fiber, coke, wood chips, sand, glass, and volcanic rock among others materials reported in literature that have been used in different Anammox bioreactors [30–32].

3.4 Molecular Evidence

Figure 6a shows the gel image of the PCR amplified Anammox bacteria DNA extracted from the ABDite® medium detected by species-specific primers. The Amx368F/Amx820R primer set targeting any members of the Anammox bacteria genera (*Candidatus – Anammoxoglobus*, *Brocadia*, *Jettenia*, *Kuenenia*, and *Scalindua*) detected the expected size of DNA bands from Anammox bacteria (AnAOB) in lanes 2, 4, 6, and 8 extracted from 4 ABDite samples between 1 and 13 months. Also, the correct size of DNA bands belonging either to *Candidatus Brocadia* or *Scalindua* was detected with the Brod541F/Amx820R primer pair from ABDite 1, 2, and 3 in lanes 3, 5, and 7, respectively (Fig. 6a).

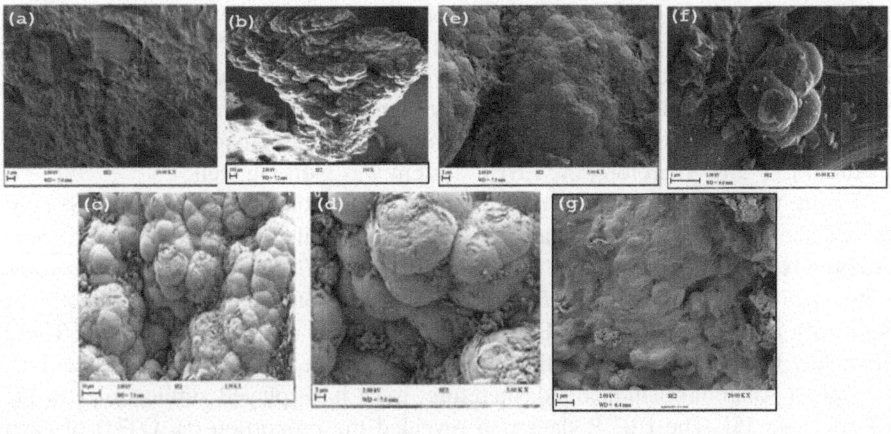

Fig. 5 EBBR SEM micrographs. (**a**) Uncolonized ABDite® surface, (**b**) ABDite® bioparticle surface, (**c**–**d**) cauliflower-like structure, (**e**–**g**) suspected Anammox bacteria cells

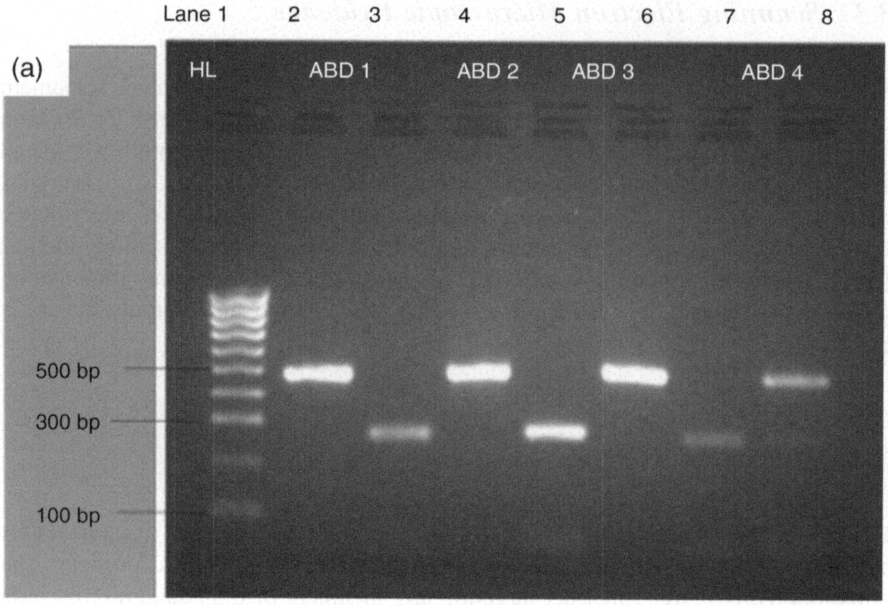

Fig. 6a Anammox bacteria DNA extracted from ABDite® in the first month (L 2 and 3), fifth month (L 4 and 5), seventh month (L 6 and 7), and 13th month (L 8), and detected with species-specific primers, Amx368F/Amx820R and Brod541F/Amx820R. Key Lane 1: 100 bp HyperLadder

Figures 6b and 6c provided evidence that key Anammox bacteria groups – ammonia-oxidizing bacteria (AOB) and anaerobic ammonia-oxidizing bacteria (AnAOB) among others – were identified by Sanger sequencing and Basic Local Alignment Search Tool (BLAST) techniques. The identification of bacteria targeted with species-specific primers using nucleotide sequences of DNA from the EBBR-ABDite® confirmed that Anammox bacteria capable of removing the nitrogenous pollutant were developed on the ABDite® medium as have been reported by other researchers that have grown Anammox biomass using different growth medium [33–35]. Therefore, the EBBR system was working well.

Figure 6d shows that the more robust next-generation sequencing (NGS) confirmed the EBBR-Anammox Sanger sequencing results by identifying the same Anammox bacteria (*Candidatus Kuenenia* and *Brocadia)* alongside *Nitrosomonas*, *Nitrosospira*, *Nitrospira*, *Nitrobacter* species, and others that were directly or indirectly involved with the EBBR-Anammox process using V3V4a primer. These same key bacteria identified in the EBBR system have been described in previous published Anammox studies and, therefore, must have played similar roles in the EBBR [34, 35]. The EBBR data also revealed the proportion (% OTU) of each community as displayed in Fig. 6d; the % OTU provided evidence that could be used to verify the contribution and influence each key bacteria group present in the EBBR

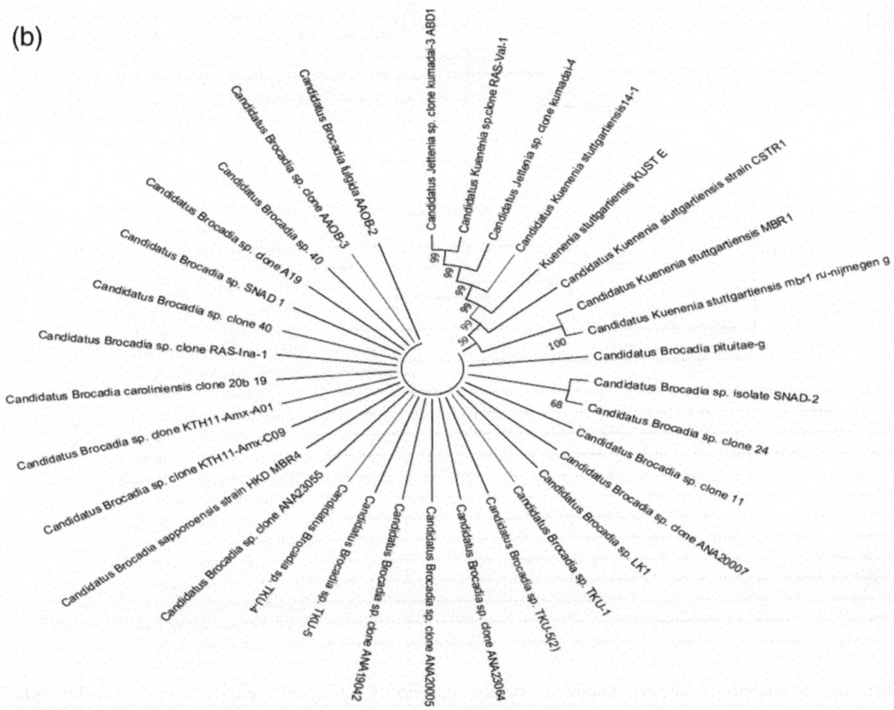

Fig. 6b Unrooted phylogenetic tree based on nucleotide sequence from targeted EBBR-Anammox bacteria. *Candidatus Brocadia* (76%), *Kuenenia stuttgartiensis* (18%), and *Jettenia* (6%). The scale bar represents 5% substitutions per 1000 bootstrap nucleotide replicates

had on the Anammox process at various stages of the research. These NGS findings could be useful in predicting the performance of the EBBR-Anammox WWT process in future scale-up operations.

4 Conclusion

The effectiveness of the Anammox process in treating ammonia-rich wastewater using different bioreactors has been reported in several laboratory-, pilot-, and industrial-scale studies. Results from this research provides strong evidence that the Anammox wastewater treatment process was established for the first time using the lab-scale EBBR by achieving a NRR of 6.1 KgN m^{-3} d^{-1} and 80–100% NRE. The ABDite® biomass medium supported the growth of three Anammox – *Candidatus Brocadia*, *Kuenenia*, and *Jettenia* among others key groups. These results suggest that EBBR technology initially designed for nitrification can be adapted successfully for the Anammox wastewater treatment process to remove ammonia and nitrite from wastewater before it is discharged into natural water bodies.

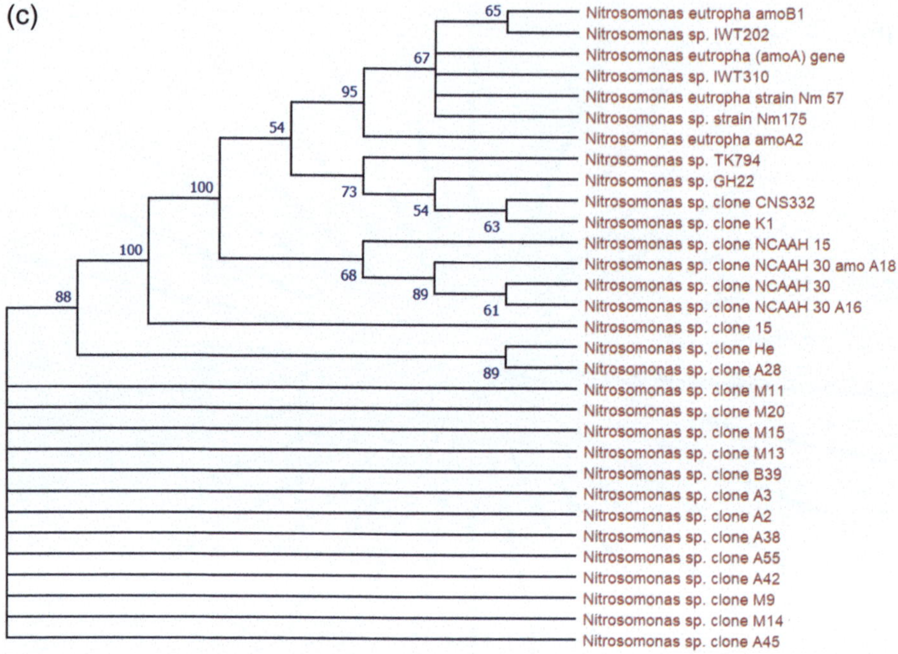

Fig. 6c Ammonia-oxidizing bacteria species that grew alongside EBBR-Anammox bacteria. *Nitrosomonas eutropha* (12%) and other species (87%). The scale bar represents 5% substitutions per 1000 bootstrap nucleotide replicates

5 Future Work

Future EBBR-Anammox study would benefit from an online sensor connected to a data logger with sufficient data storage, for N-species, temperature, pH, and dissolved oxygen (DO) measurements in real time [34, 35]. This is recommended to save time and provide additional data which could be compared with the spectrophotometric data.

In the EBBR-Anammox investigation, the identity of the suspected nitrogen gas produced was not confirmed. The advancement of research work in this area is necessary because if the gas is confirmed as nitrogen, this will further reduce the overall cost of the process. The gas could be used to deoxygenate the rig or collected and sold; gas collection will, however, require the modification of the EBBR at a cost.

Further work to evaluate the abundance and impact of different microbial communities at various stages of the EBBR-Anammox process is also recommended; this will be useful for scale-up operation.

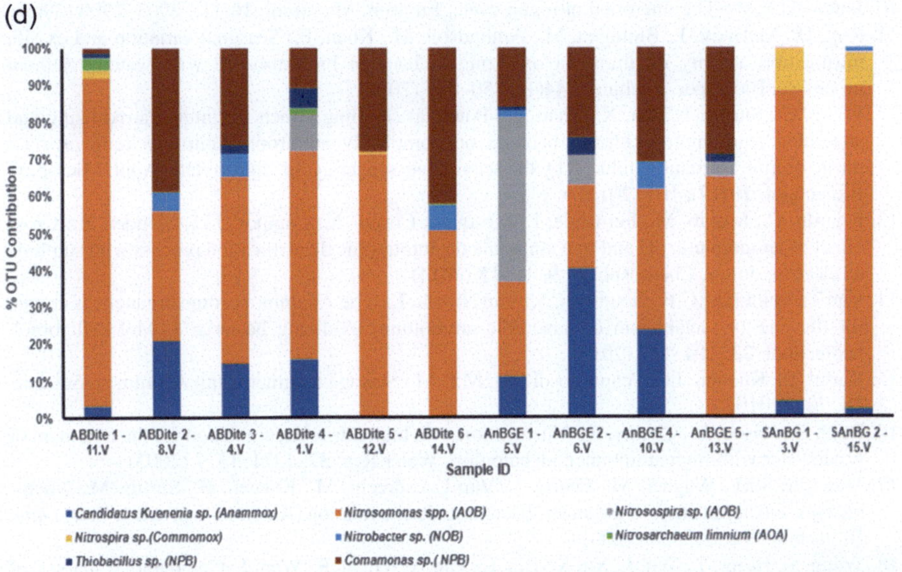

Fig. 6d Proportion (% OTUs) contributed to the EBBR-Anammox microbial communities by bacteria identified from ABDite® bioparticles 1, 2, 3, 4, 5, and 6 sampled in the 1st, 5th, 7th, 13th, 21st, and 24th month, respectively. AnAOB (14), AOB (145), complete ammonia-oxidizing bacteria (Commamox 5.2), NOB (6%), ammonia-oxidizing archaea (AOA 0.7%), and nitrifying proteobacteria (NPB 51%) respectively

References

1. World Water Development Report: UN-Water Publications. https://www.unwater.org/publications/world-water-development-report (2018)
2. Okereke, J.N., Ogidi, O.I., Obasi, K.O.: Environmental and health impact of industrial wastewater effluents in Nigeria. Int. J. Adv. Res. Biol. Sci. **3**(6), 55–67 (2016)
3. WHO, UNEP: ISBN Number: 978-92-3555-0, pp. 15–17 (2015a)
4. Canfield, D.E., Glazer, A.N., Falkowski, P.G.: The evolution and future of Earth's nitrogen cycle. Science. **330**, 192–196 (2010)
5. Rodriguez-Sanchez, A., Gonzalez-Martinez, A., Martinez-Toledo, M.V., Garcia-Ruiz, M.J., Osorio, F., Gonzalez-Lopez, J.: The effect of influent characteristics and operational conditions over the performance and microbial community structure of partial nitritation reactors. Water. **6**, 1905–1924 (2014)
6. Langone, M., Yan, J., Haaijer, S.C.M., Op den Camp, H.J.M., Jetten, M.S.M., Andreottola, G.: Coexistence of nitrifying, Anammox and denitrifying bacteria in a sequencing batch reactor. Front. Microbiol. **5**, 28 (2014)
7. Kuenen, J.G.: Anammox and beyond. Environ. Microbiol. **22**(2), 525–536 (2020)
8. Jaroszynski, L.W., Cicek, N., Sparling, R., Oleszkiewicz, J.A.: Impact of free ammonia on Anammox rates (anoxic ammonium oxidation) in a moving bed biofilm reactor. Chemosphere. **88**(2), 188–195 (2012)
9. Ward, B.B.: Nitrification: An Introduction and Overview of the State of the Field. Wiley Online Library (2011)

10. Jetten, M.S.M.: The microbial nitrogen cycle. Environ. Microbiol. **10**(11), 2903–2909 (2008)
11. Roy, D., McEvoy, J., Blonigen, M., Amundson, M., Khan, E.: Seasonal variation and ex-situ nitrification activity of ammonia oxidizing archaea in biofilm-based wastewater treatment processes. Bioresour. Technol. **244**(1), 850–859 (2017)
12. Wang, C., Sitong, L., Xu, X., Zhao, C.: Potential coupling effects of ammonia-oxidizing and anaerobic ammonium-oxidizing bacteria on completely autotrophic nitrogen removal over nitrite biofilm formation induced by the second messenger cyclic diguanylate. Appl. Microbiol. Biotechnol. **101**(9), 101 (2017)
13. Fajardo, C., Morab, M., Fernande, I., Mosquera-Corral, A., Campos, J.L., Mendez, R.: Cross effect of temperature, pH and free ammonia on autotrophic denitrification process with sulphide as electron donor. Chemosphere. **9**, 10–15 (2014)
14. Van Teeseling, M.C.F., Neumann, S., Van Niftrik, L.: The Anammoxosome organelle is crucial for the energy metabolism of anaerobic ammonium oxidizing bacteria. J. Mol. Microbiol. Biotechnol. **23**, 104–117 (2013)
15. Kartal, B., Kuenen, J.G., Van Loosdrecht, M.C.M.: Sewage treatment with Anammox. Science. **28**, 702 (2010)
16. Ruiz, G., Jeison, D., Chamy, R.: Nitrification with high nitrite accumulation for the treatment of wastewater with high ammonia concentration. Water Res. **37**, 1371–1377 (2003)
17. Jetten, M.S.M., Wagner, M., Fuerst, J., Van Loosdrecht, M., Kuenen, G., Strous, M.: Microbiology and application of the anaerobic ammonium oxidation (Anammox) process. Curr. Opin. Biotechnol. **12**, 283–288 (2001)
18. Wang, S., Hong, Y., Wu, J., Xu, X., Bin, L., Pan, Y., Guan, F., Wen, J.: Comparative analysis of two 16S rRNA gene-based PCR primer sets provide insight into the diversity distribution patterns of Anammox bacteria in different environments. Appl. Microbiol. Biotechnol. **99**, 8163–8176 (2015)
19. Castro-Barros, C.M., Jia, M., van Loosdrecht, M.C., Volcke, E.I., Winkler, M.K.: Evaluating the potential for dissimilatory nitrate reduction by Anammox bacteria for municipal wastewater treatment. Bioresour. Technol. **233**, 363–372 (2017)
20. Van Hulle, S.W.H., Vandeweyer, H.J.P., Meesschaert, B.D., Vanrolleghem, P.A., Dejans, P., Dumoulin, A.: Engineering aspects and practical application of autotrophic nitrogen removal from nitrogen rich streams. Chem. Eng. J. **162**, 1–20 (2010)
21. Cho, S., Kambey, C., Nguyen, V.K.: Performance of Anammox processes for wastewater treatment: a critical review on effects of operational conditions and environmental stresses. Water. **12**(1), 20 (2019)
22. Zhang, Z.-Z., et al.: Transient disturbance of engineered ZnO nanoparticles enhances the resistance and resilience of Anammox process in wastewater treatment. Sci. Total Environ. **622**, 402–409 (2018)
23. Kumwimba, M.N., Lotti, T., Şenel, E., Li, X., Suanon, F.: Anammox-based processes: how far have we come and what work remains? A review by bibliometric analysis. Chemosphere. **238**, 124627 (2020)
24. Akhidime, I.D., Dempsey, M.J.: Influence of biofilm thickness on ammonia oxidation rate of bioparticles from expanded bed process for tertiary nitrification. In: IWA Nutrient Removal Conference. Krakow, Poland (2009)
25. Sliekers, A.O., Derworta, N., Gomez, J.L.C., Strous, M., Kuenen, J.G., Jetten, M.S.M.: Completely autotrophic nitrogen removal over nitrite in one single reactor. Water Res. **36**, 2475–2482 (2002)
26. Hu, Z., Lotti, T., De Kreuk, M., Kleerebezem, R., van Loosdrecht, M., Kruit, J., Jetten, M.S.M., Karta, B.: Nitrogen removal by a nitritation-Anammox bioreactor at low temperature. Appl. Environ. Microbiol. **79**(8), 2807–2812 (2013)
27. Dapena-Mora, A., Campos, J.L., Mosquera-Corral, A., Jetten, M.S., Mendez, R.: Stability of the Anammox process in a gas-lift reactor and a SBR. J. Biotechnol. **110**(2), 159–170 (2004a)
28. Suneethi, S., Joseph, K.: Anammox process start up and stabilization with an anaerobic seed in Anaerobic Membrane Bioreactor (AnMBR). Bioresour. Technol. **102**(19), 8860–8867 (2011)

29. Zhang, B., Zhao, J., Zuo, J., Shi, X., Gong, J., Ren, H.: Realizing stable operation of anaerobic ammonia oxidation at low temperatures treating low strength synthetic wastewater. J. Environ. Sci. **75**, 193–200 (2019)
30. Laureni, M., Weissbrodt, D.G., Szivak, I., Robin, O., Nielsen, J.L., Morgenroth, E., Joss, A.: Activity and growth of Anammox biomass on aerobically pre-treated municipal wastewater. Water Res. **80**, 325–336 (2015)
31. Lu, Y.F., Ma, L.J., Ma, L., Shan, B., Chang, J.J.: Improvement of start-up and nitrogen removal of the Anammox process in reactors inoculated with conventional activated sludge using biofilm carrier materials. Environ. Technol. **39**(1), 59–67 (2018)
32. Zainab, A., Meraj, S., Liaquat, R.: Study on natural organic materials as biofilm carriers for the optimization of anaerobic digestion. Waste Biomass. **11**, 2521–2531 (2020)
33. Jetten, M.S.M., van Niftrik, L., Strous, M., Kartal, B., Keltjens, J.T., Op den Camp, H.J.M.: Biochemistry and molecular biology of Anammox bacteria. Biochem. Mol. Biol. **44**(2–3), 65–84 (2009)
34. Lackner, S., Gilbert, E.M., Vlaeminck, S.E., Joss, A., Horn, H., Van Loosdrecht, M.C.M.: Full-scale partial nitritation/Anammox experiences – An application survey. Water Res. **55**, 292–303 (2014)
35. Lackner, S., Horn, H.: Comparing the performance and operation stability of an SBR and MBBR for single-stage nitritation-Anammox treating wastewater with high organic load. Environ. Technol. **34**(10), 1319–1328 (2013)

Development of Compact Design Formulae for Aperture-Coupled MPA for Satellite Communication Applications

Amit Ghosh, Swarnadipto Ghosh, Dipankar Saha, Samik Chakraborty, Sunday Cookey Ekpo, and Stephen Alabi

1 Introduction

An aperture-coupled microstrip antenna [1] (Fig. 1a) consists of a rectangular patch of dimensions $a \times b$ printed on a substrate of thickness h and dielectric constant Єra. A microstrip line feeds the microstrip patch through an aperture or slot in the common ground plane. The aperture is of dimensions $la \times lw$ and centred at (Xo, Yo). The width of the microstrip line is W, and it is printed on a substrate described by thickness t and dielectric constant Єrf. The characteristic impedance of the microstrip line is denoted by Zom and that of the slot line corresponding to the coupling slot by Zos. Coupling of the slot to the dominant mode [2] of the patch and the microstrip line occurs because the slot interrupts the longitudinal current flow in them.

The coupling slot is nearly centred with respect to the patch, where the magnetic field of the patch is maximum. This is done to enhance coupling between the magnetic field of the patch and the equivalent magnetic current near the slot. The coupling amplitude can be determined [3] from the following expression:

A. Ghosh (✉) · S. Chakraborty
Department of Electronics and Communication Engineering, Regent Education and Research Foundation, Kolkata, India

S. Ghosh · D. Saha
Department of Avionics, Indian Institute of Space Science and Technology, Thiruvananthapuram, India

S. C. Ekpo · S. Alabi
Manchester Metropolitan University Manchester, Manchester, UK

S. C. Ekpo (ed.), *The Second International Adaptive and Sustainable Science, Engineering and Technology Conference*, Signals and Communication Technology,
https://doi.org/10.1007/978-3-031-53935-0_5

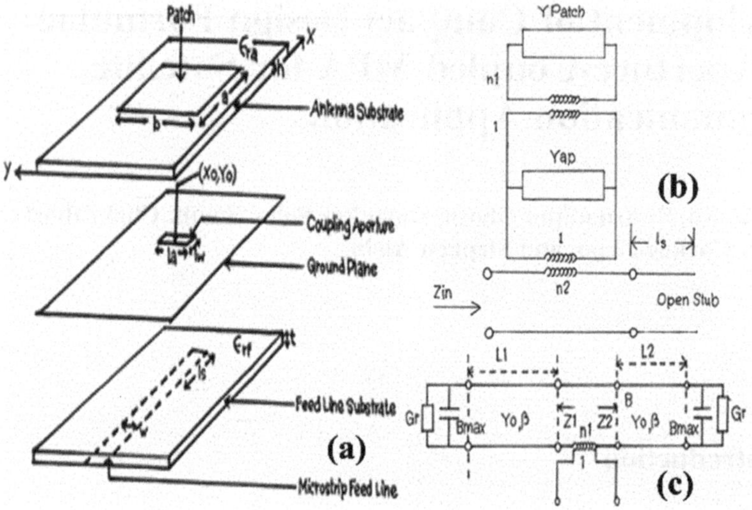

Fig. 1 (**a**) Expanded view of an aperture-coupled microstrip antenna. (**b**) Equivalent circuit of an aperture-coupled microstrip antenna. (**c**) Transmission line model of a rectangular patch fed by a slot in the ground plane

$$\text{Coupling} \cong \iiint (M.H)dv \cong \sin\left(\frac{\pi x_0}{l_a}\right) \tag{1}$$

where x_0 is the offset of the slot from the patch edge.

The biggest problem encountered by engineers while working with aperture-coupled microstrip antennas is that no closed-form design formula is available. The paper presents closed-form design expressions for the optimal design of the feed when the patch is designed according to the relations readily available.

2 Theory of Structural Analysis

In a simplified equivalent circuit of an aperture-coupled microstrip antenna (Fig. 1b, c), the patch is characterized by admittance, Y_{patch}, and the aperture by an admittance, Y_{ap}. The patch admittance is determined at the centre of the slot.

In feed configuration, the patch antenna appears in series with the feed because of slot coupling. The non-resonant slot is represented as an inductor in series with the R-L-C network representing the patch resonator. The open circuited microstrip stub of length l_s can be replaced by a shunt capacitor C_s such that $1/\omega C_s = Z_o \cot(\beta l_s)$, where Z_o is the characteristic impedance and β is the propagation constant of the microstrip feed line.

The coupling of the patch to the aperture is described by an impedance transformer of turns ratio $n_1 = l_a/b$. If Z_1 and Z_2 are the impedances looking toward the left and right of the aperture (Fig. 1c), then:

$$Y_1 = Y_0\left[\frac{(G_r + jB_{\text{open}}) + jY_0\tan(\beta L_1)}{Y_0 + j(G_r + jB_{\text{open}})\tan(\beta L_1)}\right] \tag{2}$$

where $L_1 = X_0$;

$$Y_2 = Y_0\left[\frac{(G_r + jB_{\text{open}}) + jY_0\tan(\beta L_2)}{Y_0 + j(G_r + jB_{\text{open}})\tan(\beta L_2)}\right] \tag{3}$$

where $L_2 = a - L_1$

Here (Y_o, β) characterizes the rectangular patch antenna as a microstrip line of width b, and $(G_r + jB_{\text{open}})$ is the edge admittance of the patch. The value of Y_{ap} can be obtained from the transmission line model of a slot and is given by:

$$Y_{\text{ap}} = -j2Y_{\text{os}}\cot\left(\frac{\beta_s l_a}{2}\right) \tag{4}$$

A transformer of turns ratio n_2, used to describe the coupling of the patch to the microstrip feed line, is modelled from the discontinuity ΔV in modal voltage of the feed microstrip line, that is, $n_2 = \Delta V/V_o$ where V_o is the slot voltage. Thus:

$$n_2 = \left[\frac{\left\{J_0\left(\frac{\beta_s w}{2}\right)J_0\left(\frac{\beta_m l_w}{2}\right)\right\}}{\beta_s^2\beta_m^2}\right] * \left[\frac{\beta_m^2 k_2 C_{rf}}{\{(k_2C_{rf}\cos(k_1h) - k_1\sin(k_1h))\} + \frac{\beta_m^2 k_1}{\{k_1\cos(k_1h)+k_2\sin(k_1h)\}}}\right] \tag{5}$$

where $J_0(.)$ is the zeroeth order Bessel function;

$k_1 = k_0\sqrt{C_{\text{rf}} - C_{\text{res}} - C_{\text{rem}}}$; $k_2 = k_0\sqrt{(|C_{\text{rf}} + C_{\text{rem}} - 1|)}$ and $\beta_s = k_0\sqrt{\epsilon_{\text{res}}}$ and $\beta_m = k_0\sqrt{\epsilon_{\text{rem}}}$. It can be said that $Z_{\text{patch}} = Z_1 + Z_2 = \frac{1}{Y_1} + \frac{1}{Y_2}$.

Here (w, ϵ_{rem}, β_m, Z_{om}) are microstrip line parameters and (l_w, ϵ_{res}, β_s, Y_{os}) are the corresponding slot line parameters. Taking into account the impedance of the microstrip open-circuited stub of length l_s, the input impedance [5] of the antenna at the centre of the slot is given by the following:

$$Z_{\text{in}} = \frac{n_2^2}{\left(n^2 Y_{\text{patch}} + Y_{\text{ap}}\right)} - jZ_{\text{om}} \cot(\beta_m l_s) \tag{6}$$

Setting $n_1^2 B_{\text{patch}}$ + $B_{\text{ap}} = 0$ for resonance and using Eqs. (2) and (3) yield the following condition for resonance:

$$B_{\text{patch}} = \frac{-B_{\text{ap}}}{n^2} \approx \frac{4b^2}{Z_{\text{os}} \beta_s l_a^3} \tag{7}$$

Therefore increasing l_a results in a decrease in B_{patch} and consequently the resonant frequency decreases.

3 Development of Closed-Form Design Formulae

In the analysis, series expansion of the Bessel function [4] and [6] of first kind was taken as follows:

$$J_n(x) = \frac{1}{2^n \Gamma(n+1)} \sum_{r=0}^{\alpha} (-1)^r \frac{x^{n+2r}}{2^{2r} r!(n+1)(n+2)\ldots(n+r)} \tag{8}$$

where $r = 0, 1, 2, 3, 4\ldots.$ and terms involving third or higher degrees of the argument were neglected.

Further for the inverse cotangent function, the first six terms of the corresponding Maclaurin series [7] are considered as follows:

$$\cot^{-1} z = \frac{\pi}{2} - z + \frac{z^3}{3} - \frac{z^5}{5} + \ldots \tag{9}$$

This, according to Eqs. (2)–(7), gives the following:

$$\left(\frac{l_a}{b}\right)^2 Y_{\text{patch}} - \frac{2j\cot\left(\frac{\beta_s l_a}{2}\right)}{A} = \frac{\left[\left(1 - \frac{\beta_s^2 w^2}{16} - \frac{\beta_m^2 l_w^2}{16}\right)^2 (X_1 + X_2)^2\right]}{Z_{\text{in}}} \tag{10}$$

where $X_1 = \frac{\beta_m^2 k_2 \epsilon_{rf}}{[k_2 \epsilon_{rf} \cos(k_1 h) - k_1 \sin(k_1 h)]}$ and $X_2 = \frac{\beta_s^2 k_1}{[k_1 \cos(k_1 h) + k_2 \sin(k_2 h)]}$.

The expressions for the characteristic impedance [8] and guide wavelength of a slot line on a substrate of low ϵ_r have been obtained by curve fitting of the numerical results obtained using Galerkin's method in the Fourier transform domain. These expressions are for $0.075 \leq w/\lambda_0 \leq 1.0$ and $2.22 \leq \epsilon_r \leq 3.8$.

$$\frac{\lambda_s}{\lambda_0} = 1.194 - 0.24 \ln \epsilon_r - \left[\frac{0.62\epsilon_r^{0.835}\left(\frac{w}{\lambda_0}\right)^{0.48}}{(1.344 + \frac{w}{h})} \right] - \left[0.0617 \left\{ 1.91 - \frac{(\epsilon_r + 2)}{\epsilon_r} \right\} \ln\left(\frac{h}{\lambda_0}\right) \right] \tag{11}$$

average error = 0.69%, max error = −2.6% (evaluated at two points, for $w/\lambda_0 > 0.8$), and

$$Z_{os} = 133 + 10.34(\epsilon_r - 1.8)^2 + 2.87\left\{2.96 + (\epsilon_r - 1.582)^2\right\} \times \sqrt{\left(\frac{l_w}{h} + 2.32\epsilon_r - 0.56\right)\left\{(32.5 - 6.67\epsilon_r)\left(100\frac{h}{\lambda_0}\right)^2 - 1\right\}} - \left(684.45\frac{h}{\lambda_0}\right)(\epsilon_r - 1.35)^2 + 13.23\left\{(\epsilon_r - 1.722)\frac{l_w}{\lambda_0}\right\}^2 \tag{12}$$

average error = 1.9%, $|\text{max error}| = 5.4\%$ (evaluated at three points, for $w/\lambda_0 > 0.8$). Therefore,

$$Z_{\text{in}} = \frac{\left[\left(1 - \frac{\beta_s^2 w^2}{16} - \frac{\beta_m^2 l_w^2}{16}\right)(X_1 + X_2)\right]^2}{\left(\frac{l_a}{b}\right)^2 Y_{\text{patch}} - \frac{j^2 \cot\left(\frac{\beta_s l_a}{2}\right)}{Z_{\text{os}}}} \tag{13}$$

Equating real and imaginary parts of Z_{in} as given by Eq. (13) to those of the characteristic impedance of the feed line, required design equations are obtained as follows:

$$l_a = \frac{(2n + 1)\pi}{\beta_s} \tag{14}$$

$$l_w = \frac{4}{\beta_m}\left\{\frac{\left(1 - \left(\frac{\beta_s w}{4}\right)^2\right)}{B}\right\}^2 \tag{15}$$

where Y and Z_{om} are the values of the real and imaginary parts of the microstrip feed line characteristic impedance and the other parameters are as defined earlier.

$$\text{Here, } B = \frac{\left(\frac{l_a}{b}\right)}{(X_1 + X_2)} \sqrt{Y_{\text{patch}} * Z_{\text{in}}} \quad \text{and} \quad l_s = \frac{1}{\beta_m \left(\cot^{-1}\left(\frac{Y}{Z_{\text{om}}}\right)\right)}$$

where Y and Z_{om} are the values of the real and imaginary parts of the microstrip feed line characteristic impedance and the other parameters are as defined earlier.

Thus, a set of closed-form design equations is obtained for the aperture length (l_a), width (l_w), and corresponding stub length (l_s) of the aperture-coupled microstrip patch antenna.

4 Validation of Design Procedure

In order to check the validity of the design equations developed, an aperture-coupled rectangular microstrip patch antenna was designed for operation at K_u band with the following parameters: for antenna substrate, thickness = 1.5875 mm, $\epsilon_r = 2.4$; and for feed line substrate, thickness = 0.79375 mm, $\epsilon_r = 2.4$. The corresponding resonant length and patch width are 3.70 mm and 5.60 mm, respectively.

Computer simulation using IE3D software package [9] was run for the designed antenna, which yielded a – 9.5 dB (VSWR < 2:1) bandwidth of 1.853 GHz and undistorted radiation pattern providing half power beam width of 79.284 degrees. The simulated plots for return loss vs. frequency and radiation pattern at resonance are shown in Fig. 2a, b, respectively.

To extend the validation, further investigations were carried out in an array environment. A 4 × 4 uniform array was investigated at a frequency band centred around 18.5 GHz with inter-element spacing of 12 mm along both orthogonal

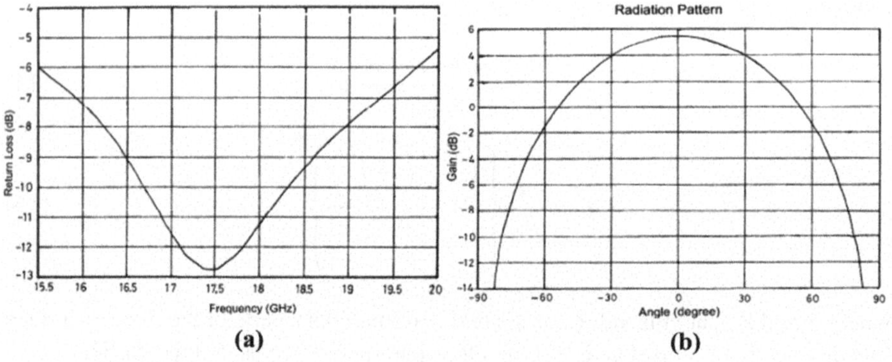

Fig. 2 (**a**) Return loss vs. frequency plot for single aperture-coupled patch (simulated). (**b**) Radiation pattern for single aperture-coupled patch (simulated)

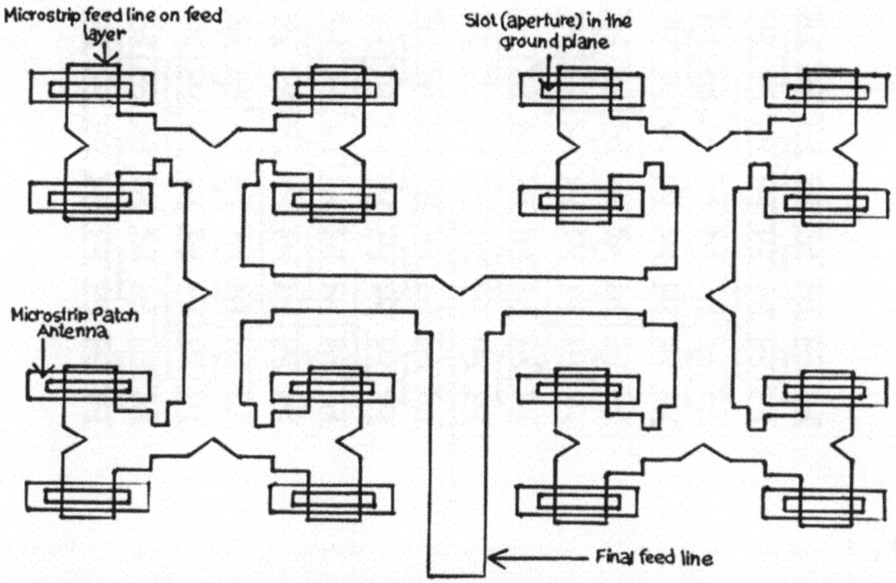

Fig. 3 4 × 4 array of aperture-coupled microstrip patches

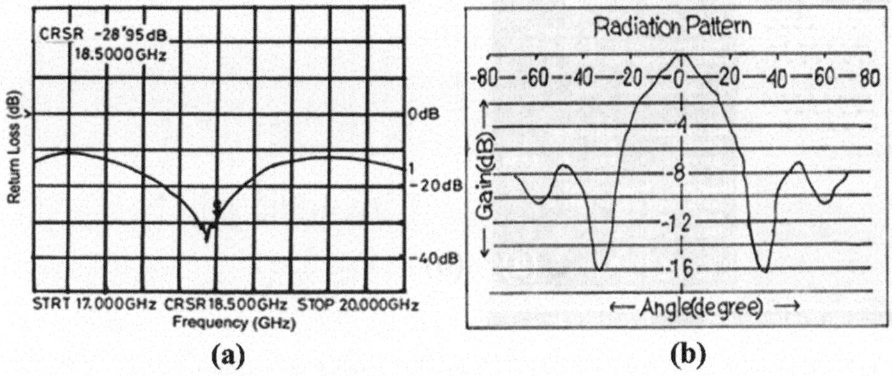

Fig. 4 (**a**) Return loss vs. frequency plot for aperture-coupled microstrip array (experimental). (**b**) Radiation pattern for aperture-coupled microstrip array (experimental)

directions (Fig. 3). The individual antennas had configuration as mentioned earlier, obtained by using the design procedure developed. Since the memory requirement for computer simulation of such a large structure is prohibitively large, the entire array was fabricated on a single substrate and tested experimentally. For return loss (Fig. 4a), measured values show very good impedance matching with the feed indicated by a less than −10 dB return loss over 17–20 GHz. The measured radiation pattern at resonance (Fig. 4b) expectedly provides half power beam width of about 25 degrees and side lobe level of −8.5 dB (Figs. 5 and 6).

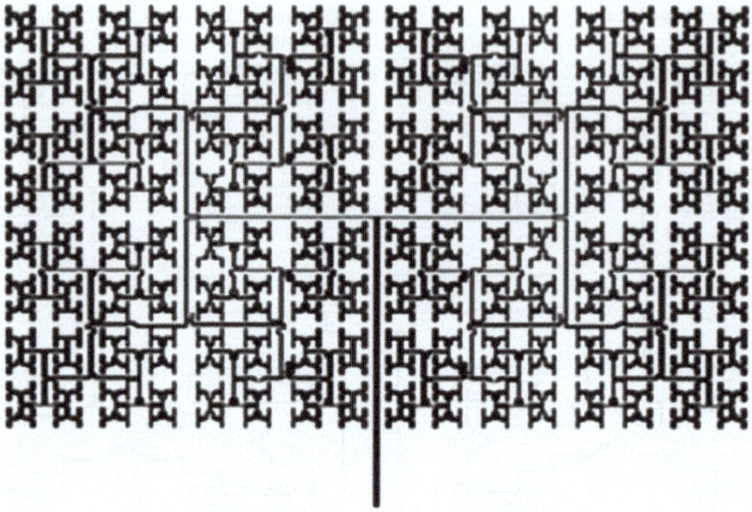

Fig. 5 Schematic drawing of 32 × 32 array with feed line (substrate: $\epsilon_r = 10.2$; height = 1/40 inch)

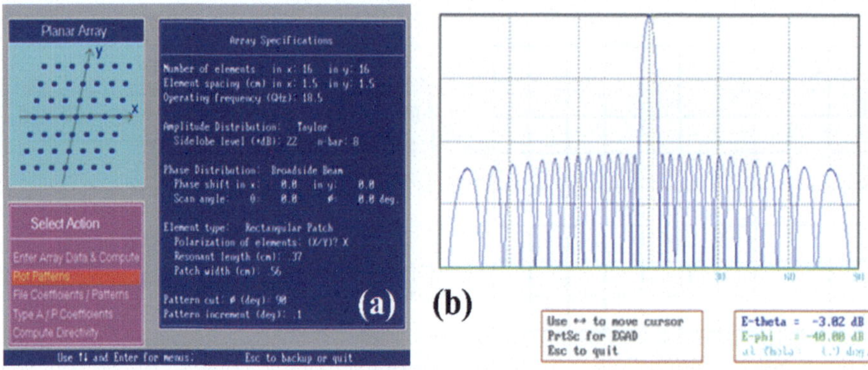

Fig. 6 (**a**) Simulation parameters of 32 × 32 array. (**b**) Elevation plane radiation pattern of 32 × 32 aperture-coupled microstrip patch antenna array

To achieve the desired satellite communication, the goal is to design, fabricate, and test an ultra-low side lobe-level microstrip antenna array to operate with the frequency range of 17.5–19.5 GHz with VSWR: 2:1 and side lobe level of −22 dB minimum with a beam width of less than 4 degree. To achieve the goal 32 × 32 planar microstrip patch antenna array, with the same aperture-coupled microstrip patch antenna explained above excited with Taylor series, current distribution and isometric phases are designed with feed line using substrate of $\epsilon_r = 10.2$ and height = 1/40 inch. And results are given below.

5 Conclusion

A set of closed-form design formulae is presented for the feed to an aperture-coupled microstrip antenna which is a novel approach in the literature on microstrip antenna. Its validity is checked through simulation of a single patch as well as fabrication and measurement in an array environment for application in SAT-COM. It is established that usage of the relations developed results in wideband performance with very low return loss, there by indicating a good design from practical point of view.

References

1. Garg, R., Bhartia, P., Bahl, I., Ittipiboon, A.: Microstrip Antenna Hand Book, pp. 28–29. Artech House, Boston (2001)
2. Das, B.N., Joshi, K.K.: Impedance of a radiating slot in the ground plane of a microstrip line. IEEE Trans. Antennas Propag. **AP-30**, 922–926 (1982)
3. Pozar, D.M.: Microstrip antennas. Proc. IEEE. **80**, 79–91 (1992)
4. Ryder, J.D.: Networks, Lines and Fields, p. 264. Prentice Hall, New Delhi (1998)
5. Garg, R., Bhartia, P., Bahl, I., Ittipiboon, A.: Microstrip Antenna Hand Book, p. 542. Artech House, Boston (2001)
6. Rajput, B.S.: Mathematical Physics, p. 454. Pragati Prakashan, Meerut (1996)
7. http://mathworld.wolfram.com/inversecotangent.htm
8. Janaswamy, R., Shaubert, D.H.: Characteristic impedance of a wide slotline on low permittivity substrates. IEEE Trans. Microw. Theory Tech. MTT. **34**, 900–902 (1986)
9. IE3D, release 10.1, Zeland Software Inc, Fremont, CA 94538, USA

A Comparative Analysis of Different Polygon-Shaped Slotted Anisotropic Unit Cell for Holographic Metasurface Application

Swarnadipto Ghosh, Dipankar Saha, Ayona Chakraborty, Samik Chakraborty, Sunday Cookey Ekpo, and Fanuel Elias

1 Introduction

Holographic beamformer metasurface (HBM) and leaky wave antennas (LWAs) are very popular and suitable candidates now in the domain of multi-user wireless communications [1], wideband beam-scanning applications [2], frequency scanning with multiple polarizations [3], holographic radar, space applications [4–8], and 5G communications. HBM and LWAs have structural compactness, low power consumption, and light weight, in addition to their unique ability to beam form with high gain and narrow beam width. To enable integrated satellite-cellular fifth-generation (5G)/6G wireless intelligent information networks, massive multiple-input multiple-output (mMIMO) RF front-end subsystems are expected to provide us the revolutionary mobile connectivity, extremely high-throughput data service, and energy efficient, wide coverage, and low-cost operations. mMIMO will utilize metamaterials and reconfigurable intelligent metasurfaces to accomplish energy-efficient networks with simultaneous wireless data transmission and autonomous RF energy harvesting capabilities. The inherent limitations of the phased array systems (such as the requirement of numerous phase shifters, power amplifiers, complex phase shifting circuits) make them unable to meet these challenging demands of the 6G networks for beamforming [4]. For the design considerations of the holographic surface, the effective surface impedance of each unit cell needs to

S. Ghosh (✉) · D. Saha
Indian Institute of Space Science and Technology, Thiruvananthapuram, India

A. Chakraborty · S. Chakraborty
Jadavpur University, Kolkata, India

S. C. Ekpo · F. Elias
Manchester Metropolitan University, Manchester, UK

S. C. Ekpo (ed.), *The Second International Adaptive and Sustainable Science, Engineering and Technology Conference*, Signals and Communication Technology,
https://doi.org/10.1007/978-3-031-53935-0_6

be modeled to get directive beamforming. To resolve the issue, a systematic study has been provided in this paper to get the desired Z_surf as per the phase angle of the beam. A comparative study and equivalent closed-form mathematical relationship for the different polygon-shaped anisotropic unit cells are presented in the following section.

2 Mathematical Analysis of Z_surf

In a holographic or impedance-modulated metasurface design, the surface current distribution (*J*_surf) of the meta unit cell needs to be characterized. To characterize the effective *J*_surf of an anisotropic meta unit cell, the effective surface impedance (Z_surf) needs to be extracted. Here, a systematic way to extract the Z_surf for a particular anisotropic unit cell is presented.

By applying the periodic boundary conditions and the Floquet port solutions, the equivalent Z_{sheet} has been extracted with the help of different EM solvers. By applying the transverse resonance method to solve the equivalent Z_{surf}, we have thus:

$$\frac{1}{Z_{\text{sheet}}} = \frac{1}{Z_{\text{in}}} - \frac{1}{j\eta_1 \tan(k_1 d)} = \frac{1}{Z_{\text{in}}} - \frac{1}{j\frac{\epsilon_0}{\sqrt{\epsilon_{r1}}} \tan\left(k_0\sqrt{\epsilon_{r1}} d\right)} \tag{1}$$

From Fig. 1, to compute the transcendental Eq. (1) by using the transmission line concept and the transverse resonance method, it can be stated that:

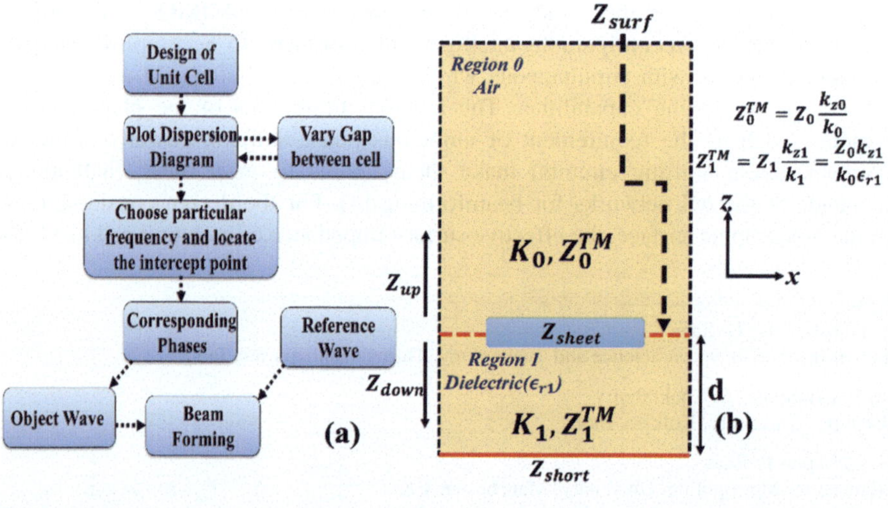

Fig. 1 (**a**) Block diagram representation of systematic study of Z_{surf}. (**b**) Schematic diagram for equivalent transverse resonance method solution

$$\frac{1}{Z_{\text{up}}(x)} + \frac{1}{Z_{\text{down}}(x)} = 0 \tag{2}$$

Incorporating the transverse Eqs. (2) in (1) yields thus:

$$\frac{1}{Z_{\text{surf}}} = \frac{1}{Z_{\text{sheet}}} + \frac{1}{j\frac{Z_0 k_{x1}}{k_0 \epsilon_{r1}} \tan(k_{x1} d)} \tag{3}$$

where

$$k_{x1} = \sqrt{k_0^2(\epsilon_r - 1) + \left(\frac{Z_{\text{surf}} k_0}{Z_0}\right)^2} \tag{4}$$

To extract the Z_{surf} from (3), polynomial numerical techniques need to be implemented. From an EM simulation, we can extract the Z_{sheet}. The extracted Z_{surf} is now utilized for designing the holographic metasurface through the following equation thus:

$$Z_{\text{surf}}(x_t) = j\left[X + M\,\text{Re}\left(\psi_{\text{rad}}\psi^*_{\text{surf}}\right)\right] \tag{5}$$

where $\psi_{\text{surf}} = e^{-jk_0 nr}$ is the surface wave vector component generated by the source antenna; $\psi_{\text{rad}} = e^{-jk_0 x \sin(\theta_L)}$, the corresponding incident target wave vector component; r, the distance (m) from the source antenna; and θ_L (deg), the angle deviated from the normal direction of the plane wave. The interference of the surface wave (ψ_{surf}) and the object wave (ψ_{rad}) is realized by the inner product ($\langle \psi_{\text{rad}}, \psi_{\text{surf}} \rangle = \psi_{\text{rad}}\psi^*_{\text{surf}}$) between the two complex vectors, where ψ^*_{surf} is the Hermition format of surface wave vector [2]. Due to the illumination of interference pattern by the reference wave (Eq. 6), the original wave pattern has been reconstructed thus:

$$\left(\psi_{\text{rad}}\psi^*_{\text{surf}}\right)\psi_{\text{surf}} = \psi_{\text{rad}}\|\psi_{\text{surf}}\|^2 \tag{6}$$

In (5), X usually sets the average impedance value for the geometry of the surface where M defines the entire available impedance range as the modulation depth.

3 Mathematical Modeling for Each Polygon Architecture

In this paper, the single- and the cross-slotted circular, square, and hexagonal unit cells have been analyzed to extract the Z_{surf} by expressing the equivalent J_{surf} (which is extracted from a perfect boundary (PB) and Floquet port simulation).

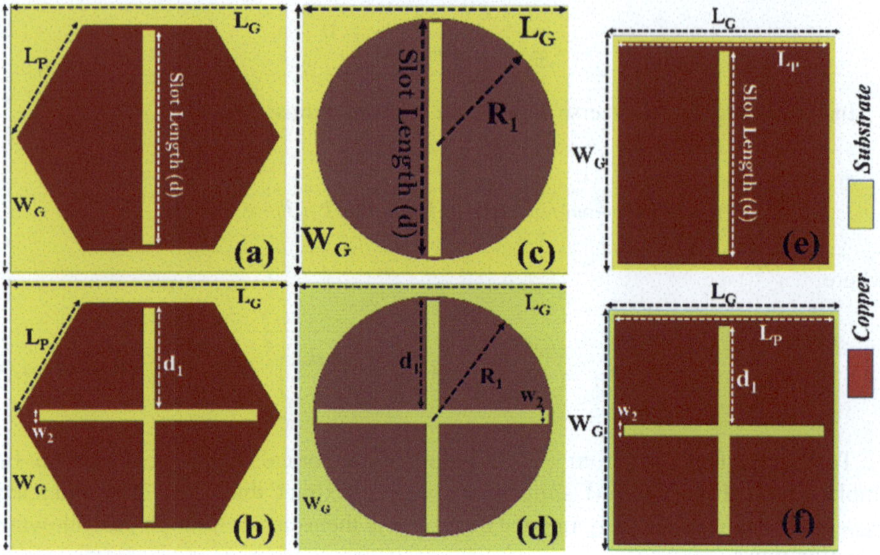

Fig. 2 Schematic of hexagonal unit cell, (**a**) single and (**b**) cross slotted; circular unit cell, (**c**) single and (**d**) cross slotted; square unit cell, (**e**) single and (**f**) cross slotted

Figure 2 expresses the single- and the cross-slotted hexagonal, circular, and square meta unit cells. To express the comparative analysis of the surface impedance (Z_{surf}) variation with respect to the parameter optimization of the slots for different polygon-shaped unit cells, a comparative 2D plot for Z_{surf} vs slot length (d) and angular rotation (θ) are presented (Figs.1 and 2). A detailed analysis of the Z_{surf} variation is discussed in the following section.

3.1 Single- and Cross-Slotted Meta Unit Cell wrt. Slot Length (d) Variation

To make a comparative study between the surface impedance variations for each type of unit meta-atom structure, a surface impedance vs slot length (d) variation plot is presented (Fig. 3). For the single- and cross-slotted circular and hexagonal meta-atoms, the surface impedance (Im(z)) has a range of 225–425 Ω, whereas both the single- and cross-slotted square meta-atoms have the surface impedance within a range spanning -800 Ω to -100 Ω.

From the surface impedance vs slot length (d) plot (Fig. 3), the surface impedances for both single- and cross-slotted square meta-atoms exhibit highly capacitive reactance. By increasing the slot length, the effect of capacitive reactance is gradually suppressed by magnetic coupling between the cross slots. The suppression of capacitive reactance through magnetic coupling is distributed through a higher-order

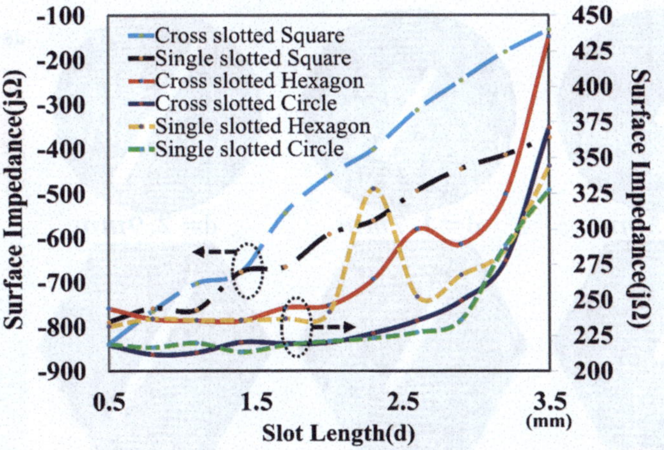

Fig. 3 Surface impedance vs slot length (d) for single- and cross-slotted circular, hexagonal, and square meta-atom for particular angular rotation of 30°

polynomial over the slot length (d). The surface impedance growth (Fig. 3) for the cross-slotted hexagonal meta-atom structure has much rapidness rather than single-slotted hexagonal meta-atoms over the increment of slot length (d). Similarly, the cross-slotted circular meta-atom has a higher impedance growth than the single-slotted circular meta-atom. By incorporating the single and cross slots upon the meta-atom surface, the diversification of impedance has been clearly visible from the surface current distribution plots (Fig. 4). Both the single- and the cross-slotted circular, hexagonal, and square meta-atom surfaces have been analyzed (Figs. 4 and 5). The single-slotted square meta-atom structure exhibits a highly capacitive reactance for the lower slot length (d) (Fig. 4). The current distribution plots for circular, hexagonal, and square meta-atoms are giving a strong explanation for the surface impedance (Im(z)) vs slot length (d) plot. According to Ohm's law (electrostatics), the surface impedance has a direct proportional effect on the surface current density (J_{surf}). For lower slot dimensions, the square-shaped meta-atom exhibits a higher surface current distribution (J_{surf}) due to a high capacitive reactance (jX_c). For the circular and the hexagonal meta-atoms, a gradual increment of the slot dimension yields a rapid increment of the surface impedance. This causes the structure to exhibit a higher surface current density gradually. Also, the hexagonal meta-atom has a denser current density through the sharp edges for the higher slot dimensions. The circular meta-atom makes a smooth-edged surface current density (J_{surf}) to reduce the probability of leaky current density through sharp edges.

Similarly, the current distribution plots for cross-slotted circular, hexagonal, and square meta-atoms (Fig. 5) exhibit a strong correlation with the surface impedance (Im(z)) vs slot length (d) plot. Hence, the surface impedance plays a vital role for the current distribution on the surface. In comparison with the single-slotted meta-atom structures, the cross-slotted meta-atom structures exhibit a surface current declassification by integrating the cross-slotted arrangement. Due to the cross-slot

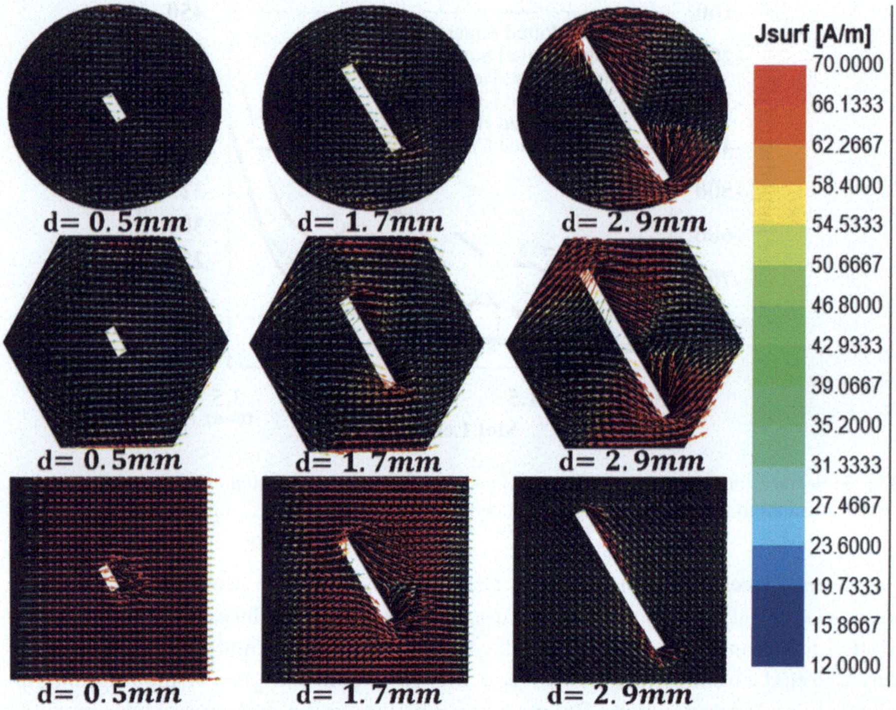

Fig. 4 Current distribution of single-slotted circular, hexagonal, and square meta-atom for different slot length (d) at angular rotation (θ) = 30°

arrangement, for both circular and hexagonal meta-atoms, the surface current density has been increased along the edge of orthogonally arranged cross slots. Some of the current basis vectors are also mutually coupled to each other. Similarly, we have encountered the reverse behavior of the surface current distribution for the cross-slotted square meta-atoms. But, due to decoupling more, the rate of increment of the capacitive-type surface impedance nature is more rapid than the single-slotted square meta-atom.

3.2 Single- and Cross-Slotted Meta Unit Cell wrt. Angular Rotation (θ) Variation

This section contains an analysis of the surface impedance with respect to the variation of the angular rotation of the center aligned cross slot for each meta-atom structures. According to the surface current distribution (Figs. 7 and 8) and for the surface impedance (Im(z)) vs angular rotation (θ) plot, the maximum field variation

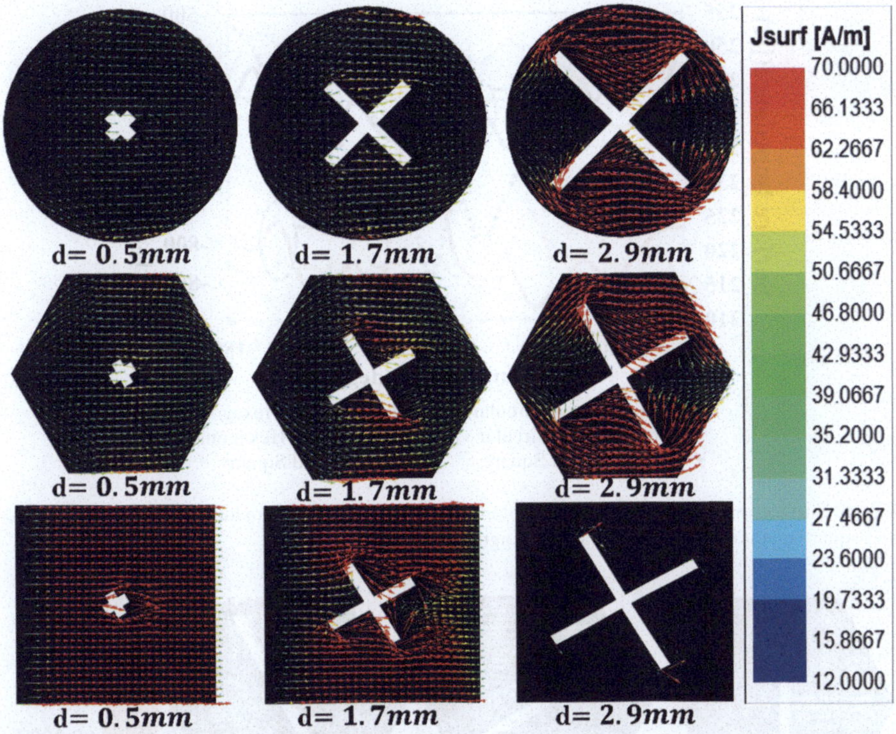

Fig. 5 Current distribution of cross-slotted circular, hexagonal, and square meta-atom for different slot length (d) at angular rotation (θ) = 30°

has occurred for slot length (d) = 2.9 mm. For a comparative study of each type of unit meta-atom structure, a surface impedance (Im(z)) vs angular rotation (θ) plot is presented (Fig. 6).

Figure 6 shows both the surface impedances of the cross- and the single-slotted square meta-atoms behave as capacitive reactances. The surface impedance for circular and hexagonal cross-slotted meta-atoms is behaving as an inductive reactance. The single-slotted square meta-atom has a significant variation of the surface impedance (Im(z)), whereas the cross-slotted square meta-atom has a less variation over the angular variation of slot. However, in the point of view of the current distribution (Fig. 7), the single-slotted meta-atom has created an impedance symmetry about the angular rotation (θ) = 90°. This impedance variation symmetry is the cause for the inverted symmetric current distribution when the slots rotated in a complementary angle. Due to higher current distribution throughout the hexagonal edges, single-slotted hexagonal meta-atom has a higher surface impedance through the edges which has been highly declassified by the rotation of the slots. By rotating the slots, less current vectors will be mutually coupled through the parallel edges of the slots. Due to the absence of sharp edges in the circular meta-atom, fewer current vectors are degenerating by inserting the slot than hexagonal meta-atom. Similarly,

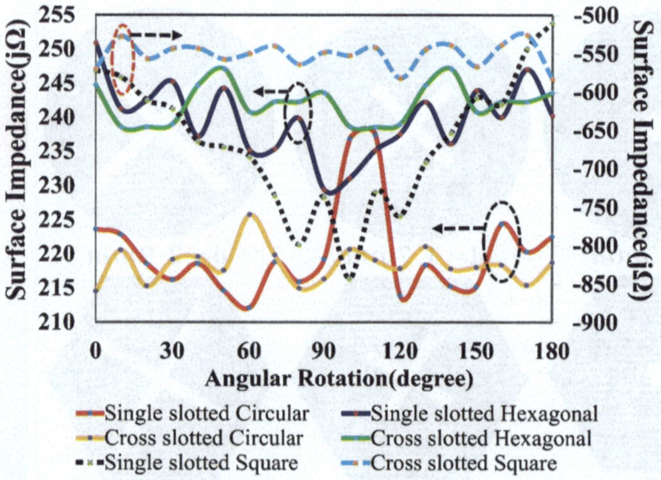

Fig. 6 Surface impedance vs angular rotation (θ) for single- and cross-slotted circular, hexagonal, and square meta-atom for particular slot length (d) of 2.9 mm

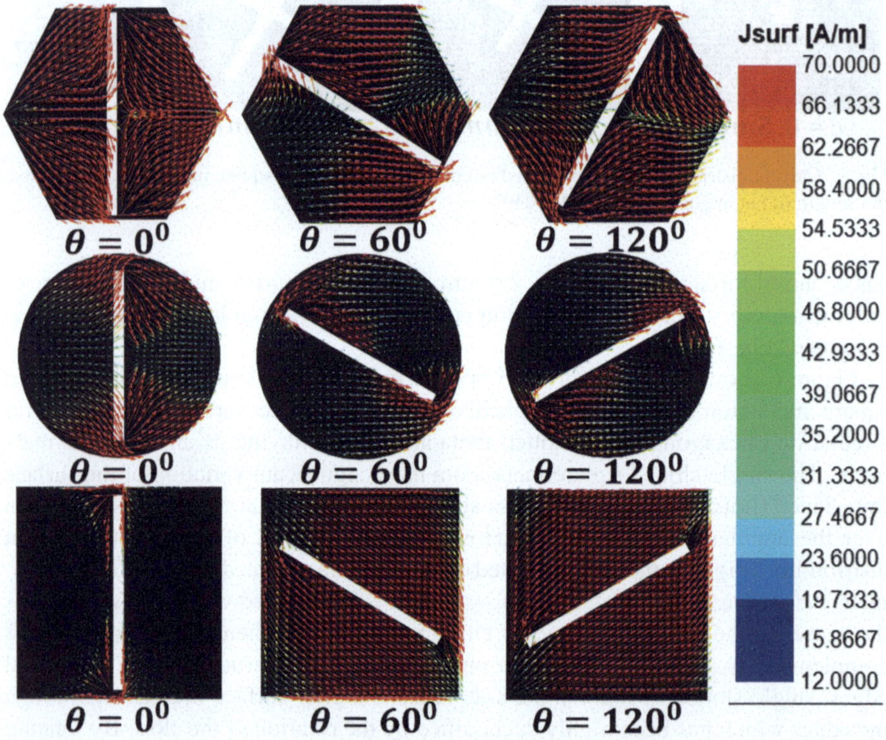

Fig. 7 Current distribution of single-slotted circular, hexagonal, and square meta-atom for different angular rotation at slot length (d) = 2.9 mm

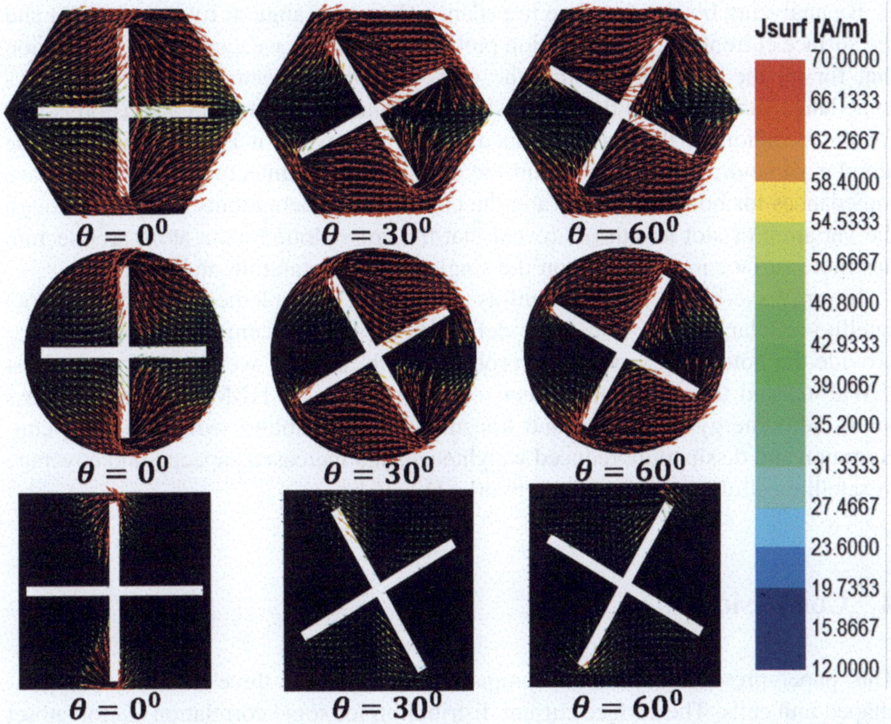

Fig. 8 Current distribution of cross-slotted circular, hexagonal, and square meta-atom for different angular rotation at slot length (d) = 2.9 mm

the cross-slotted meta-atoms presented in Fig. 8 can be analyzed by the correlation between the surface impedance vs angular rotation and the phase current distribution. Similarly, the surface impedance of the cross-slotted square meta-atom has a capacitive reactance response (ranging from 575 to 598 Ω), whereas the circular meta-atom has a low-inductive reactance response (ranging from 215 to 220 Ω). The hexagonal meta-atom has a higher inductive surface impedance (ranging from 245 to 250 Ω) than the cross-slotted circular surface impedance. For the cross-slotted square meta-atom, the range of impedance variation is very less over the angular orientation of the cross slots. That happens for the less amount of current distribution over the edges. For the different angular orientations of the cross-slotted circular and hexagonal meta-atoms, no huge variations of the surface impedances have been encountered. The less amount of variation can be perfectly correlated with the surface current distribution (Fig. 8). Though the hexagonal meta-atom has more surface impedances than the circular meta-atom, it is evident from both the surface impedance (Im(z)) vs angular rotation (θ) plot and the surface current distribution (J_{surf}) plot that the variation of impedance is very less achieved due to the particular slot length (d) = 2.9 mm.

By analyzing both the surface impedance (Im(z)) vs angular rotation (θ) plot and the surface current (J_{surf}) distribution plot (Figs. 7 and 8), we can make a conclusion that for all the unit meta-atoms, the cross-slotted architecture [9–11] has a less impedance variation than the single-slotted architecture. But this conclusion is true for the variation of the surface impedance (Im(z)) when it is with respect to the angular variation of the single- and the cross-slotted architectures. But the surface impedances for both the single- and the cross-slotted meta-atoms analyzed through the variation of slot length (d) reveal that the cross-slotted meta-atom architecture has more surface impedance than the single-slotted meta-atom architecture.

In mmWave-based vehicle-to-infrastructure (V2X) implementations and 5G/6G satellite-cellular networks, software-defined holographic beamforming metasurfaces provide the potential to overcome problems with the size, weight, and power-cost (SWaP-C) and thermal management requirements [5–8]. HBMs enhance wireless multi-radio energy harvesting and transmission by enabling reduced energy consumption and dissipation, reduced weight/size, and increased capacity and coverage in satellite-cellular convergence networks [12–15].

4 Conclusion

This paper presents a detailed comparative analysis of three different polygon-shaped unit cells. The surface current distribution for some correlation vector subset of the slot parameters has been portrayed to visualize the variation of imaginary Z_surf along with the anisotropic variation for each cell. Both the inductive and the capacitive surface impedances are validated. The Z_surf can be applied to design the impedance-modulated metasurface structure. Also, the single- and cross-slotted architecture can modulate the Z_surf. The presented research findings will support future-generation multi-site and multi-static holographic metasurfaces in wireless power transfer, energy harvesting, and holographic radar applications.

References

1. Liu, Y., Li, N., Jia, Y., Zhang, W., Zhou, Z.: Low RCS and high-gain patch antenna based on a holographic metasurface. IEEE Antennas Wirel. Propag. Lett. **18**(3), 492–496 (2019). https://doi.org/10.1109/LAWP.2019.2895117
2. Pandi, S., Balanis, C.A., Birtcher, C.R.: Design of scalar impedance holographic metasurfaces for antenna beam formation with desired polarization. IEEE Trans. Antennas Propag. **63**(7), 3016–3024 (2015). https://doi.org/10.1109/TAP.2015.2426832
3. Kampouridou, D., Feresidis, A.: Tunable multibeam holographic metasurface antenna. IEEE Antennas Wirel. Propag. Lett. **21**(11), 2264–2267 (2022). https://doi.org/10.1109/LAWP.2022.3192977
4. Sowande, O., Idachaba, F., Ekpo, S., Faruk, N., Uko, M., Ogunmodimu, O.: Sub-6 GHz 5G spectrum for satellite-cellular convergence broadband internet access in Nigeria. Int. Rev. Aerosp. Eng. J. **15**(2), 85 (2022)

5. Uko, M., Ekpo, S.: A 23-28 GHz pHEMT MMIC low-noise amplifier for satellite-cellular convergence applications. Int. Rev. Aerosp. Eng. J. **14**(5), 1–10 (2021)
6. Ekpo, S.C., Adebisi, B., Wells, A.: Regulated-element frost beamformer for vehicular multimedia sound enhancement and noise reduction applications. IEEE Access. **5**, 27254–27262 (2017)
7. Uko, M., Ekpo, S.: 8-12 GHz pHEMT MMIC low-noise amplifier for 5G and fiber-integrated satellite applications. Int. Rev. Aerosp. Eng. J. **13**(3), 99–107 (2020)
8. Ekpo, S.C.: Parametric system engineering analysis of capability-based small satellite missions. IEEE Syst. J. **13**(3), 3546–3555 (2019). https://doi.org/10.1109/JSYST.2019.2919526
9. Ghosh, S., Saha, D., Prahannathan, V., Saha, C.: Mathematical modelling of slotted circular meta-atom for holographic meta-surface applications. In: 2023 International Conference on Microwave, Optical, and Communication Engineering (ICMOCE), pp. 1–4. Bhubaneswar, India (2023). https://doi.org/10.1109/ICMOCE57812.2023.10167286
10. Chakraborty, A., Ghosh, S., Roy, S.K., Chakraborty, S., Chatterjee, S., Gupta, B.: Loaded multi loop hexagram shaped printed antennas for multifrequency communication. In: 2022 Microwave Mediterranean Symposium (MMS), pp. 1–6. Pizzo Calabro, Italy (2022). https://doi.org/10.1109/MMS55062.2022.9825536
11. Saha, D., Ghosh, S., Chakrabarti, M., Chattopadhyay, A.: Prism-shaped nonuniformly distributed slotted SIW array with highly reduced SLL for mm-wave ground surveillance application. In: 2023 First International Conference on Microwave, Antenna and Communication (MAC), pp. 1–5. Prayagraj, India (2023). https://doi.org/10.1109/MAC58191.2023.10177066
12. Zafar, M., Ekpo, S., George, J., Sheedy, P., Uko, M., Gibson, A.: Hybrid power divider and combiner for passive RFID tag wireless energy harvesting. IEEE Access. **10**, 502–515 (2022). https://doi.org/10.1109/ACCESS.2021.3138070
13. Lau, I., Ekpo, S.C., Zafar, M., Ijaz, M., Gibson, A.: Hybrid mmWave-Li-Fi 5G architecture for reconfigurable variable latency and data rate communications. IEEE Access. **11**, 42850–42861 (2023). https://doi.org/10.1109/ACCESS.2023.3270777
14. Ekpo, S.C., George, D.: Impact of noise figure on a satellite link performance. IEEE Commun. Lett. **15**(9), 977–979 (2011). https://doi.org/10.1109/LCOMM.2011.072011.111073
15. Ekpo, S., George, D.: A system engineering analysis of highly adaptive small satellites. IEEE Syst. J. **7**(4), 642–648 (2013). https://doi.org/10.1109/JSYST.2012.2198138

Theoretical Analysis and Design Equation of Rotated Dual-Split Elliptical Split Ring Resonator (RDS-ESRR)

Dipankar Saha, Swarnadipto Ghosh, Ayona Chakraborty, Samik Chakraborty, Sunday Cookey Ekpo, Fanuel Elias, and Stephen Alabi

1 Introduction

Split ring resonators (SRR) having various structural geometry have been extensively studied and implemented in different interesting electromagnetic and communication applications by the different research groups across the globe [1–5] since it was originally proposed by Pendry [1] in 1999. Most of the applications of SRRs reveal their usefulness to realize the practicality of double-negative (DNG) metamaterial and as one of the integral subsystems of narrow and wideband microwave filters, novel compact-phase shifter, etc. A unique advantage of using an SRR for such application is that as it is a sub-wavelength structure, the size of these structures is very compact at resonance [6]. For all such applications the proper calculation and formulation for estimating the resonant frequency of the SRR become very important [7–11].

This article presents an appropriate mathematical closed-form expression for the resonance frequency calculation of the elliptical and double-split elliptical SRRs (DS-ESRRs) and examined the effect of the relative angular rotation of the rings of the double-split elliptical SRR on its resonance.

The rest of the article is discussed as follows: Sect. 2 provides the theoretical background and calculation steps of the resonance frequency of the ESRR and DS-ESRR and the rotational effect of the rings of the SRR and Sect. 3 concludes the article.

D. Saha (✉) · S. Ghosh
Indian Institute of Space Science and Technology, Thiruvananthapuram, India

A. Chakraborty · S. Chakraborty
Jadavpur University, Kolkata, India

S. C. Ekpo · F. Elias · S. Alabi
Manchester Metropolitan University, Manchester, UK

S. C. Ekpo (ed.), *The Second International Adaptive and Sustainable Science, Engineering and Technology Conference*, Signals and Communication Technology,
https://doi.org/10.1007/978-3-031-53935-0_7

2 Theoretical Analysis of E-SRR and DS-ESRR

Figure 1a, b reveal the schematic diagrams of elliptical and double-split elliptical split ring resonators, respectively. The rings are designed on the top of a substrate having the relative permittivity, εr, and dielectric height of h_t backed by the ground metallic plate on the other side. The other design parameters are as follows: a_{ext} and b_{ext} are the semimajor and semiminor axes of the external elliptical ring, c is the ring width, d is the gap between the outer and inner elliptical ring, and g is the gap width of the two splits facing opposite to each other. In Fig. 1b the four split widths of double-split elliptical SRR are taken as $g1$, $g2$, $g3$, and $g4$ where we usually consider the $g1 = g2 = g3 = g4 = g$.

2.1 *Accurate Calculation of Equivalent Capacitance*

The proper calculation of the realizable equivalent capacitance is very important to estimate the resonance frequency of the SRR. To obtain the ring capacitance, we take C_{pul} times the average ring length ($\pi(a_{\text{avg}} + b_{\text{avg}})$), where C_{pul} is the distributed capacitance per unit length. C_{pul}, a_{avg}, b_{avg} can be expressed as follows (Fig. 2):

$$a_{\text{avg}} = a_{ext} - c - \frac{d}{2}, \quad b_{\text{avg}} = b_{\text{ext}} - c - \frac{d}{2} \tag{1}$$

And,

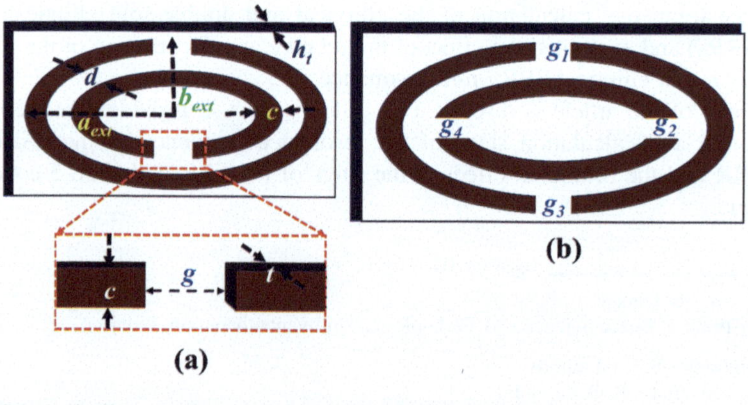

Fig. 1 Schematic diagram of the proposed (**a**) ESRR and (**b**) DS-ESRR

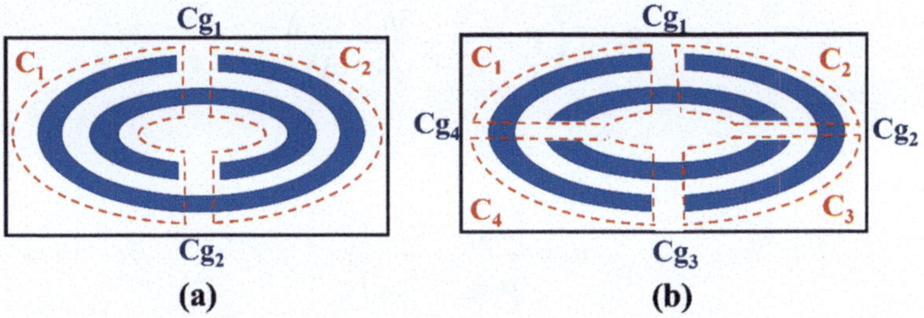

Fig. 2 (**a**) Visualization of the realizable capacitances in ESRR and (**b**) visualization of realizable capacitances in DS-ESRR

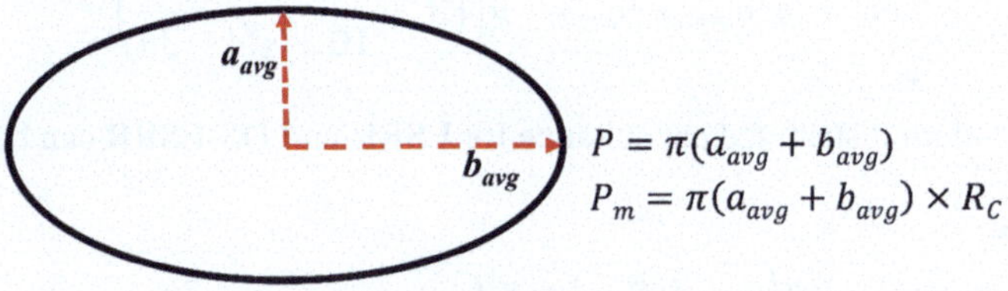

Fig. 3 Schematic of an elliptical geometry to estimate the perimeter

$$C_{\text{pul}} = \sqrt{\varepsilon_e}/c_0 Z_0 \tag{2}$$

Figure 3 shows a schematic diagram of an ellipse having the semimajor axis a_{avg} and semiminor axis b_{avg}.

The commonly used formula for estimating the perimeter of any ellipse is as follows:

$$P = \pi\left(a_{\text{avg}} + b_{\text{avg}}\right) \tag{3}$$

where P is the perimeter of the ellipse and a_{avg}, b_{avg} are the semimajor and semiminor axes of the ellipse.

But Eq. (1) is a round-off approximation formula for the perimeter calculation which may cause a significant error in high-frequency applications. As a result, Ramanujan's correction factor has been introduced along with it to calculate the perimeter more accurately using Eq. (2):

$$P_m = \pi\left(a_{\text{avg}} + b_{\text{avg}}\right) R_C \tag{4}$$

where the P_m and R_c are the modified perimeter formula and Ramanujan's correction factor, respectively.

$$R_C = \left(1 + \frac{3h}{10 + \sqrt{4 - 3h}}\right) \tag{5}$$

and

$$h = \frac{\left(a_{\text{avg}} - b_{\text{avg}}\right)^2}{\left(a_{\text{avg}} + b_{\text{avg}}\right)^2} \tag{6}$$

Hence,

$$P_m = \left[\pi\left(a_{\text{avg}} + b_{\text{avg}}\right) - g\right]\left(1 + \frac{3h}{10 + \sqrt{4 - 3h}}\right) \tag{7}$$

The equivalent distributed capacitance for ESRR and DS-ESRR can be expressed as follows:

$$C^{\text{ESRR}} = \frac{\left[\pi\left(a_{\text{avg}} + b_{\text{avg}}\right) - g\right]C_{\text{pul}}}{2}\left(1 + \frac{3h}{10 + \sqrt{4 - 3h}}\right) \tag{8}$$

and

$$C^{\text{DS}-\text{ESRR}} = \frac{\left[\pi\left(a_{\text{avg}} + b_{\text{avg}}\right) - g\right]C_{\text{pul}}}{4}\left(1 + \frac{3h}{10 + \sqrt{4 - 3h}}\right) \tag{9}$$

Hence, $C^{\text{DSESRR}} = C^{\text{ESRR}}/2$.

The gap capacitance (C_g) experienced between the splits can be derived as follows:

$$C_g = \frac{\varepsilon_0 t}{g} \tag{10}$$

Hence, the equivalent capacitance expression for ESRR:

$$C_{\text{eq}}^{\text{ESRR}} = \frac{\left(C^{\text{ESRR}} + C_g\right)}{2} \tag{11}$$

The equivalent capacitance expression for DS-ESRR can be written as follows:

$$C_{\text{eq}}^{\text{DS}-\text{ESRR}} = \frac{\left(C^{\text{DS}-\text{ESRR}} + C_g\right)}{4} \tag{12}$$

2.2 Resonant Frequency Calculation

Figure 4a, b are the equivalent circuital realization of the ESRR and DS-ESRR, respectively. As shown from Fig. 2, the expressions for resonance frequencies for ESRR and DS-ESRR can be written as follows:

$$\omega_0 = \frac{1}{\sqrt{L_T C_{\text{eq}}}} \tag{13}$$

The expression of the resonant frequencies of different SRRs are given as follows.

For E-SRR:

$$f_0^{\text{ESRR}} = \frac{1}{2\pi\sqrt{L_T^{\text{ES}}\left\{\frac{\left[\pi\left(a_{\text{avg}}+b_{\text{avg}}\right)-g\right]C_{\text{pul}}}{4}\left(1+\frac{3h}{10+\sqrt{4-3h}}\right)+\frac{\varepsilon_0 t}{2g}\right\}}} \tag{14}$$

For DS-ESRR:

$$f_0^{\text{DS}-\text{ESRR}} = \frac{1}{2\pi\sqrt{L_T^{\text{DS}}\left\{\frac{\left[\pi\left(a_{\text{avg}}+b_{\text{avg}}\right)-g\right]C_{\text{pul}}}{16}\left(1+\frac{3h}{10+\sqrt{4-3h}}\right)+\frac{\varepsilon_0 t}{4g}\right\}}} \tag{15}$$

From, Eqs. (14) and (15), we can conclude that:

$$f_0^{\text{ESRR}} = f_0^{\text{DS}-\text{ESRR}}/2$$

where L_T^{ES} and L_T^{DS} are the equivalent distributed inductances of the ESRR and DS-ESRR, respectively. The general expression for the equivalent distributed inductance is given as follows:

$$L_T = 0.0002l\left(\frac{4l}{c} - \theta\right) \quad \text{nH} \tag{16}$$

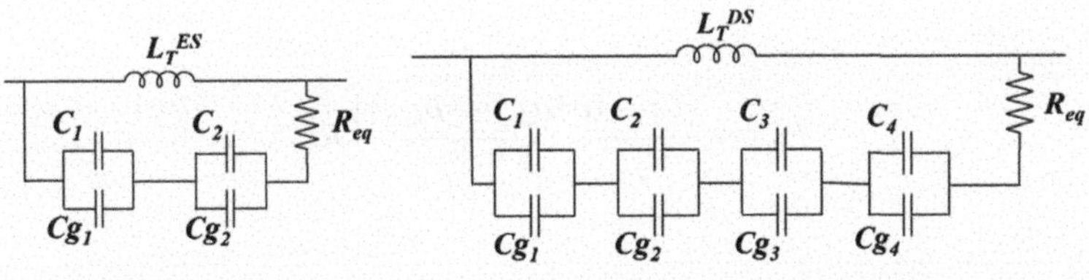

Fig. 4 (a) and (b) are the equivalent circuital realization of the ESRR and DS-ESRR

Table 1 Resonance frequency calculation for DS-ESRR

a_{ext} (mm)	b_{ext} (mm)	c (mm)	d (mm)	g (mm)	freq. (calculated)	freq. (Eigen mode)	error (%)
8.5	6	0.5	0.7	0.7	4.66	4.30	8.3
5.5	3	0.5	0.8	0.6	8.11	8.36	1.6
6	3	0.5	0.7	0.3	7.34	7.5	2.1
6	2.5	0.4	0.6	0.8	7.18	7.76	2.4
4	2.5	0.4	0.6	0.5	11.30	10.82	4.4
3.5	1.9	0.4	0.5	0.6	13.12	12.44	5.05
3	1.8	0.3	0.6	0.6	14.2	14.12	0.5

where l and c are the length and width of the metallic ring shape and $\theta = 2.451$ for elliptical shape.

For ESRR:

$$l^{\mathrm{ESRR}} = \pi(a_{\mathrm{ext}} + b_{\mathrm{ext}})\left(1 + \frac{3h}{10 + \sqrt{4 - 3h}}\right) - g \tag{17}$$

For DS-ESRR:

$$l^{\mathrm{DSESRR}} = \pi(a_{\mathrm{ext}} + b_{\mathrm{ext}})\left(1 + \frac{3h}{10 + \sqrt{4 - 3h}}\right) - g \tag{18}$$

The resonance frequencies of the DS-ESRR are calculated using Eq. (15), respectively, for different parameters and compared with the eigenmode simulation results in Table 1.

2.3 Rotation Effect on Resonance Frequency

As shown in Fig. 5, the length of the sections (1, 3) and (2, 4) can be expressed as P_{m1} and P_{m2}, respectively, through Eqs. (19) and (20), respectively:

$$P_{m1} = \frac{[(\pi + \theta)(a_{\mathrm{avg}} + b_{\mathrm{avg}})]}{4} Rc \tag{19}$$

$$P_{m2} = \frac{[(\pi - \theta)(a_{\mathrm{avg}} + b_{\mathrm{avg}})]}{4} Rc \tag{20}$$

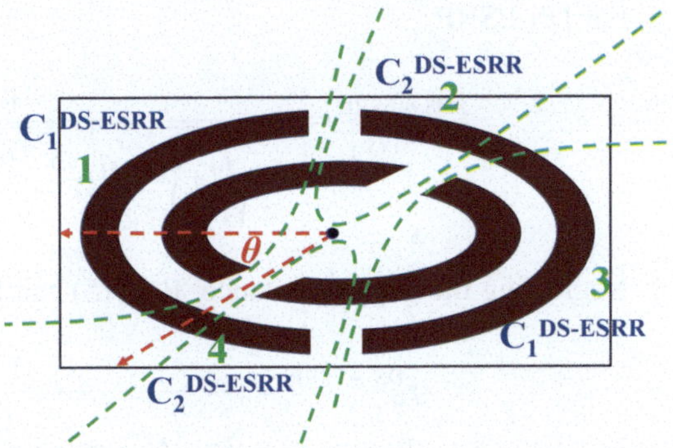

Fig. 5 Visualization of realizable capacitances in DS-ESRR

The equivalent distributed capacitance for each section of the rotated DS-ESRR can be written in Eqs. (21) and (22), respectively, as follows:

$$C_1^{\mathrm{DS-ESRR}} = \frac{\left[(\pi + \theta)\left(a_{\mathrm{avg}} + b_{\mathrm{avg}}\right) - g\right] R_C C_{\mathrm{pul}}}{4} \tag{21}$$

$$C_2^{\mathrm{DS-ESRR}} = \frac{\left[(\pi - \theta)\left(a_{\mathrm{avg}} + b_{\mathrm{avg}}\right) - g\right] R_C C_{\mathrm{pul}}}{4} \tag{22}$$

The equivalent capacitance expression for DS-ESRR can be written as follows:

$$C_r^{\mathrm{DS-ESRR}} = \frac{\left(C_1^{\mathrm{DS-ESRR}} + C_g\right)\left(C_2^{\mathrm{DS-ESRR}} + C_g\right)}{2\left(C_1^{\mathrm{DS-ESRR}} + C_2^{\mathrm{DS-ESRR}} + 2C_g\right)} \tag{23}$$

Substituting $C_1^{\text{DS-ESRR}}$ and $C_2^{\text{DS-ESRR}}$ from Eqs. (21) and (22), respectively, the simplified $C_r^{\text{DS-ESRR}}$ can be expressed as follows:

$$C_r^{\mathrm{DS-ESRR}} = \frac{(\pi + x)^2 - \theta^2}{4(\pi + x)} \left[\frac{\left(a_{\mathrm{avg}} + b_{\mathrm{avg}}\right) R_c C_{\mathrm{pul}}}{4}\right] \tag{24}$$

where $x = {}^{4C_g}\!/\!{\left(a_{\mathrm{avg}} + b_{\mathrm{avg}}\right) R_c C_{\mathrm{pul}}}$.

The equivalent capacitance expression for DS-ESRR can be written as follows:

$$C_{\mathrm{req}}^{\mathrm{DS-ESRR}} = \frac{\left(C_r^{\mathrm{DS-ESRR}} + C_g\right)}{4} \tag{25}$$

As the C_g term is very small, it can be neglected and the expression of the resonant frequencies of rotational DS-ESRR is given as follows.

For DS-ESRR:

$$f_{r0}^{\text{DS-ESRR}} = \frac{1}{2\pi\sqrt{L_T\left\{\frac{(\pi+x)^2-\theta^2}{4(\pi+x)}\left[\frac{(a_{\text{avg}}+b_{\text{avg}})R_c C_{\text{pul}}}{16}\right]R_c\right\}}} \tag{26}$$

Neglecting the gap capacitance, Eq. (15) can be approximated as follows:

$$f_0^{\text{DS-ESRR}} = \frac{1}{2\pi\sqrt{L_T^{\text{DS}}(\pi+x)\left[\frac{(a_{\text{avg}}+b_{\text{avg}})R_c C_{\text{pul}}}{16}\right]}} \tag{27}$$

Comparing Eqs. (26) and (27), we can come up with the relation between the resonance frequency of DS-ESRR and that of the rotated DS-ESRR as follows:

$$\frac{f_{r0}^{\text{DS-ESRR}}}{f_0^{\text{DS-ESRR}}} = \frac{2(\pi+x)}{\sqrt{(\pi+x)^2-\theta^2}} \tag{28}$$

For the very small value of x, Eq. (28) can be further compressed as follows:

$$\frac{f_{r0}^{\text{DS-ESRR}}}{f_0^{\text{DS-ESRR}}} = \frac{2\pi}{\sqrt{\pi^2-\theta^2}} \tag{29}$$

2.4 *Mathematical Modeling Between* f_0 *and the Axes of the Ellipse*

In this section, different empirical relations have been proposed for the design of the double-split ring resonator (DS-SRR) and rotational. Depending upon the different electrical parameters of the structure, the effect on the resonance frequency of the SRR structure has been modeled as a customized empirical nonlinear relationship. Two cases are taken in consideration here. In first case, the semimajor and semiminor axes of the outer elliptical ring are sequentially varied to estimate the effects on the resonating frequency. Due to the nonlinear variation of semimajor, different semiminor axis values have been correlated and returns to resonance frequencies of the DS-ESRR. From the parametric data taken from the simulation environment, due to the nonconvex nature and some nonlinear dependency of column basis vectors, linear regression becomes inefficient to correlate different basis values. Therefore, customized curve fitting tool is used to achieve the estimation of the optimum correlation between basis data and final empirical equation using MATLAB computational platform.

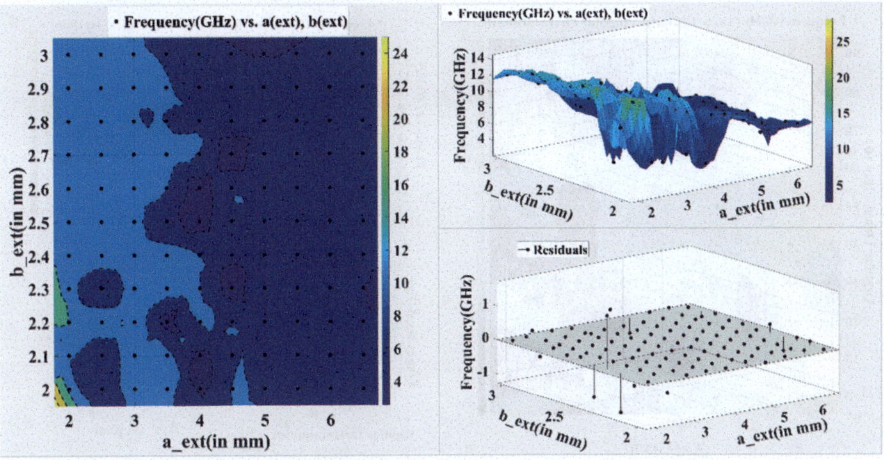

Fig. 6 Contour plots and residuals for each (a_{ext}, b_{ext}, f_0) basis vector mapping

The abovementioned double-split elliptical SRR (DS-ESRR) can be nonlinearly mapped into the jointly distributed vector span of semimajor axis (a_{ext}) and semiminor axis (b_{ext}). The cross-correlated vector basis can be effectively mapped into resonating frequency of the SRR through Loess regression process [12–14] by obtaining multivariate Gaussian distribution. The multivariate Gaussian distribution can be expressed as follows:

$$p(x|\mu,\Sigma)=\frac{1}{(2\pi)^{\frac{p+q}{2}}|\Sigma|^{0.5}}\times\exp\left\{-\frac{1}{2}\begin{pmatrix}x_1-\mu_1\\x_2-\mu_2\end{pmatrix}^T\begin{bmatrix}\Sigma_{11}&\Sigma_{12}\\\Sigma_{21}&\Sigma_{22}\end{bmatrix}^{-1}\begin{pmatrix}x_1-\mu_1\\x_2-\mu_2\end{pmatrix}\right\} \quad (30)$$

where $\mu=\begin{pmatrix}1.44\\0.31\end{pmatrix}$ and $\Sigma_{11}=4.25$, $\Sigma_{22}=2.5$, Σ_{12}, $\Sigma_{21}=0$.

Based on the above distribution, the jointly distributed X_1 and X_2 vector has been mapped to semimajor axis (a_{ext}) and semiminor axis (b_{ext}). The respective contour plots and residuals have been mentioned in Fig. 6 for each basis vector.

For, the resonance frequency of the rotational double-split elliptical SRR (DS-ESRR) can be again nonlinearly mapped into the jointly distributed vector span of semimajor axis (a_{ext}) and relative angular orientation (θ) of the inner splits. The cross-correlated vector basis can be effectively mapped into resonating frequency of the SRR through Loess regression process by obtaining multivariate Gaussian distribution. The multivariate Gaussian distribution can be expressed this time as follows:

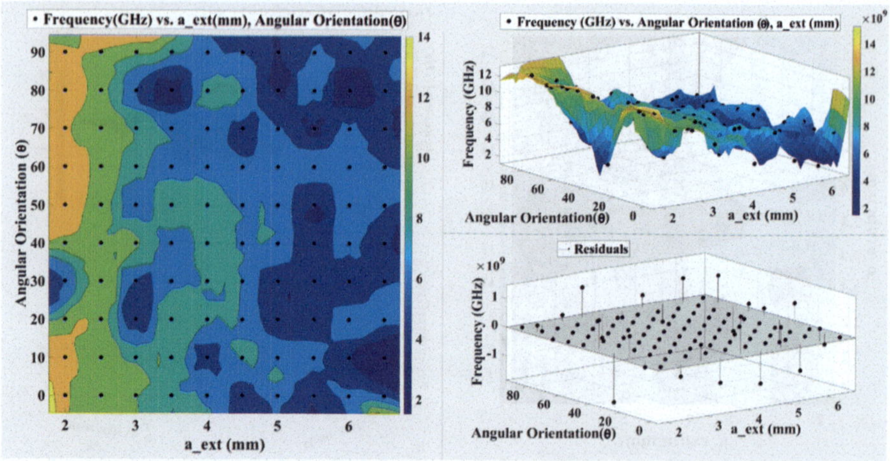

Fig. 7 Contour plots and residuals for each (a_{ext}, b_{ext}, f_0) basis vector mapping

$$p(x|\mu,\Sigma)=\frac{1}{(2\pi)^{\frac{p+q}{2}}|\Sigma|^{0.5}}\times\exp\left\{-\frac{1}{2}\begin{pmatrix}x_1-\mu_1\\x_2-\mu_2\end{pmatrix}^T\begin{bmatrix}\Sigma_{11}&\Sigma_{12}\\\Sigma_{21}&\Sigma_{22}\end{bmatrix}^{-1}\begin{pmatrix}x_1-\mu_1\\x_2-\mu_2\end{pmatrix}\right\} \quad (31)$$

where $\mu=\begin{pmatrix}4.25\\45\end{pmatrix}$ and $\Sigma_{11}=1.44$, $\Sigma_{22}=28.87$, Σ_{12}, $\Sigma_{21}=0$.

Based on the above distribution, the jointly distributed X_1 and X_2 vector has been mapped to semimajor axis (a_{ext}) and relative angular orientation (θ) of the inner splits. The respective contour plots and residuals have been mentioned in Fig. 7 for each basis vector.

3 Conclusion

From this article, the overall numerical analysis for both single and dual-split SRR has been analyzed. The effect of rotation in the inner annular ring has been analyzed by the point of view of comparative analysis between rotated and unrotated DS-ESRR. This analysis of the concepts we can apply for different strong band notching characteristics and jamming applications. The stopband characteristics of the double-split elliptical SRR (DS-ESRR) are presented with the help of an UWB filtenna structure in detail. The proposed idea is very simple and can be scaled to any other futuristic applications.

References

1. Zhou, J., Koschny, T., Kafesaki, M., Economou, E.N., Pendry, J.B., Soukoulis, C.M.: Saturation of the magnetic response of split-ring resonators at optical frequencies. Phys. Rev. Lett. **95**(22), 223902 (2005)
2. Marqués, R., Martel, J., Mesa, F., Medina, F.: Left-handed-media simulation and transmission of EM waves in subwavelength split-ring-resonator-loaded metallic waveguides. Phys. Rev. Lett. **89**(18), 183901 (2002)
3. Martın, F., Bonache, J., Falcone, F.A., Sorolla, M., Marqués, R.: Split ring resonator-based left-handed coplanar waveguide. Appl. Phys. Lett. **83**(22), 4652–4654 (2003)
4. Movchan, A.B., Guenneau, S.: Split-ring resonators and localized modes. Phys. Rev. B. **70**(12), 125116 (2004)
5. Zhang, X.C., Yu, Z.Y., Xu, J.: Novel band-pass substrate integrated waveguide (SIW) filter based on complementary split ring resonators (CSRRs). Prog. Electromagn. Res. **72**, 39–46 (2007)
6. Siddiqui, J.Y., Saha, C., Antar, Y.M.M.: Compact SRR loaded UWB circular monopole antenna with frequency notch characteristics. IEEE Trans. Antennas Propag. **62**(8), 4015–4020 (2014). https://doi.org/10.1109/TAP.2014.2327124
7. Shaik, L.A., Saha, C., Antar, Y.M.M., Siddiqui, J.Y.: An antenna advance for cognitive radio: introducing a multilayered split ring resonator-loaded printed ultrawideband antenna with multifunctional characteristics. IEEE Antennas Propag. Magaz. **60**(2), 20–33 (2018). https://doi.org/10.1109/MAP.2018.2796027
8. Wang, Y., Huang, T., Ma, D., Shen, P., Hu, J., Wu, W.: Ultra-wideband (UWB) monopole antenna with dual notched bands by combining electromagnetic-bandgap (EBG) and slot structures. 2019 IEEE MTT-S international microwave biomedical conference (IMBioC), Nanjing, China, 2019, pp. 1–3. https://doi.org/10.1109/IMBIOC.2019.8777856
9. Saha, C., Siddiqui, J.Y., Freundorfer, A.P., Shaik, L.A., Antar, Y.M.M.: Active reconfigurable ultra-wideband antenna with complementary frequency notched and narrowband response. IEEE Access. **8**, 100802–100809 (2020). https://doi.org/10.1109/ACCESS.2020.2997933
10. Basu, S., Pahadsingh, S., Sahu, S.: Compact elliptical slotted antenna with multi bands notched for UWB applications. 2021 IEEE 2nd international conference on applied electromagnetics, signal processing, & communication (AESPC), Bhubaneswar, India, 2021, pp. 1–6. https://doi.org/10.1109/AESPC52704.2021.9708519
11. Villarino, M.B.: Ramanujan's perimeter of an ellipse. arXiv preprint math/0506384. (2005)
12. Ghosh, S., Saha, D., Prahannathan, V., Saha, C.: Mathematical modelling of slotted circular meta-atom for holographic meta-surface applications. 2023 international conference on microwave, optical, and communication engineering (ICMOCE), Bhubaneswar, India, 2023, pp. 1–4. https://doi.org/10.1109/ICMOCE57812.2023.10167286
13. Chakraborty, A., Ghosh, S., Roy, S.K., Chakraborty, S., Chatterjee, S., Gupta, B.: Loaded multi loop hexagram shaped printed antennas for multifrequency communication. 2022 microwave mediterranean symposium (MMS), Pizzo Calabro, Italy, 2022, pp. 1–6. https://doi.org/10.1109/MMS55062.2022.9825536
14. Saha, D., Ghosh, S., Chakrabarti, M., Chattopadhyay, A.: Prism-shaped nonuniformly distributed slotted SIW Array with highly reduced SLL for mm-wave ground surveillance application. 2023 first international conference on microwave, antenna and communication (MAC), Prayagraj, India, 2023, pp. 1–5. https://doi.org/10.1109/MAC58191.2023.10177066

Rectifier and Reconfigurable Impedance-Matching Network Analysis for Wireless Sub-6 GHz 5G/Wi-Fi 6/6E Energy Harvester

Fanuel Elias, Sunday Cookey Ekpo, Stephen Alabi, Dipankar Saha, Samik Chakraborty, Swarnadipto Ghosh, Mfonobong Charles Uko, Muhammad Ijaz, and Umar Raza

1 Introduction

Sustainability is one of the world's significant challenges today, and the increasing use of fossil fuels to generate electricity exacerbates the situation. As a result, carbon-free energy generation methods have been investigated. Energy harvesting is a modern technology that collects and converts energy from the environment, such as wind turbines, solar cells, and electromagnetic waves in the ambient environment [1, 2]. Energy harvesting has gained significant attention in recent years as a promising solution to address the ever-increasing demand for power in wireless devices and systems. Among the various ambient energy sources, wireless energy harvesting technology has risen rapidly due to the popularity of wireless communications, such as TV, radio, cellular, satellite, and Wi-Fi signals, particularly since the early 1990s [3, 4]. One avenue of exploration within this domain is RF energy harvesting, which involves the conversion of ambient electromagnetic radiation into usable electrical energy [5, 6].

F. Elias (✉) · S. C. Ekpo · M. C. Uko · M. Ijaz · U. Raza
Manchester Metropolitan University, Manchester, UK
e-mail: Fanuel.Elias@stu.mmu.ac.uk

S. Alabi
Wilsons Park, Manchester, UK

D. Saha · S. Ghosh
Indian Institute of Space Science and Tech, Thiruvananthapuram, India

S. Chakraborty
Regent Education and Research Foundation, Kolkata, India

S. C. Ekpo (ed.), *The Second International Adaptive and Sustainable Science, Engineering and Technology Conference*, Signals and Communication Technology,
https://doi.org/10.1007/978-3-031-53935-0_8

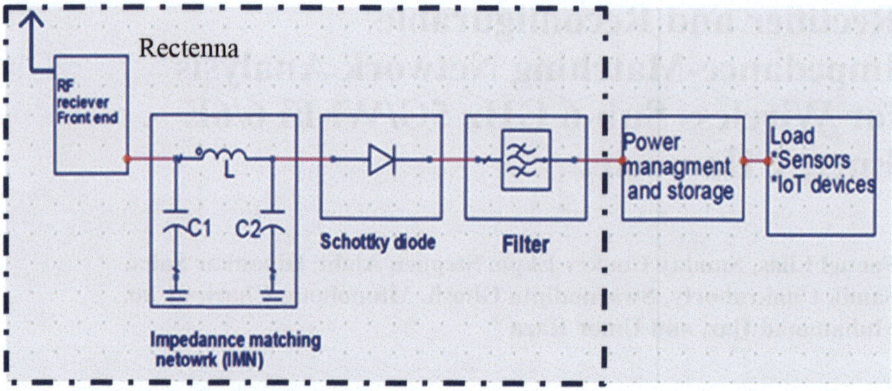

Fig. 1 RF energy-harvesting block diagram

The rapid population growth and the emergence of new communication technologies increased the demand for modern wireless communication technology, particularly mobile and television, which expanded base stations. As a result, RF energy distribution in the environment is becoming more widespread. The promising features of 5G/6G technology, such as high speed, low latency, enhanced capacity, and increased bandwidth, have advanced the use of indoor low-power devices such as IoT and sensors [7].

Rectenna, an essential component in energy-harvesting systems, consists of a receiving antenna, bandpass filter, rectifying diode, and DC-pass filter, allowing for the reception and conversion of RF power into usable DC power [3, 8]. Their significance lies in their ability to capture ambient electromagnetic energy, particularly in the radio frequency spectrum, and convert it into electrical power for various applications. Despite the widespread availability of radio waves for broadcasting and wireless communication, their power output could be improved, making it challenging to generate high voltage. To overcome this limitation, rectifiers can be employed in the RF energy-harvesting circuits to increase the output power.

Figure 1 depicts the block diagram of an RF energy-harvesting system comprising the antenna, matching network, rectifier, and load. The matching network maximises power transfer, while the rectifier efficiently converts RF to DC. Consequently, the performance and efficiency of the system, represented by the obtained DC voltage value and the conversion efficiency (the ratio between DC power and RF power), are directly influenced by the non-linear devices present in the rectifier, matching structure, and load. Moreover, the characteristics of the RF signal also impact the performance of the RF/DC conversion system.

Reconfigurable matching circuits provide several advantages, including expanding the frequency bands beyond those achievable with passive components and providing a broader available bandwidth [4–8]. When subjected to varied hostile environments that affect the frequency response of the antenna, reconfigurability is also vital for tuning the frequency bands of operation [5–9].

2 Modelling and Selecting RF Schottky Diode

The main challenge in designing an RF energy harvesting is operating with low RF power. For a typical 50 Ω antenna, the average RF power can be as low as −25 dBm (3.2uW) and generate a peak AC voltage of 5.7 mV; this requires a typical diode with a threshold much higher than the peak voltage. Additionally, these diodes must have a faster switching time than traditional ones, as RF energy harvesting operates at high frequencies. Therefore, Schottky diodes are a good candidate for these designs [4, 10].

RF Schottky diodes are high-frequency semiconductor devices widely used in microwave circuits, signal detection, and mixing. These diodes are ideal for RF applications due to their low forward voltage drop and fast switching speed. The maximum frequency of operation, forward voltage drops, reverse breakdown voltage, and noise characteristics are all important considerations when choosing an RF Schottky diode. Schottky diode efficiency can vary from 30% to 80%, depending on the device and operating conditions [9–14].

The most used RF Schottky diodes for this application are listed below, based on the application's unique criteria, such as frequency range, power level, and sensitivity. Table 1 summarises the determining characteristics of RF Schottky diodes. Despite having a relatively high forward voltage drop, the SMS7630 diode from Skyworks Solutions Inc. was used in all designs in this paper due to its market availability and wide operation frequency (up to 24 GHz). This diode has a peak current of 150 mA and a low turn-on voltage, and it is conveniently available in surface mount 0201 packages.

Table 1 Parameters of RF Schottky diodes

Parameters	Unit	MA4E2054	SMS7630	SMS3922
Reverse leakage current, I_s	A	3E-8	5E-6	3E-8
Series resistance, R_s	W	11	20	9
Emission coefficient, N	–	1.05	1.05	1.08
TT	sec	0	1E-11	8E-11
Zero bias junction capacitance, C_{JO}	pF	0.13	0.14	0.7
M	–	0.5	0.40	0.26
EG	eV	0.69	0.69	0.69
Breakdown voltage, B_v	V	5	2	20
Reverse bias current, IBV	A	1E-5	1E-4	1E-5
Junction voltage, V_j	V	0.4	0.34	0.595

3 Rectifier Design and Analysis

When selecting a diode for RF rectification, the most important criterion is its forward voltage drop (V_f) at the expected operating frequency and current level. The forward voltage drop is the voltage that the diode must overcome before the current can flow forward through it. As a result, limiting the forward voltage drop is critical since it decreases the amount of power wasted as heat in the diode and boosts the rectifier's efficiency. Furthermore, a low forward voltage drop reduces the diode's impact on the RF signal being rectified.

The SMS7630 RF Schottky diode was modelled in ADS using the parameters provided in Table 1. Several rectifiers, such as voltage doubler, Delon doubler, Delon tripler, Delon quadrupler, two-stage charge pump, three-stage charge pump, and five-stage charge pump rectifiers, have been designed and analysed on ADS using the selected diode model (SMS7630). The output voltage and efficiency of each rectifier were compared, and it was found that the voltage doubler rectifier (Fig. 2) had the highest values. This can be attributed to the voltage doubler utilising only two diodes for rectification. On the other hand, the five-stage charge pump design employed ten diodes, resulting in the lowest output voltage and efficiency among the other designs. A diode was added in series to the load (50 Ω) in the rectifier circuit to improve the sensitivity of the design by acting as a voltage booster, rectifying and stacking multiple RF cycles to increase the output voltage. In Fig. 2, a power generator tool from ADS (P_TONE) replaced the antenna and was set to generate input power.

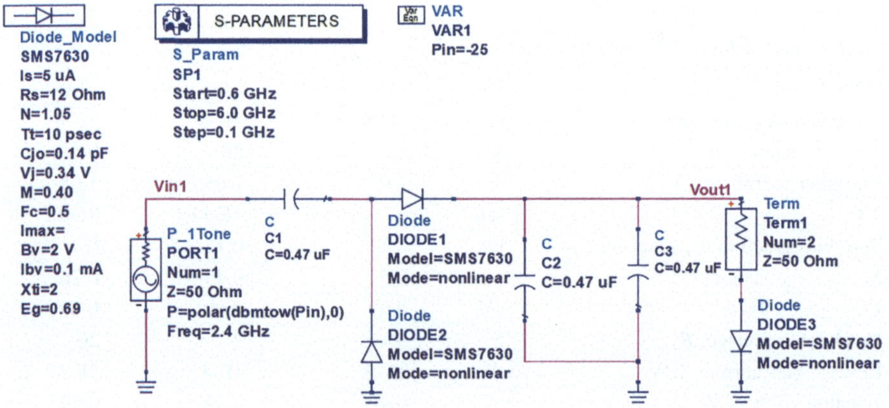

Fig. 2 The selected voltage doubler rectifier

$$\text{Output Voltage, } V_{\text{out}} = \sum_{n=0}^{N} C_n\,(P)^n \tag{1}$$

Here the coefficient bounds are as follows: $C_1 = 1.77\text{e}{-05}$, $C_2 = 0.001$, $C_3 = 0.025$, $C_4 = 0.233$.

$$\text{Efficiency, } \varepsilon\ (\%) = \sum_{n=0}^{N} K_n(P)^n \tag{2}$$

Here, the coefficient bounds are as follows: $K_1 = -1.73\text{e}{-05}$, $K_2 = -0.0002$, $K_3 = 0.013$, $K_4 = 0.01$, $K_5 = -2.92$, $K_6 = 52.75$.

The output voltage and efficiency of the voltage doubler rectifier can also be expressed using polynomials (1) and (2), respectively. These polynomial approximations were based on the data presented in Figs. 3 and 4, respectively, along with their corresponding residual plots.

The voltage doubler rectifier's output voltage and output current are given in Figs. 5 and 6, respectively.

The RF-DC conversion efficiency of a rectifier was calculated using the ratio of the output DC power (which depends on the output voltage and load) to the input power. In this report, the efficiency of each rectifier was calculated using (3).

$$\eta_{\text{RF}-\text{DC}} = \frac{V_{\text{cap}}^2}{2 * R_{\text{L}} * P_{\text{in}}} * 100 \tag{3}$$

where P_{in} = input power, V_{cap} = output voltage, and R_{L} = load resistance.

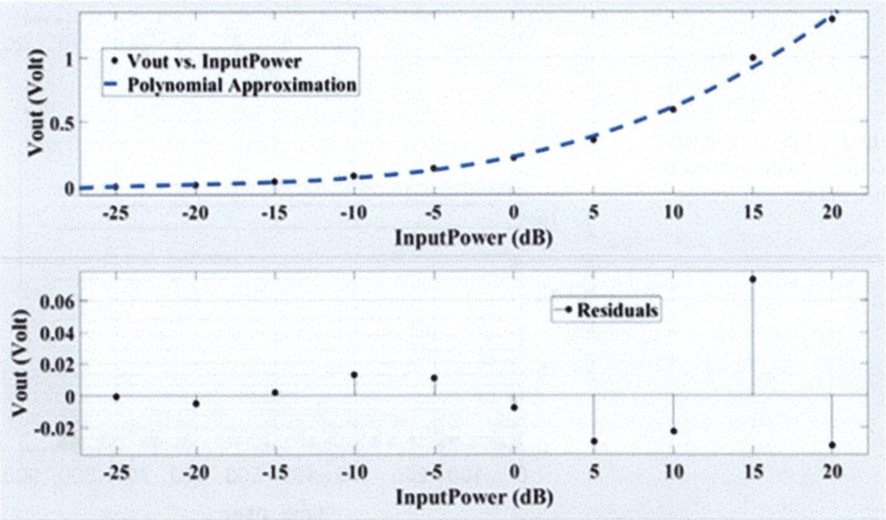

Fig. 3 Residual plot of output voltage (V) against the input power (dB)

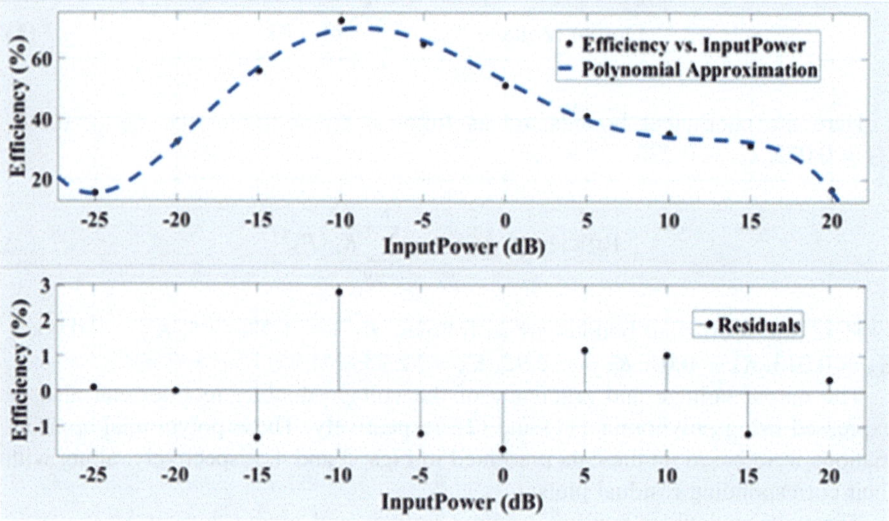

Fig. 4 Residual plot of efficiency (%) against the input power (dB)

Fig. 5 Voltage doubler, output voltage (V_{out1}) in V vs input power (P_{in}) in dBm

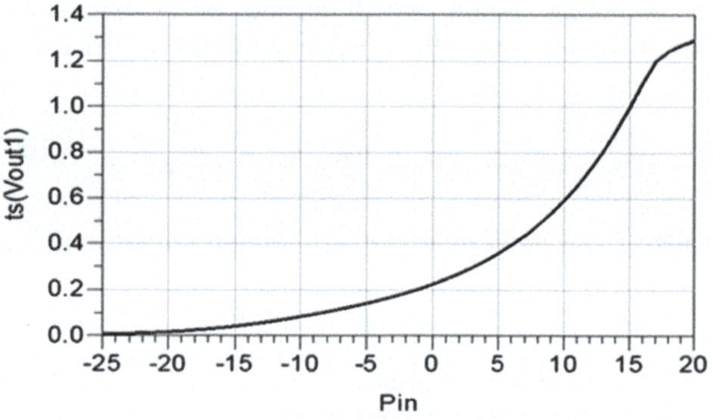

Fig. 6 Voltage doubler rectifier, output current in mA vs time

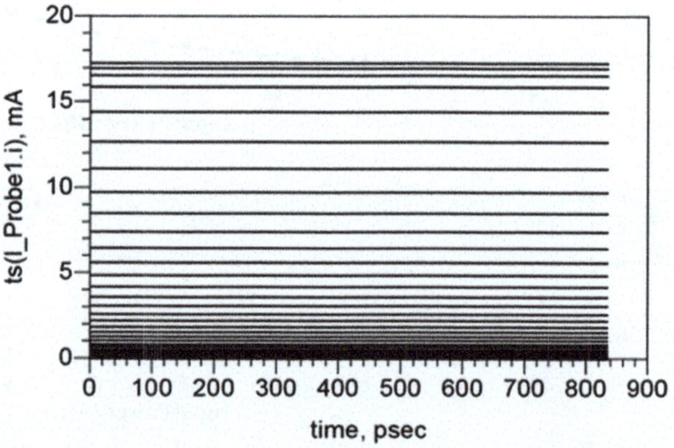

4 Matching Circuit Topologies and Reconfigurability

Impedance-matching circuits are essential in RF energy-harvesting technologies to ensure efficient power transfer and minimise signal reflection. These circuits are used to match the complex impedance of the antenna with the rectifier at specific frequencies. Different matching circuits can be implemented using distributed elements like transmission lines or discrete lumped elements such as capacitors and inductors. Each matching circuit must be designed and tested individually for a specific antenna and frequency range. While discrete lumped element circuits are more practical in size, have low noise figure, and have good power-handling capabilities, they are commonly used in RF and microwave applications up to 20GHz [7, 15, 16]. On the other hand, distributed element circuits are considered more feasible for matching networks in microwave applications, although they may pose challenges in predicting reflection for short wires [4, 16].

The choice of network topology significantly impacts the reconfigurable network's effectiveness. Possible options include L, Π, and T structures (Fig. 7) and transmission line topology. Π and T structures offer similar reconfigurability across different frequencies; however, Π structure proves to be less sensitive and facilitates the implementation of tuneable elements [8, 15]. Hence, Π topology in series to an inductor was practised in this paper.

Various virtual antennas have been researched and investigated; however, a non-resonant cellular-embedded IoT antenna from Ignion, RUN mXTEND™, has been selected for this work. This antenna is compact (12.0 × 3.0 × 2.4 mm), has a 698–5875 MHz operating frequency, and can be used for applications such as wearable, IoT sensors and medical devices. It also has an omnidirectional radiation pattern, linear polarisation, and less VSWR [17]. This antenna is a reliable, off-the-shelf product and compatible with multiple regionals. The library file of the antenna was downloaded from the Ignion website [17], and the S-parameter file (. s1p) was imported into the ADS. The S-parameter of the selected antenna and rectifier circuit was extracted at various frequencies using ADS with the aid of the SP-Probe tool. The Smith chart (ADS tool) was used to analyse and adjust the impedance values of the antenna and rectifier at 2.4 GHz. Then, the impedance-matching circuit was adjusted to match different frequencies by varying the variable capacitor (C2), 0.3–6.2 pF. The reconfigurable impedance-matching (RIM) circuit diagram with the value of the capacitors and inductors is given in Fig. 8.

Fig. 7 Impedance-matching topology (**a**) L, (**b**) Π, and (**c**) T

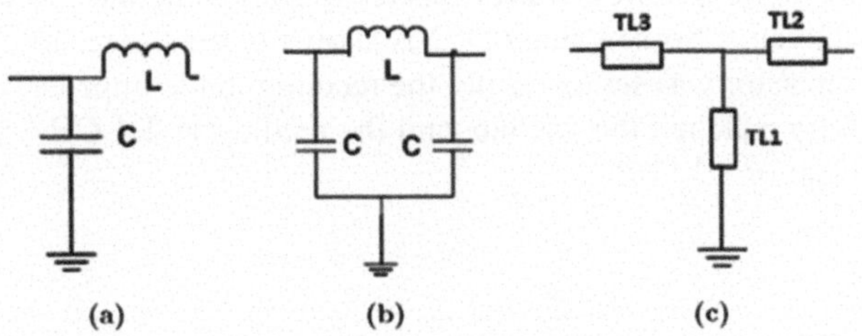

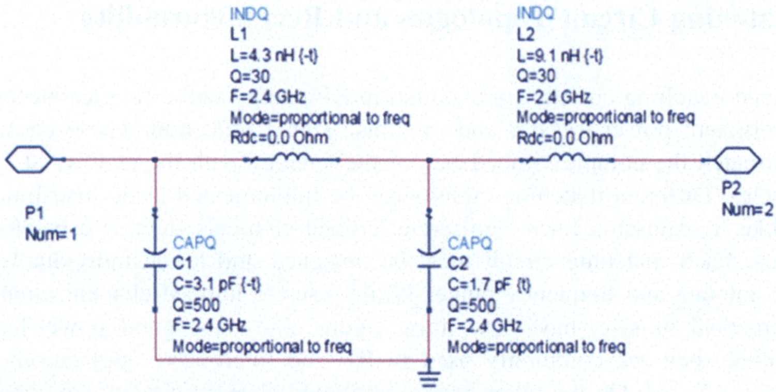

Fig. 8 Reconfigurable impedance-matching (RIM) circuit diagram

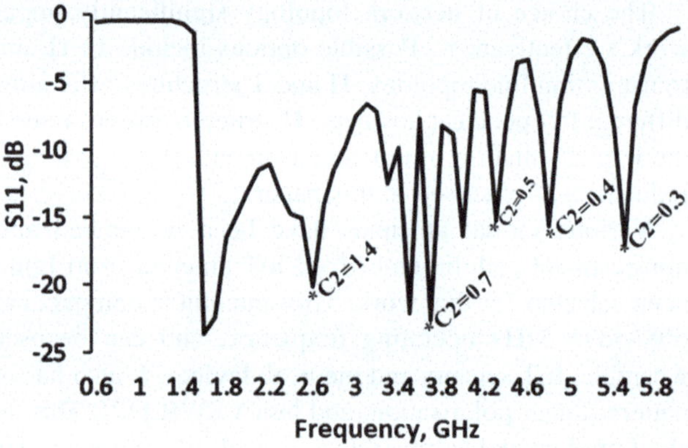

Fig. 9 The design's reflection coefficient (S11) at different frequencies by manually tuning the capacitor, C2, (pF) of the designed RIM network

Figure 9 describes the simulation results of S11 when the RIM circuit was adjusted by varying the variable capacitor's value. It demonstrates that the RIM circuit effectively matched the impedance of both the antenna and the rectifier at multiple frequencies. The horizontal dashed line represents the maximum allowable value (−10 dB) for the reflection coefficient (S11), and the graph indicates that the design achieved a satisfactory match at frequencies where S11 was below the dashed line (−10 dB). Figure 9 illustrates that the design achieved a good match (S11 $\leq -$ 10 dB) across different frequency ranges (1.5–5.6 GHz) by adjusting the variable capacitor. Figure 9 was produced by combining the S11 data below −10 dB for each C2 value. Some value of C2 is shown in the figure; for instance, when the variable capacitor was set to 1.7 pF, the reconfigurable impedance-matching circuit successfully matched the antenna and the rectifier at 2.4 GHz, yielding a reflection coefficient of −14 dB.

5 Conclusion

The proposed architecture comprises an antenna, a reconfigurable impedance-matching network, and a rectifier. All the rectifier designs have been designed and evaluated in Keysight ADS software. However, only the selected rectifier is presented in this paper due to commercial sensitivity and page limit. The proposed design had an output voltage of 1.3 V and an efficiency of 70%. It also had an S11 ≤ -10 dB and VSWR ≤ 2 for a wide range of frequencies. The proposed RF energy-harvesting design had a sensitivity of -25 dBm.

References

1. IEEE: RF energy harvesting circuit for vital sensing platform. In: *2021 IEEE Asia-Pacific Conference on Applied Electromagnetics* (APACE), pp. 1–3. IEEE, Penang, Malaysia (2021)
2. Lau, I., Ekpo, S.C., Zafar, M., Ijaz, M., Gibson, A.: Hybrid mmWave-Li-Fi 5G architecture for reconfigurable variable latency and data rate communications. IEEE Acess. **11**, 42850–42861 (2023)
3. Kim, S., et al.: Ambient RF energy-harvesting technologies for self-sustainable standalone wireless sensor platforms. Proc. IEEE. **102**(11), 1649–1666 (2014)
4. Ekpo, S., George, D.: Impact of noise fig on a satellite link performance. IEEE Commun. Lett. **15**(9), 977–979 (2011)
5. Ekpo, S., Adebisi, B., Wells, A.: Regulated-element frost Beamformer for vehicular multimedia sound enhancement and noise reduction applications. IEEE Access J. **5**, 27254–27262 (2017)
6. Uko, M., Ekpo, S.: 8-12 GHz pHEMT MMIC low-noise amplifier for 5G and fiber-integrated satellite applications. Int. Rev. Aerospace Eng J. **13**(3), 99–107 (2020)
7. Wu, P., Huang, S.Y., Zhou, W.: High-efficient rectifier with extended input power range based on self-tuning impedance matching[J]. IEEE Microw. Wirel. Compon. Lett. **28**(12), 1116–1118 (2018)
8. Sowande, O., Idachaba, F., Ekpo, S., Faruk, N., Uko, M., Ogunmodimu, O.: Sub-6 GHz 5G spectrum for satellite-cellular convergence broadband internet access in Nigeria. Int. Rev. Aerospace Eng. J. **15**(2), 85–96 (2022)
9. Anguera, J., Andújar, A., Leiva, J., Massó, O., Tonnesen, J., Rindalsholt, E., Brandsegg, R., Gaddi, R.: Reconfigurable multiband operation for wireless devices embedding antenna boosters. Electronics. **10**(7), 808 (2021)
10. Wang, X., Yang, L., Lee, T., Su, J.K., Zhu, B., Yao, X.F.: RF modeling of integrated RF CMOS Schottky diodes for rectifier designs. In: 9th International Conference on Solid-State and Integrated-Circuit Technology, pp. 305–308. IEEE, Beijing (2008)
11. Assogba, O., Mbodji, A.K., Karim, D.A.: Efficiency in RF energy harvesting systems: a comprehensive review. In: 2020 IEEE International Conf on Natural and Engineering Sciences for Sahel's Sustainable Development - Impact of Big Data Application on Society and Environment (IBASE-BF), pp. 1–10. IEEE, Ouagadougou, Burkina Faso (2020)
12. Rahola, J.: Estimating the performance of matching circuits for antennas'. European conference on antennas and propagation. In: Proceedings of the Fourth European Conference on Antennas and Propagation, Barcelona, Spain (2010)
13. Scucchia, L., Palomba, M., Cleriti, R., Ciccognani, W., Limiti, E.: Reconfigurable matching network for RF energy harvesting circuits. In: *2014 International Workshop on Integrated Nonlinear Microwave and Millimetre-wave Circuits* (INMMiC), pp. 1–3, Leuven, Belgium (2014)

14. Zafar, M., Ekpo, S., George, J., Sheedy, P., Uko, M., Gibson, A.: Hybrid power divider and combiner for passive RFID tag wireless energy harvesting. IEEE Access. **10**, 502–515 (2022)
15. Ekpo, S., George, D.: 4–8 GHz LNA design for an adaptive small satellite transponder using InGaAs PHEMT technology. In: Proc. 11th IEEE Wireless & Microwave Conference, pp. 1–4, Melbourne, FL, USA (2010)
16. Ekpo, S.C.: Parametric system engineering analysis of capability based small satellite missions. IEEE Syst. J. **13**(3), 3546–3555 (2019)
17. Ignion: Design Center: Libraries. Ignion.com.Ignion - IoT Antenna Manufacturers (2023)

A DC-50 GHz SPDT Switch for 5G Wireless Applications

Mfonobong Uko, Sunday Cookey Ekpo, Fanuel Elias, Dipankar Saha, Swarnadipto Ghosh, Muhammad Ijaz, Samik Chakraborty, and Andrew Gibson

1 Introduction

The demand for broadband RF switches is significant in microwave and millimetre-wave transceiver system design [1, 2]. These RF switches are vital building blocks utilised in a wide variety of wireless applications, including space-based applications, for the control of RF signal input and output [3–7]. These switches are key components in wireless communication systems because of their low loss, high isolation, and high bandwidth which captures a wide range of frequencies up to millimetre-wave range [8–10].

Figure 1 shows a typical single pole double throw (SPDT) switch integrated with a transceiver sub-system. The transceiver sub-system consists of the receiver front end made up of the low noise amplifier (LNA) as its primary building block and the transmitter front end made up of the power amplifier (PA).

SPDT switch is required for switching between the receiver front end and the transmitter side at different frequency bands. It is used in RF transceiver design for transmit and receive operation. Several SPDT switches have been reported utilising stacked FET series-shunt GaAs pHEMT and GaN processes [8, 9, 11], since III–V semiconductor technologies have shown outstanding performances for high-power wireless communication applications. The use of FETs for switch realisation is of

M. Uko (✉) · S. C. Ekpo · F. Elias · M. Ijaz · A. Gibson
Manchester Metropolitan University, Manchester, UK
e-mail: m.uko@mmu.ac.uk; s.ekpo@mmu.ac.uk; fanuel.elias@stu.mmu.ac.uk; M.Ijaz@mmu.ac.uk; A.Gibson@mmu.ac.uk

D. Saha · S. Ghosh
Indian Institute of Space Science and Tech, Thiruvananthapuram, India

S. Chakraborty
Regent Education and Research Foundation, Kolkata, India

S. C. Ekpo (ed.), *The Second International Adaptive and Sustainable Science, Engineering and Technology Conference*, Signals and Communication Technology,
https://doi.org/10.1007/978-3-031-53935-0_9

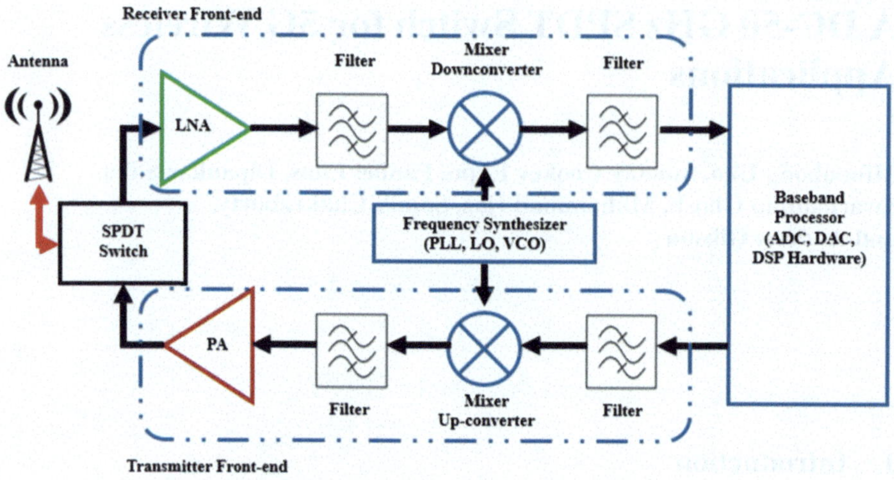

Fig. 1 A typical SPDT switch integrated with a transceiver sub-system

advantage owing to its drain-source resistance behaving as a voltage variable resistor [6, 12].

For millimetre-wave applications, FET switches are preferred over PIN diode owing to its reliable integration capability with other transceiver components, fast switching speed, and small power dissipation [4, 8]. The major shortcomings of FET switches are their low isolation at high frequencies, low breakdown voltage, and limited power-handling capability due to their gate-drain parasitic capacitance C_{gd} [8].

In this paper, a 0.15 μm pHEMT process provided by WIN Semiconductors Corporation is used to design a broadband SPDT switch for satellite and 5G communication applications. The well-matched single pole double throw (SPDT) switch is designed using four shunt-stacked FET units on each branch with two FETs in series for increased voltage-handling capability and operation over large bandwidth [13–15]. The series FETs allow the SPDT switch to operate down to DC. Section 2 reviews device technologies with regard to evolving satellite and 5G communication standards [16, 17]. In Sect. 3, we present design for two broadband low noise amplifiers for satellites and 5G applications. The results of the simulated LNAs are presented in Sect. 4. Section 5 concludes the paper.

2 Switch FET Model Technology

The right process technology is vital based on the required operational specification of the intended application. For a successful design to be achieved, a combination of active (diode, transistors) and passive (inductor, capacitors) devices must be selected, using a well-defined process technology to meet the design goals and expected level of performance of the system.

3 Broadband DC-50 GHz SPDT Switch Design

A broadband SPDT switch is designed over the frequency range from DC-50 GHz for 5G communication applications. The series and shunt FET switching elements make use of a depletion mode scalable PL15-10 0.15 µm InGaAs pseudomorphic high electron mobility transistor process technology provided by WIN Semiconductors.

3.1 FET Device Biasing

When used as switches, FET devices such as pHEMT operate at 0 V DC. In the "on" state, the control voltage for the gate is biased at 0 V, while in the "off" state, the control voltage for the gate is biased beyond pinch-off voltage (−1.5 V for this work). At 0 V on the gate, there is a low resistance signal path, and the transistor is on.

As the gate voltage reduces, the resistance increases and the transistor is off. The RF signal path is drain to source and the gate is the control terminal where the gate voltages are applied through large resistors. The drain-source voltage is swept around 0 V for different gate voltages. Figure 2 shows the typical I–V characteristics of a depletion mode FET for different negative gate-source bias voltages about the 0 V operating point.

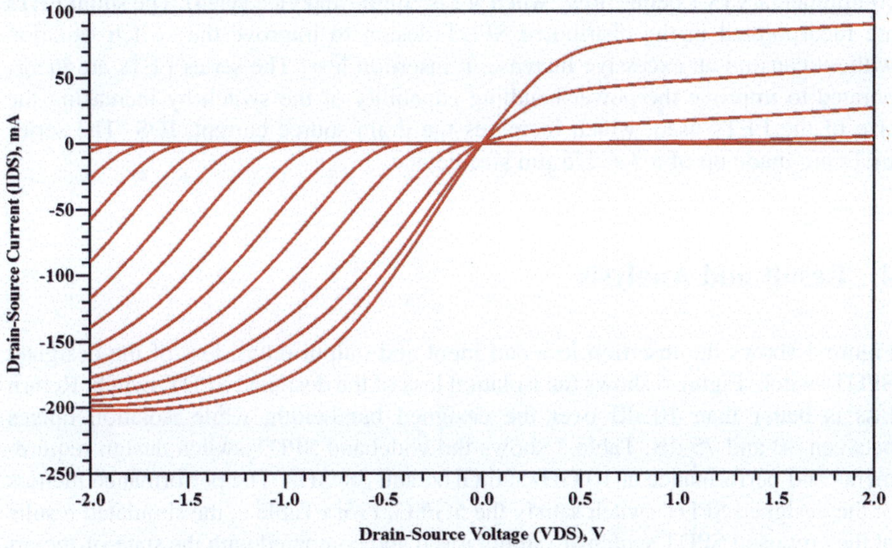

Fig. 2 DC characteristics of a FET switch about 0 V region of operation

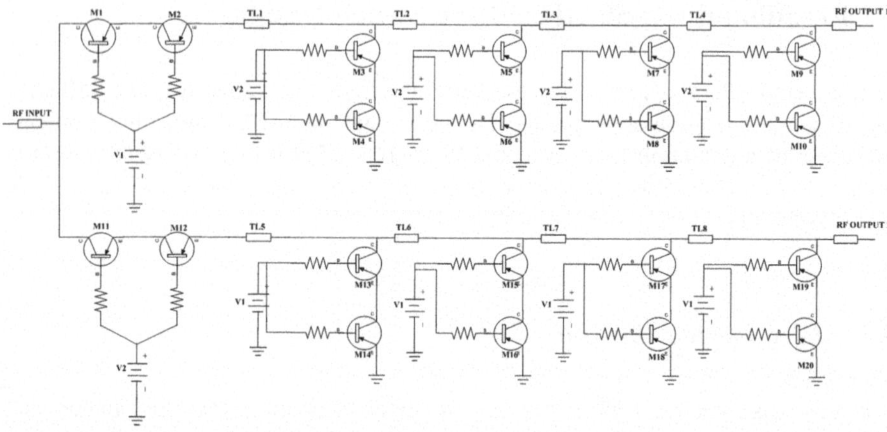

Fig. 3 Schematic of the distributed SPDT switch

3.2 *Switch Design Topology*

A distributed topology is adopted for the SPDT switch design, as shown in Fig. 3. The distributed topology made up of two series FETs and four cascaded stacks of shunt FETs is used to extend the operating frequency of the switch by absorbing the off-state capacitance of the shunt FETs into a low pass filter.

The shunt FETs in the "on"-state arm are at pinched-off voltage when the series FET is switched on, while the FETs in "off"-state arm are switched on when the series FETs are at pinched-off voltage. In Fig. 3, the control voltages V1 and V2 are complimentary (V1 being "low" when V2 is "high" and vice versa). The shunt FETs are incorporated in the distributed SPDT design to improve the switch isolation without causing an excessive increase in insertion loss. The series FETs are incorporated to improve the power-handling capability of the switch by increasing the size of the FETs used, which increases the drain-source current, IDS. The series FETs are made up of a 3 × 275 μm size device.

4 Result and Analysis

Figure 4 shows the insertion loss and input and output return loss of the designed SPDT switch. Figure 5 shows the isolation loss of the designed SPDT switch. Return loss is better than 10 dB over the designed bandwidth, while isolation ranges between 40 and 75 dB. Table 1 shows the wideband SPDT switch design requirements and performance at 10 GHz, 26 GHz, and 38 GHz. The performance metrics of the designed SPDT switch satisfy the 5G/6G. From Table 2, the simulated results of the proposed SPDT switch are summarised and compared with the state-of-the-art designs at 10 GHz. The excellent input and output reflection coefficients of −16 dB and −70 dB, respectively, compensate for the low forward transmission gain

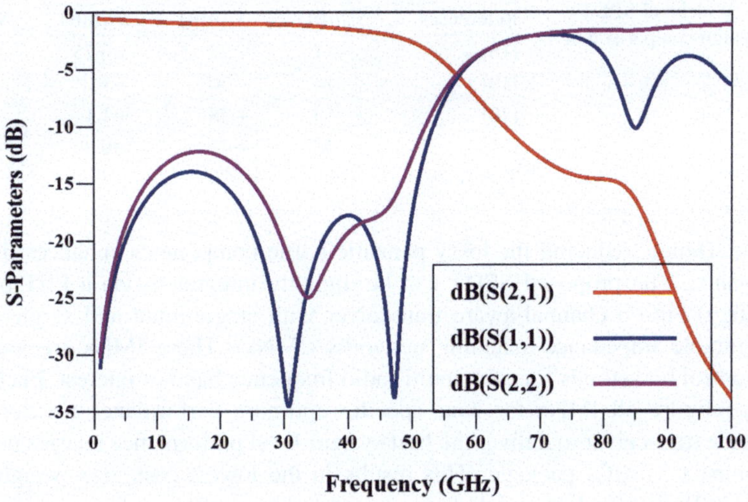

Fig. 4 Insertion loss, input and output return loss of distributed SPDT switch

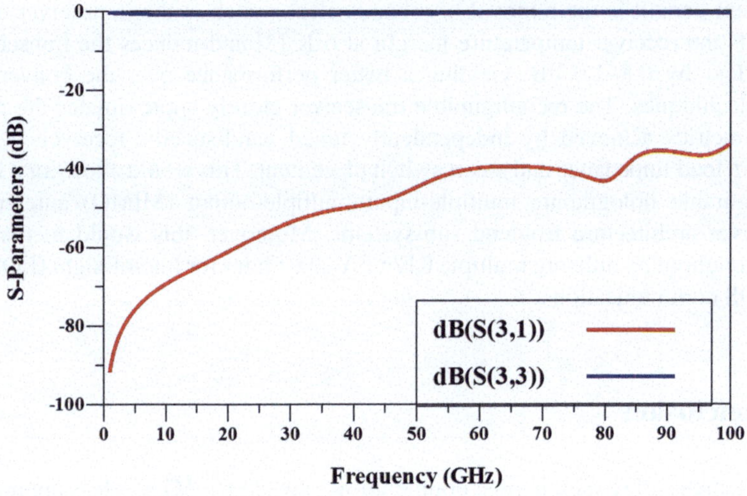

Fig. 5 Isolation loss of the SPDT switch

Table 1 Wideband SPDT switch design requirements and performance at 10 GHz, 26 GHz, and 38 GHz frequencies

Design parameter	Requirement	10 GHz	26 GHz	38 GHz
S_{11}(dB)	≤−10	−21	−21	−21
S_{12}(dB)	≤−40	−60	−21	−21
S_{21}(dB)	≥25	40	−21	−21
S_{22}(dB)	≤−10	−40	−21	−21
In-band ripple (dB)	≤3	1	−21	−21

Table 2 Wideband SPDT switch design comparison at 10 GHz

References	S_{11}(dB)	S_{12}(dB)	S_{21}(dB)	S_{22}(dB)
[18]	≤−11	−25	−0.75	–
[19]	≤−14	−30	−1	–
[20]	≥−13	−28	−1.1	–
[21]	≤−13	−44	−0.8	–
This work	−16	−70	−3	−15

response. Hence, reducing the lossy parasitic-borne components enhances the S21 performance. The proposed SPDT can be digitally integrated (via a FET process technology) onto a channel-aware transceiver with single-input and single-output reconfigurable impedance-matching networks (IMNs). These IMNs are varactor-based control bit settings for each multi-radio frequency band of interest. Each bit is set to configure all IMNs for band-specific optimum performance metrics. This enables the transceiver to deliver the best system-level performance metrics possible in a compact MMICs package. This results in the lowest cost, size, weight, and power (C-SWaP) adaptive and dynamic integrated transmitter and receiver frontend architecture possible. This novel reconfigurable receiver design approach obviates multiple shortcomings of conventional multiband receiver frontend designs. The prominent benefit is the removal of a band-select switch at the transceiver output. This sets the receiver temperature margin at 6 K [3] and reduces the transceiver's output loss by 0.8–1.0 dB, yielding a better performance over the conventional design techniques. The reconfigurable transceiver closely approximates the performance metrics achieved by independently tuned band-specific receivers with an optimum load impedance and smart switch placement. This would allow for a hybrid reconfigurable holographic multiple-input multiple-output (MIMO) antenna and transceiver architecture frontend sub-systems. Moreover, this would increase the data throughput by utilising multiple RF/mmWave nodes for line-of-sight (LOS) and non-LOS communications.

5 Conclusion

The designed SPDT (single pole double throw) switch for 5G wireless applications represents a significant advancement in radio frequency (RF) technology, providing crucial functionality in the rapidly evolving landscape of wireless communication. The switch is designed to operate at the frequency range of DC-50 GHz with less than 3 dB insertion loss and more than 40 dB isolation, for X-band and K/Ka band frequency applications. This switch offers numerous benefits and addresses critical challenges faced in 5G networks. The switch's compact and power-efficient design contributes to reducing the overall footprint and power consumption of 5G wireless access points, thereby enhancing network scalability and energy efficiency. Additionally, its robust and durable construction ensures reliable operation even in challenging environmental conditions, contributing to the longevity and stability of 5G infrastructure.

References

1. Zafar, M., Ekpo, S., George, J., Sheedy, P., Uko, M., Gibson, A.: Hybrid power divider and combiner for passive RFID tag wireless energy harvesting. IEEE Access. **10**, 502–515 (2021)
2. Lin, K.-Y., Wang, Y.-J., Niu, D.-C., Wang, H.: Millimeterwave MMIC single-pole-double-throw passive HEMT switches using impedance transformation networks. IEEE Trans. Microw. Theory Tech. **51**, 1076–1085 (2003)
3. Ekpo, S.C., George, D.: Impact of noise figure on a satellite link performance. IEEE Commun. Lett. **15**, 977–979 (2011)
4. Ayasli, Y.: Microwave switching with GaAs FETs. Microw. J. **25**, 61–64 (1982)
5. Kim, J., Park, J., Kim, J.: 28 GHz common-leg T/R IC in 65 nm CMOS technology. Electron. Lett. **54**(10), 616–618 (2018)
6. Fang, W., Chen, C., Lin, Y.: 2.4-GHz absorptive MMIC switch for switched beamformer application. IEEE Trans. Microw. Theory Tech. **65**, 3950–3961 (2017)
7. Zobilah, A.M., Zakaria, Z., Shairi, N.A.: Selectable multiband isolation of single pole double throw switch using transmission line stub resonator for WiMAX and LTE applications. IET Microw. Antennas Propag. **11**(6), 844–851 (2017)
8. Nguyen, D.P., Pham, A., Aryanfar, F.: A k-band high power and high isolation stacked-FET single pole double throw MMIC switch using resonating capacitor. IEEE Microw. Wirel. Comp. Lett. **26**, 696–698 (2016)
9. Uko, M., Ekpo, S.: A 23-28 GHz pHEMT MMIC low-noise amplifier for satellite-cellular convergence applications. Int. Rev. Aerospace Eng. J. **14**(5), 1–10 (2021)
10. Ekpo, S.C., Adebisi, B., Wells, A.: Regulated-element frost beamformer for vehicular multimedia sound enhancement and noise reduction applications. IEEE Access. **5**, 27254–27262 (2017)
11. Chang, H., Chan, C.: A low loss high isolation DC-60 GHz SPDT traveling wave switch with a body bias technique in 90 nm CMOS process. IEEE Microw. Wirel. Comp. Lett. **20**, 82–84 (2010)
12. Uko, M., Ekpo, S.: 8-12 GHz pHEMT MMIC low-noise amplifier for 5G and fiber-integrated satellite applications. Int. Rev. Aerospace Eng. **13**(3), 99–107 (2020)
13. Rao, C.V.N., Ghodgaonkar, D.K., Sharma, N.: GaAs MMIC low noise amplifier with integrated high-power absorptive receive protection switch. IEEE Microw. Wirel. Comp. Lett. **28**, 1128–1130 (2018)
14. Hieda, M., Nakahara, K., Miyaguchi, K., Kurusu, H., Iyama, Y., Takagi, T., Urasaki, S.: High-isolation series-shunt FET SPDT switch with a capacitor cancelling fet parasitic inductance. IEEE Trans. Microw. Theory Techn. **49**, 2453–2458 (2001)
15. Ekpo, S.: Thermal subsystem operational times analysis for ubiquitous small satellites relay in LEO. Int. Rev. Aerospace Eng. **11**(2), 48–57 (2018)
16. Ekpo, S.C.: Parametric system engineering analysis of capability-based small satellite missions. IEEE Syst. J. **13**(3), 3546–3555 (2019)
17. Lau, I., Ekpo, S., Zafar, M., Ijaz, M., Gibson, A.: Hybrid mmWave-Li-Fi 5G architecture for reconfigurable variable latency and data rate communications. IEEE Access. **11**, 42850 (2023)
18. Erturk, V., Gurdal, A., Ozbay, E.: A high-power and broadband GaN SPDT MMIC switch using gate-optimized HEMTs. IEEE Microw. Wirel. Technol. Lett. **33**, 1207 (2023)
19. Campbell, C.F., Dumka, D.C.: Wideband high power GaN on SiC SPDT switch MMICs. In: 2010 IEEE MTT-S International Microwave Symposium, pp. 145–148. IEEE (2010)
20. D'Angelo, S., Biondi, A., Scappaviva, F., Resca, D., Monaco, V.A.: A GaN MMIC chipset suitable for integration in future X-band spaceborne radar T/R module frontends. In: 2016 21st International Conference on Microwave, Radar and Wireless Communications (MIKON), pp. 1–4. IEEE (2016)
21. Memioglu, O., Turan, D.I., Kocer, F., Civi, O.A.: A high power, GaN, quarter-wavelength switch for X-band applications. In: 2018 18th Mediterranean Microwave Symposium (MMS), pp. 195–197. IEEE (2018)

Multi-Radio Frequency Antenna for Sub-6 GHz 5G Carrier and Data Link Margin Enhancement

Sunday Enahoro, Sunday Cookey Ekpo, Andy Gibson, Kin Kee Chow, Helen Ji, and Khaled Rabie

1 Introduction

Wireless technologies and communication systems are nowadays increasing at a fast pace, and the biggest challenge is designing and optimizing ultralow-power, low-cost, miniature wireless communication devices without degrading the performance matrices such as data rate and range characteristics.

To exploit the system's full capabilities, two common questions arise when designing high-speed data links. What distance the signal can cover and at what throughput? Wireless radio communication systems' performance depends to a large extent on these two characteristics. Wireless system radio performance is affected by many factors such as the quality of the equipment used, antenna gain, receiver sensitivity, environmental conditions, and available output power. It is therefore imperative that a detailed site analysis and network layout design be undertaken before deploying the network on a large scale [2].

To improve throughput and directivity, multiple antennas can be integrated in a device. The integration of multiple antennas in a single device causes mutual coupling between them due to limited space [6–8]. These couplings limit antenna characteristics and performance metrics. Hence, multi-broadband design trade-offs such as radiation pattern, bandwidth, efficiency, beamforming, beamwidth, and input impedance must be carefully implemented [3–5, 9, 10, 11].

The challenge is to design and optimize the performance of an antenna for a limited form factor/volume. To address some of these issues, a novel multi-radio solution standard planar monopole antenna that covers the sub-6 GHz 5G mobile wireless radio access technology (RAT) is presented.

S. Enahoro (✉) · S. C. Ekpo · A. Gibson · K. K. Chow · H. Ji · K. Rabie
Department of Engineering, Manchester Metropolitan University, Manchester, UK
e-mail: Sunday.enahoro@stu.mmu.ac.uk

S. C. Ekpo (ed.), *The Second International Adaptive and Sustainable Science, Engineering and Technology Conference*, Signals and Communication Technology,
https://doi.org/10.1007/978-3-031-53935-0_10

2 Radio Communication System Network Model

To design a robust network, a balanced trade-off is essential [1]. The radio link is the first step in communication network design. It is needed to establish radio coverage and density of base stations in various operating environments.

The mobile station (MS) (uplink) power budget is more critical than base station (BS) (downlink) link budget since the BS transmitted power is much greater than that of the MS (due to MS size and space restrictions).

The flux density at the receiving antenna, resulting from the transmitting antenna gain, can be determined when considering a transmitter equipped with a hypothetical isotropic antenna. This isotropic antenna radiates energy uniformly in all directions. The expression for the flux density at the receiving antenna incorporates the transmitting antenna gain and is given by:

$$F = \frac{G_t P_t}{4\pi R^2}\ \text{W/m} \tag{1}$$

The power received by a receiving antenna with an effective aperture area located at distance d from a transmitting antenna with power P_t and antenna gain G_t is given by the following:

$$P_r = FA_e = \frac{P_t G_t}{4\pi d^2} A_e = \frac{P_t G_t}{4\pi d^2} . G_r\ \ \text{(Watts)} \tag{2}$$

where A_e is the antenna effective area in m^2, and it can be expressed as follows:

$$A_e = \frac{G\lambda^2}{4\pi} \tag{3}$$

And $\frac{P_t G_t}{4\pi d^2}$ = power flux density in (w/m^2) and $\frac{\lambda^2}{4\pi}$ is the spreading loss, s in m^2. Rearranging Eq. (2) gives the following:

$$P_r = P_t G_t G_r \left[\left(\frac{\lambda}{4\pi d}\right)^2\right] \tag{4}$$

The inverse form of the term in bracket represents the isotropic radiator free space path loss, L_{PL}, and it accounts for the inverse square loss.

$$L_{PL} = \left(\frac{4\pi d}{\lambda}\right)^2 \tag{5}$$

$$L_{dB} = 10\frac{P_t}{P_r} = 20\log\left(\frac{4\pi d}{\lambda}\right) = -20\log(\lambda) + 20\log(d) + 21.98\ \text{dB} \tag{6}$$

Different environments can be accommodated by modifying the Friis equation to account for decrease in signal power as the distance between the transmitter and receiver increases with nth distance power factor, i.e.:

$$P_{r(d)} = P_t \frac{G_t G_r \lambda^2}{(4\pi)^2 d^n L_0} \tag{7}$$

where n is the path loss exponent and takes constant empirical values depending on the modelled environment.

3 Broadband Monopole Antenna Design Considerations

The planar antenna consisting of four different shape monopoles, each designed to operate at 900 MHz, 1900 MHz, 3500 MHz, and 5500 MHz frequency bands, combined to form a single antenna with a common microstrip line feeding point.

Using the quarter wavelength of each resonance frequencies as the starting total length, CST Microwave Studio Suite 2021 was used to obtain the optimized length as shown in Table 1. The combination of folded monopoles 1, 2, and 3 are used to generate Ant1 with four resonance frequencies (3.09, 3.68, 4.72, and 5.87 GHz) with bandwidths of 1.064 GHz (2.943–4.007 GHz), 1.064 GHz (2.943–4.007 GHz), 0.253 GHz (4.611–4.864 GHz), and 0.331 GHz (5.737–6.058 GHz). Folded monopole four is then added to Ant1 which then retunes the antenna from four resonance frequencies to three resonance frequencies and generates the 5.38 GHz bands as shown in S11 plot of Fig. 2 and Table 2. The proposed multiband monopole antenna structure is shown in Fig. 1a with the antennas' detailed dimensions shown in Fig. 1c, d. The antenna is printed on a commercially available low-loss FR4 substrate with dielectric constant ? $r = 4.4$, loss tangent = 0.02, and substrate height, $h = 0.8$. The overall antenna occupies an area of 110 × 60 mm^2 with the radiating (monopole)

Table 1 Optimized geometry parameters of proposed antenna

Part	Value (mm)	Part	Value (mm)	Part	Value (mm)	Part	Value (mm)
W	60	b	29.5	m	6.5	Wp2	0.5
L	100	c	6	n	9	a	6
h	0.8	wgg	13	p	2	j	20
ly	10	g1	9.5	r	4	k	1.5
gslot	1	g3	2	s	5	wx	49
feedp	9.75	d	26	t	4	g	3
fl	10	e	5	slotW	58	g2	8.5
fw	1.5	f	2	Wgg2	6	Wp	1

Table 2 Simulation results at various antenna frequency stages

Antenna	Fr (GHz)	BW (GHz)	S11 (dB)	GAIN (dBi)	VSWR
Ant1	3.09	2.943–4.007 (34.4%)	−15.83	3.29	1.37
	3.68	2.943–4.007 (28.9%)	−13.35	4.22	1.55
	4.72	4.611–4.864 (5.40%)	−6.62	6.00	2.76
	5.87	5.727–6.058 (5.60%)	−21.90	2.66	1.23
Proposed	3.05	2.932–3.800 (28.5%)	−17.35	3.16	1.09
	3.42	2.932–3.800 (25.4%)	−9.76	4.52	1.90
	5.38	5.115–5.819 (12.3%)	−12.22	5.43	1.71

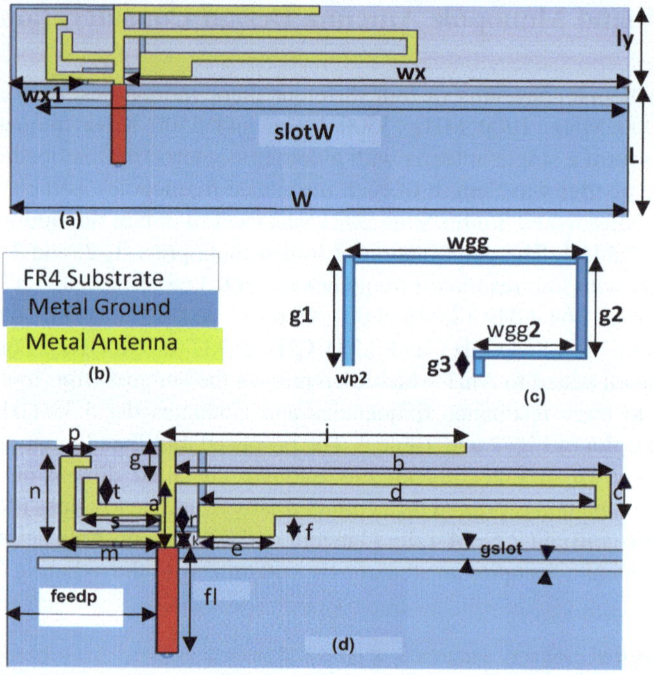

Fig. 1 Proposed antenna structure: (**a**) structure, (**b**) material definition, (**c**) ground patch dimensions, (**d**) detailed dimension of the proposed antenna

antenna occupying an area of 10 × 60 mm^2 with an antenna metal height of 0.035 mm. The PCB has a bottom ground layer dimension of 100 × 60 mm and a thickness of 0.035 mm. The antenna is fed through a 50 Ohm printed microstrip line (width = 1.5 mm) connected to a sub-miniature version A (SMA) through a ground via hole. The monopole antenna and the ground patch have 1 mm and 0.5 mm width, respectively.

4 Result and Discussion

This section presents the realization and analysis of the proposed antenna in terms of its operational performance metrics.

4.1 *Scattering Parameters*

Figure 2 presents the antenna reflection coefficient (S11) simulation result. From the figure, Ant1 resonates at four different frequency bands of 3.09, 3.68, 4.72, and 5.87 GHz with corresponding S11 responses of −15.83, −13.35, −6.62, and −21.90 dB. From the same plot, it is observed that the proposed antenna has three functional bands at −6 dB bandwidth with three resonant frequencies of 3.05, 3.42, and 5.38 GHz and corresponding return loss (S11) responses of −17.35, −9.76, and −12.22 dB. Table 2 presents the simulated results of the proposed antenna at different stages (Fig. 3).

4.2 *Surface Current Distribution*

Figure 4 displays the simulated electric field surface current distribution of the proposed antenna at the designed resonant frequencies. From the responses, it is observed that the current is highly concentrated at the lower section of the antenna

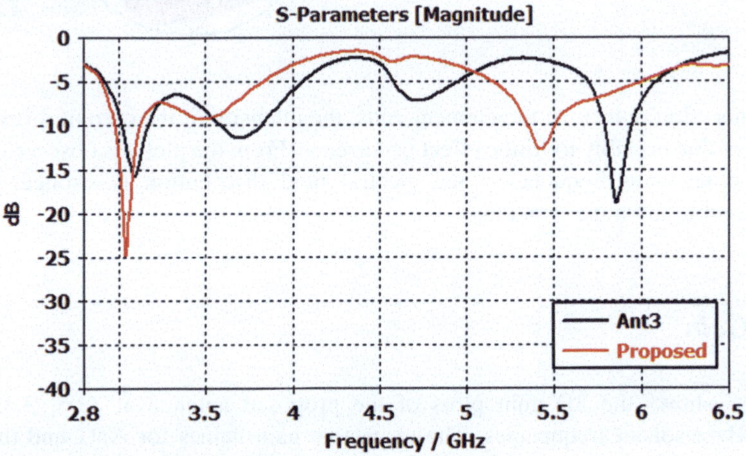

Fig. 2 S11 comparison of Ant3 and the proposed antenna

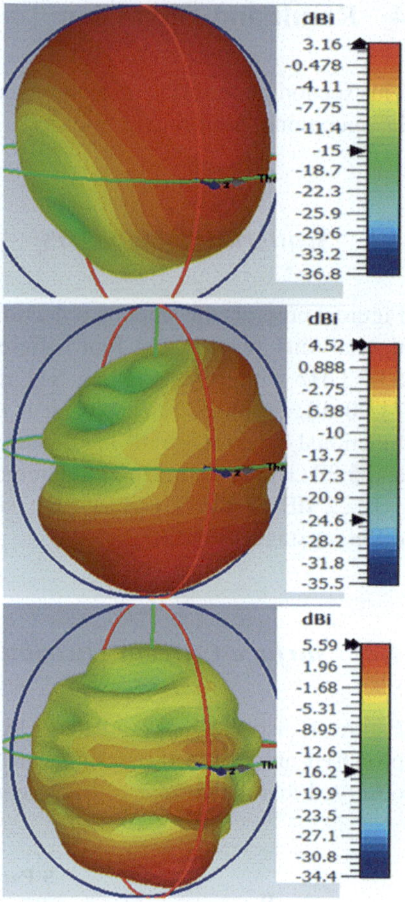

Fig. 3 3D gain plot of proposed antenna at 3.05, 3.42, and 5.38 GHz

(the connection point). At the element ends, the intensity of the current distribution decreases due possibly to fringe effect occurrence. From the plot it is observed that at specified resonant frequencies, the electric field distribution is stronger at the corresponding monopole antenna.

4.3 Gain

Figure 3 shows the 3D gain plots of the proposed antenna at 3.05, 3.42, and 5.38 GHz resonant frequencies. The respective gain values for Ant1 and the proposed antenna are shown in Table 2.

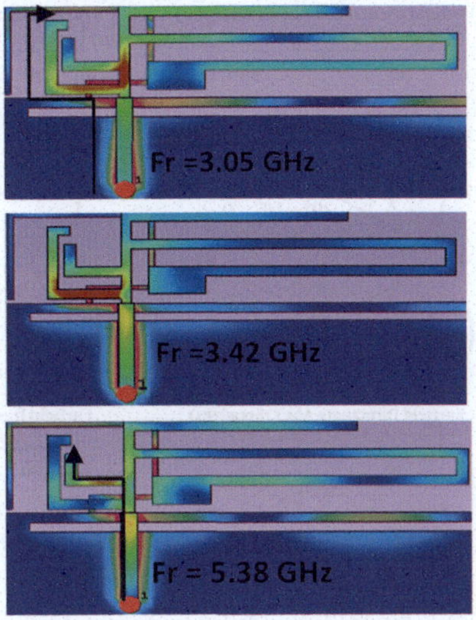

Fig. 4 Surface current distribution of proposed antenna at designed resonant frequencies

4.4 Radiation Pattern

Simulated radiation pattern plots of the proposed antenna at 3.05, 3.42, and 5.38 GHz resonance frequencies, respectively, are shown in Fig. 5a–i. The plot displays the antennas' far-field region at phi = 90° (*xy*-plane) and phi = 0° (*yz*-plane). The radiation patterns (gain) of the proposed antenna frequency bands measured at 3.05, 3.42, and 5.38 GHz are displayed in Fig. 3. The antenna peak values are 3.16, 4.52, and 5.59 dBi at 3.05, 3.42, and 5.38 GHz, respectively. The variations of the gain at upper bands are less than 0.26 dB, while the overall gain variation is less than 1.44 dB.

4.5 Input Impedance

Figure 6 shows the input impedance and the variation of real and imaginary part of the proposed antenna. It is observed from the plot that the first and third resonance occurring at around 3.05 and 5.38 GHz have slowly changed impedance values versus frequency, indicating that they are of the parallel types. Below the 3.05 and 5.38 GHz resonant frequencies, the antennas' impedance is inductive (positive imaginary parts) and capacitive for frequencies above 3.05 and 5.38 GHz (negative imaginary) (Fig. 7).

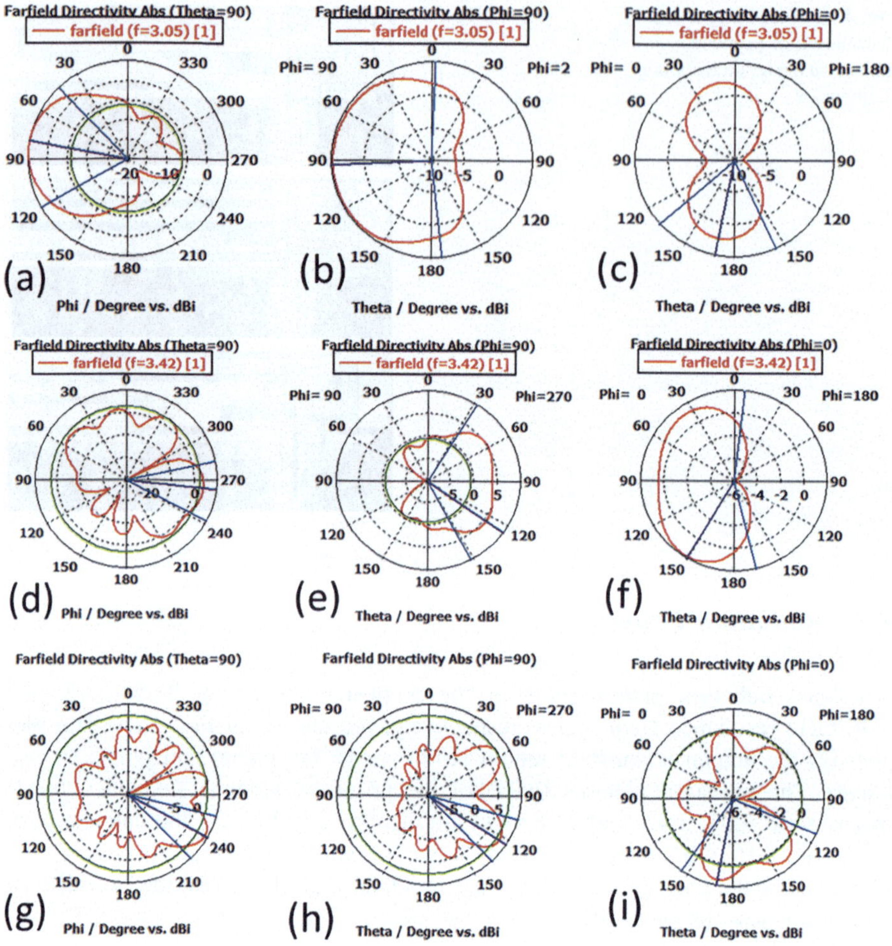

Fig. 5 Radiation patterns of projected antenna at 3.05 GHz (**a–c**), 3.42 GHz (**d–f**), and 5.38 GHz (**g–i**) at *xy*-, *yz*-, and *xz*-planes

5 System Sensitivity and the Link Budget Calculation

Urban area cellular radio (NLOS urban macro, UMa) 5G propagation model defined in [10] applicable to 0.5–100 GHz frequency bands is considered for the link budget calculation. Figure 8 shows the antennas' received and transmitted (radiated) power in dBm at the designed resonance frequencies. The figure shows the variation between the received and transmitted power due to transmission media and other losses. The antenna does not radiate all the received power due of these losses, thereby reducing the antennas' efficiency (Fig. 7). Table 3 displays the calculated values of the link budget at the receiving antenna (sensitivity) for the system path loss, effective isotropic radiated power (EIRP), and the receive signal power level.

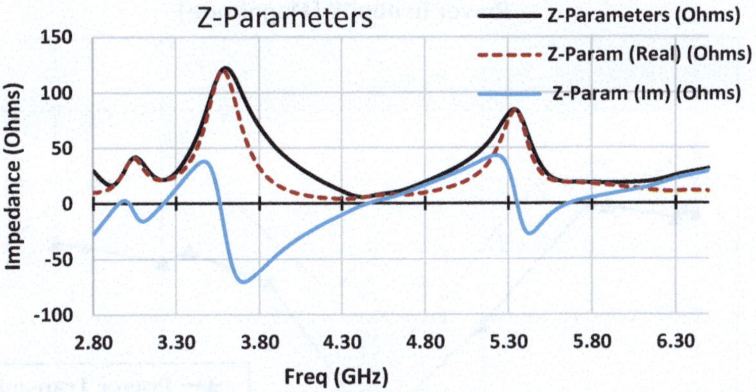

Fig. 6 Input impedance of the proposed communication system

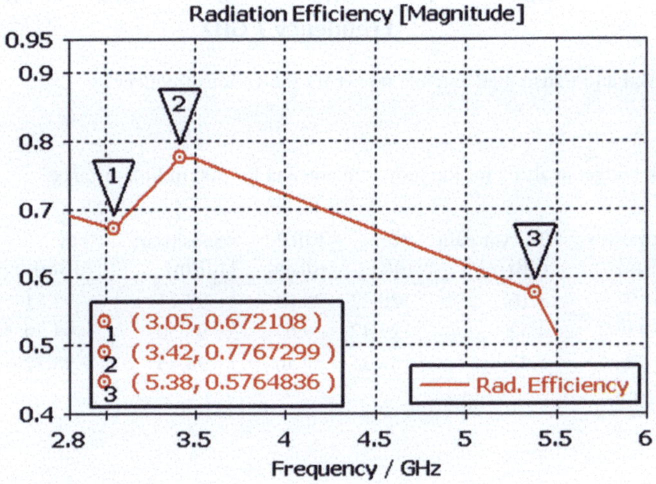

Fig. 7 Radiation efficiencies of the proposed antenna at frequencies 3.05, 3.42, and 5.38 GHz

(i) *The path loss*, PL, of a RF link is given by the following:

$$PL_{(UMA,NLOS)} = 13.54 + 39.08 \log_{10}(d) + 20 \log_{10}(fc) - 0.6 * (h_{UT} - 1.5) \quad (8)$$

From Eq. 8, we have the path loss for a 500 m distance at 3.05 GHz, PL (3.05 GHz), given by the following:

$$\begin{aligned} PL_{(UMA,NLOS)} &= 13.54 + 39.08 \log_{10}(d) + 20 \log_{10}(f_{3.05GHz}) - 0.6(h_{UT} - 1.5) \\ &= 13.54 + 39.08* \log_{10}(500) + 20* \log_{10}(3.05) - 0.6(2 - 1.5) \\ &= 128.42 \text{ dB} \end{aligned}$$

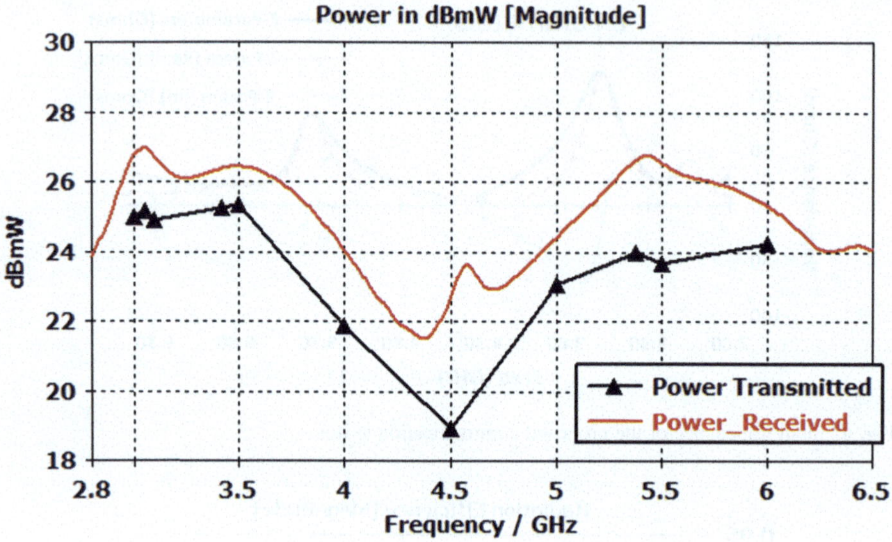

Fig. 8 Received and transmitted antenna power for the various frequencies

Table 3 Link budget analysis for the proposed antenna for 500 m link distance

Freq (GHz)	Tx power (dBm)	Tx Ant gain (dBi)	PL (dB)	EIRP (dBm)	Sensitivity (dBm)	TIS (dBm)	Link margin (dB)
3.05	25.19	3.16	128.42	28.04	−97.21	−95.51	3.45
3.42	25.41	4.52	129.42	29.63	−95.50	−94.39	0.53
5.38	24.33	5.43	133.35	29.46	−98.45	−95.72	0.29

(ii) *Effective isotropic radiated power* (EIRP):

$$\mathrm{EIRP}_{\mathrm{dB}} = P_{\mathrm{Tx(dBm)}} + G_{\mathrm{Tx(dBi)}} - L_{\mathrm{wg(dB)}} \tag{9}$$

$$\mathrm{EIRP}_{(3.05\mathrm{GHz})\mathrm{dB}} = 25.19 + 3.16 - 0.3 = 28.05\ \mathrm{dB}$$

(iii) Receiver sensitivity (minimum receive power signals):

$$\mathrm{RSL}_{\mathrm{dBm}} = \mathrm{EIRP}_{\mathrm{dB}} - \mathrm{PL} - G_{\mathrm{Rx(dBi)}} \tag{10}$$

$$\mathrm{RSL}_{(3.05\ \mathrm{GHz})\mathrm{dBm}} = 28.05 - 128.42 + 3.16 = 97.21\ \mathrm{dBm}$$

(iv) *Total isotropic sensitivity:*

$$\mathrm{TIS}_{[\mathrm{dBm},3.05\mathrm{GHz}]} = \text{Receiver Sensitivity}_{[\mathrm{dBm},3.05]} - \text{Antenna Efficiency}_{[\mathrm{dB},3.05]} \quad (11)$$

$$\mathrm{TIS}_{[\mathrm{dBm},3.05\mathrm{GHz}]} = -97.21 - 10 * \log 10\left(\frac{67.76}{100}\right) = 95.51\ \mathrm{dBm}$$

(v) *Received power* (dBm):

$$P_{\mathrm{Rx}} = T_{\mathrm{x}} + G_{\mathrm{Tx}} + G_{\mathrm{Rx}} - \mathrm{FSPL} \quad (12)$$

$$P_{\mathrm{Rx}[\mathrm{dBm},3.05\ \mathrm{GHz}} = T_{\mathrm{x}[\mathrm{dbm},3.05\mathrm{GHz}]} + G_{\mathrm{Tx}[\mathrm{dBi},3.05\ \mathrm{GHz}]} + G_{\mathrm{Rx}[\mathrm{dBi},3.05} - \mathrm{FSPL}_{[\mathrm{dB},3.05]}$$
$$= 25.19 + 3.16 + 3.16 - 128.42 = -96.92\ \mathrm{dBm}$$

(vi) *Link Margin* (DB)*:*

$$\mathrm{LM\ (dB)} = \text{Received Power (dBm)} - \text{Receiver Sensitivity (dBm)}$$
$$= P_{\mathrm{Rx}} - \text{sensitivity} \quad (13)$$
$$\mathrm{LM_{dBm}} = -96.92 - (-100.37) = 3.45\ \mathrm{dB}$$

Figure 9 shows plot of Friis equation for received power in dBm for a range of distances in metres. From the plot, the received power decreases as the distance increases. Figure 10 shows plot of the antenna propagation path loss as a function of

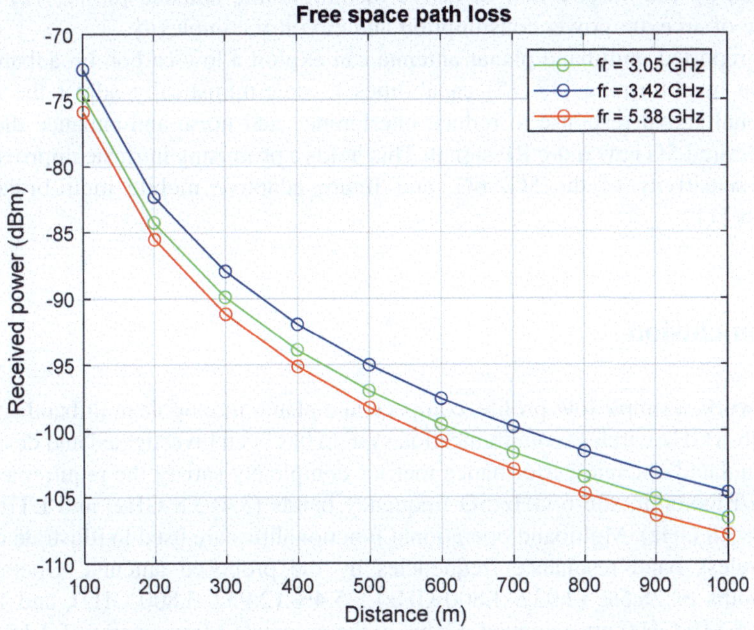

Fig. 9 Received signal power using Friis model for the proposed antenna transmission at $f = 3.05$, 3.42, and 5.38 GHz

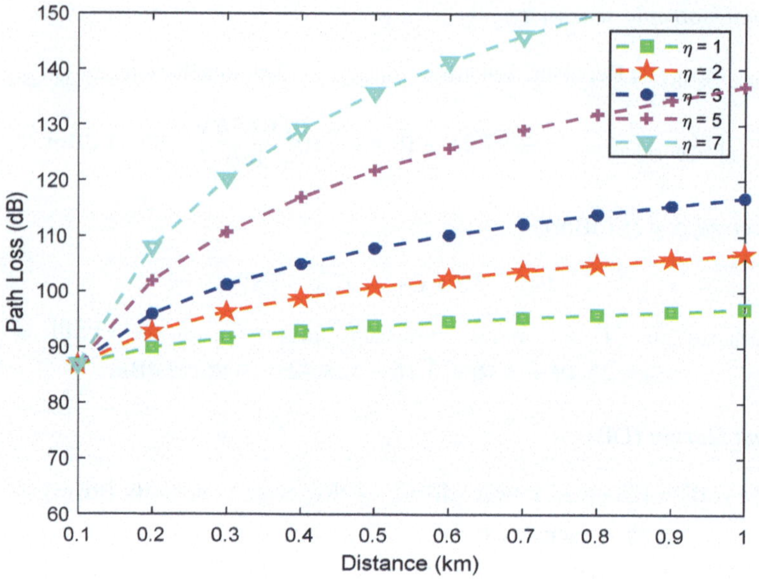

Fig. 10 Path loss vs distance of the proposed antenna

distance and path loss exponent n (Eq. 7). The figure also shows how large-scale fading increases as distance and path loss exponent n increases. The loss can be mitigated by the introduction of active reconfigurable antenna [6, 11, 12] at the expense of an extra power consumption and circuitry complexity.

The reported multiband planar antenna can exploit a low-carbon broadband and adaptive beamforming [13–15] capabilities to deterministically adjust the arrays directional signal response to reduce interference and noise and enhance the gain of the desired 5G new radio RF signal. This holds a promising intrinsic improvement of the sensitivity of the 5G, 6G, and future adaptive mobile multi-broadband handsets [11].

6 Conclusion

In this work, a simple low-profile, compact, and planar monopole multiband antenna for a sub-6 GHz wireless communication system has been investigated and designed. The simulated antenna performance metrics completely satisfy the requirements of WLAN/Bluetooth, sub-6 GHz 5G frequency bands (3.4–3.8 GHz) and LTE band 46 (5.2–5.8 GHz). Multiband operational functionalities are used to illustrate different wireless band resonance frequencies by the proposed antenna. Operational bandwidths of 28.5% (2.932–3.800 GHz), 25.4% (2.932–3.800 GHz), and 12.3% (5.115–5.819 GHz) are achieved. Gains in the associated frequencies of 3.16, 4.52, and 5.43 dBi, respectively, are also achieved with the proposed antenna. The proposed antenna also achieves good VSWR (<2) in all resonance frequency bands, i.e. 1.09, 1.90, and 1.71. The investigated and proposed antenna has the capability of fulfilling performance of a modern handheld communication system device in the sub-6 GHz FR1 bands. Moreover, it demonstrates good radiation characteristics, optimal gain, and high radiation efficiency that meet small satellite (CubeSats and nanosatellites) and smart internet of things sensor connectivity requirements.

References

1. Garg, V.K.: Wireless Communications and Networking. Elsevier Morgan Kaufmann, Amsterdam (2007)
2. *Wireless Link Budget Analysis - Tranzeo* (no date). Available at: http://www.tranzeo.com/allowed/Tranzeo_Link_Budget_Whitepaper.pdf. Accessed Feb 20 2023
3. Sim, C., Cheng, P., Lee, C.: Low-profile multiband planar hybrid antenna design for mobile handset applications. Microw. Opt. Technol. Lett. **53**(4), 910–915 (2011). https://doi.org/10.1002/mop.25889
4. Ekpo, S.C.: Parametric system engineering analysis of capability-based small satellite missions. IEEE Syst. J. **13**(3), 3546–3555 (2019). https://doi.org/10.1109/JSYST.2019.2919526
5. Ekpo, S., George, D.: A deterministic multifunctional architecture for highly adaptive small satellites. Int. J. Satell. Commun. Policy Manag. **1**(2–3), 174–194 (2012). https://doi.org/10.1504/IJSCPM.2012.049543
6. Ekpo, S.C., George, D.: A system engineering analysis of highly adaptive small satellites. IEEE Syst. J. **7**(4), 642–648 (2013). https://doi.org/10.1109/JSYST.2012.2198138
7. Ekpo, S.C., George, D.: Reconfigurable cooperative intelligent control design for space missions. Recent Patents Sp. Technol. **2**(1), 2–11 (2012). https://doi.org/10.2174/1877611611202010002
8. Jalali Khalilabadi, A.: Planar multi-broadband antenna for LTE/5G/GPS/GSM/UMTS and WLAN/WiMAX wireless applications. Wirel. Pers. Commun. **118**(4), 2611–2620 (2021). https://doi.org/10.1007/s11277-021-08145-4
9. Millimeter-wave microstrip patch antenna design for 5G. Int. J. Innov. Technol. Explor. Eng. **8**(12), 1183–1187 (2019). https://doi.org/10.35940/ijitee.l3896.1081219
10. 3GPP TR 38.901 version 15.0.0 Release 15
11. Ekpo, S.C., George, D.: Impact of noise figure on a satellite link performance. IEEE Commun. Lett. **15**(9), 977–979 (2011). https://doi.org/10.1109/LCOMM.2011.072011.111073
12. Ekpo, S.C.: Thermal subsystem operational times analysis for ubiquitous small satellites relay in LEO. Int. Rev. Aerosp. Eng. **11**(2), 48–57 (2018). https://doi.org/10.15866/irease.v11i2.13663
13. Enahoro, S., Ekpo, S., Uko, M.C., Ansari, U., Altaf, A., Zafar, M.: Adaptive beamforming for mmWave 5G MIMO antennas. In: Proc. 21st IEEE Annual Wireless & Microwave Conference, pp. 1–4, Sand Key, Clearwater Beach, Florida, USA. (Online/Virtual) (28–29 April 2021). https://doi.org/10.1109/WAMICON47156.2021.9443616
14. Ekpo, S.C., Adebisi, B., Wells, A.: Regulated-element frost Beamformer for vehicular multimedia sound enhancement and noise reduction applications. IEEE Access. **5**, 27254–27262 (2017). https://doi.org/10.1109/ACCESS.2017.2775707
15. Enahoro, S., Ekpo, S., Gibson, A.: Massive multiple-input multiple-output antenna architecture for multiband 5G adaptive beamforming applications. In: 2022 IEEE22nd Annual Wireless and Microwave Technology Conference (WAMICON), p. 1–4 (2022). https://doi.org/10.1109/WAMICON53991.2022.9786100

Multiband Monopole Antenna Design for Sub-6 GHz 5G Internet of Things Applications

Sunday Enahoro, Sunday Cookey Ekpo, Andy Gibson, Kin Kee Chow, Helen Ji, and Khaled Rabie

1 Introduction

Narrowband IoT (NB-IoT), a new radio (NR) interface [1] introduced in Release 13 by 3GPP is a technology designed to enable the connection of a large number of devices with diverse application areas. The application of the Internet has extended to include surveillance, social media, banking, machine-type communication, and connected vehicles (vehicle to everything). The RF antenna front end must sustainably achieve good isolation between elements respecting the lower size, weight, power, and cost (SWaP-C) design considerations which present new design challenges for smaller form factor devices [2–4]. Hence, multi-broadband radio design trade-offs such as radiation pattern, radiation efficiency, gain, beamforming, and input impedance must be carefully implemented [5–8] to reduce antennas' complexities and improve the performance metrics. Multiband antennas capable of covering wide bandwidths and functionalities are being integrated with communication devices such as smartphones and CubeSats [4–7].

The trend is to implement an antenna system characteristic for a low-component footprint form factor/volume for mobile IoT applications. To tackle some of these design issues, this work presents a novel multi-radio antenna solution standard that covers the sub-6 GHz 5G-IoTs mobile wireless radio access technology (RAT).

S. Enahoro (✉) · S. C. Ekpo · A. Gibson · K. K. Chow · H. Ji · K. Rabie
Department of Engineering, Manchester Metropolitan University, Manchester, UK
e-mail: Sunday.enahoro@stu.mmu.ac.uk

S. C. Ekpo (ed.), *The Second International Adaptive and Sustainable Science, Engineering and Technology Conference*, Signals and Communication Technology,
https://doi.org/10.1007/978-3-031-53935-0_11

2 Antenna Design Consideration for IoT Applications

Four monopoles each folded to reduce antenna size are designed to operate at 900 MHz, 1900 MHz, and 3500 MHz frequency bands with initial simulation quarter-wavelength (83.33, 41, and 21) values combined to form a single antenna with a common microstrip feed point. The proposed antenna is shown in Fig. 1a and the detailed dimensions in Fig. 1c, d. A high-performance 3D EM solver and analysis software package (i.e., CST Studio Suite 2021) is utilized for the antenna analysis and optimization. Table 1 shows the optimized antenna design parameters. The antenna is realized on a low-loss FR4 substrate (110 × 60 × 0.8 mm^3) with a dielectric constant of 4.4 and a loss tangent of 0.02.

Various resonance modes (0.25λ, 0.5λ, 0.75λ) are observed in the antenna's output. The quarter-wavelength mode of Ant1 (type 1, Fig. 2) resonates at 1000 MHz and the open-ground slot at 850 MHz. A tuning pad (type 3) added to Ant1 (type 2) tunes the frequency band to GSM bands (820–1050 MHz). The 0.75λ mode of an inverted-L antenna (Ant2, type 1) (Fig. 3) coupled with the 0.75λ mode of Ant1 is used to generate and tune the frequency bandwidth to 650 MHz

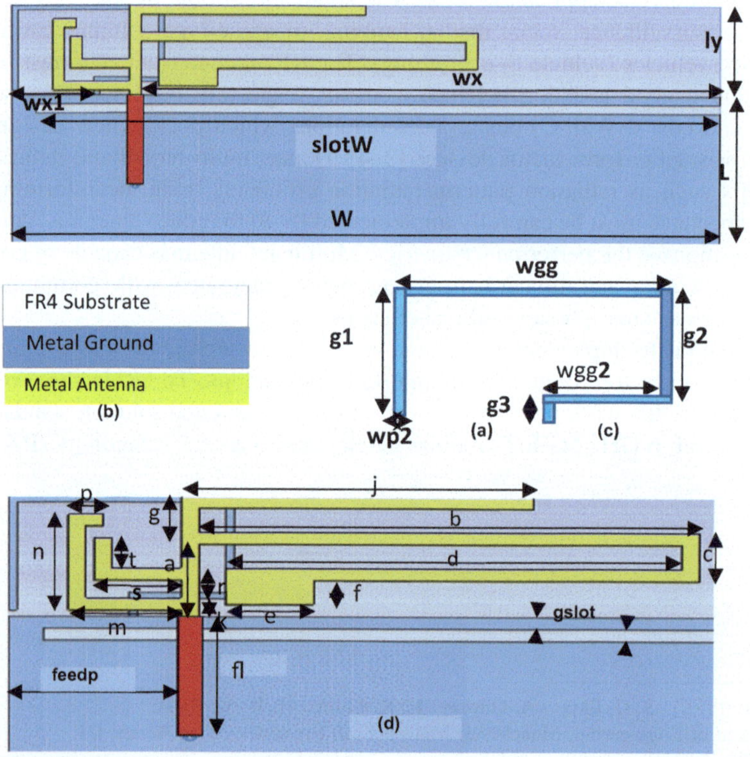

Fig. 1 Proposed antenna structure: (**a**) structure, (**b**) material definition, (**c**) ground patch dimensions, (**d**) detailed dimension of the proposed antenna

Table 1 Optimized geometry parameters of the proposed antenna

Part	Value (mm)	Part	Value (mm)	Part	Value (mm)	Part	Value (mm)
W	60	b	29.5	m	6.5	Wp2	0.5
L	100	c	6	n	9	a	6
h	0.8	wgg	13	p	2	j	20
ly	10	g1	9.5	r	4	k	1.5
gslot	1	g3	2	s	5	wx	49
feedp	9.75	d	26	t	4	g	3
fl	10	e	5	slotW	58	g2	8.5
fw	1.5	f	2	Wgg2	6	Wp	1

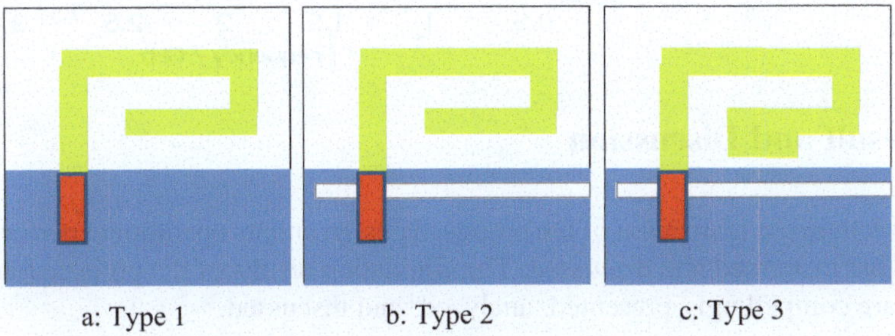

Fig. 2 Antenna 1 (Ant1, GSM) band). (**a**) Type 1. (**b**) Type 2. (**c**) Type 3

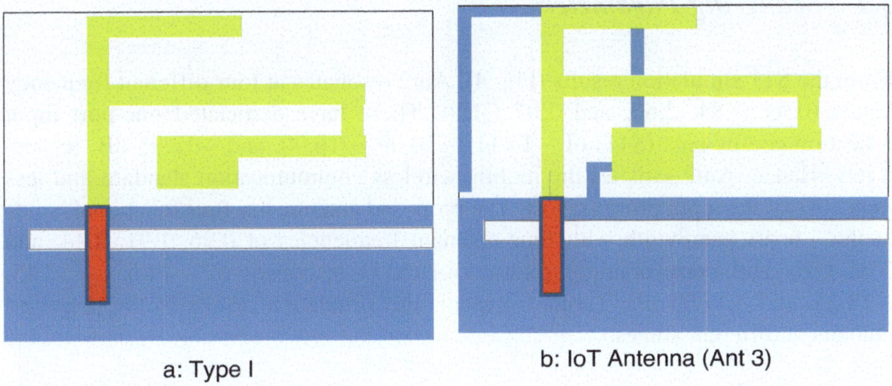

Fig. 3 Antenna 2 (Ant2) variants. (**a**) Type I. (**b**) IoT Antenna (Ant 3)

(1900–2550 MHz) from 400 MHz (2100–2550 MHz) of Ant1. A ground patch added to Ant2 (type 2) further tunes and improves on Ant2 frequency bandwidth to 175 MHz (850–1025 MHz), 1277 MHz (1540–2666 MHz), and 288 MHz (2920–3210 MHz). The simulated input reflection coefficients (S11 responses) are shown in Fig. 4.

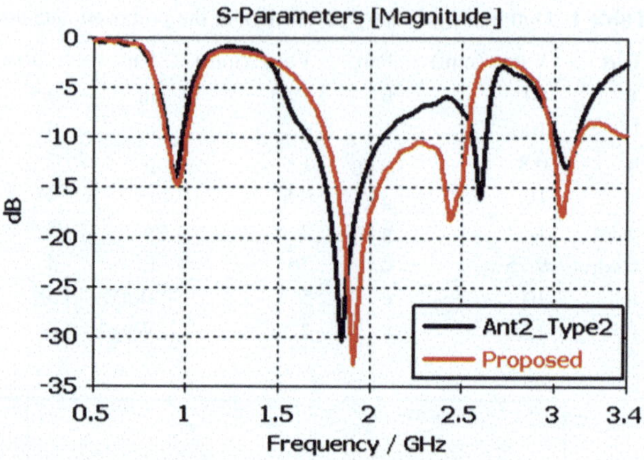

Fig. 4 S11 comparison of Ant2 and proposed IoT antenna

3 Result and Discussion

In this section, the realization of the proposed design and its operational performance metrics are examined and discussed. The simulation results of the proposed antenna model are compellingly presented, analyzed, and discussed.

3.1 *Scattering Parameters*

From the S11 simulation results (Fig. 4), Ant2 resonates at four different frequency bands (0.95, 1.84, 2.60, and 3.07 GHz). These have associated one-port input reflection coefficients (S11) of −14.11, −30.09, −16.00, and −12.95 dB, respectively. Hence, Ant2 satisfies the mobile wireless communication standard and less than 10% of the input power is lost. The proposed antenna has four functional bands at the −6 dB bandwidth with four resonant frequencies of 0.96, 1.91, 2.44, and 3.05 GHz and corresponding return loss (S11) responses of −14.65, −32.45, −18.25, and −17.35 dB. Table 2 presents the simulation results of the proposed antenna at different stages.

3.2 *Gain*

Gain is another antenna performance characteristic that represents an antenna's capability to concentrate its radiated power in a particular direction compared to an isotropic radiator. Figure 5 shows 3D gain plots of the projected antenna at 0.96, 1.91, 2.44, and 3.05 GHz resonant frequencies. The respective gain values for Ant2, Ant3, and the proposed antenna are shown in Table 2.

Table 2 Simulation results at various antenna frequency stages

Antenna	Fr (GHz)	BW (GHz)	S11 (dB)	GAIN (dBi)	VSWR
Antenna1 (Ant3)	0.95	0.857–1.031 (18.3%)	−15.48	1.80	1.43
	1.88	1.646–2.586 (50.0%)	−27.37	3.65	1.15
	2.45	1.646–2.586 (38.4%)	−19.71	1.64	1.23
	3.09	2.943–4.007 (34.4%)	−15.83	3.29	1.37
Proposed	0.96	0.854–1.043 (19.6%)	−14.65	1.86	1.43
	1.91	1.673–2.559 (46.4%)	−32.45	3.60	1.02
	2.44	1.673–2.559 (36.3%)	−18.25	1.74	1.22
	3.05	2.932–3.800 (28.5%)	−17.35	3.16	1.09

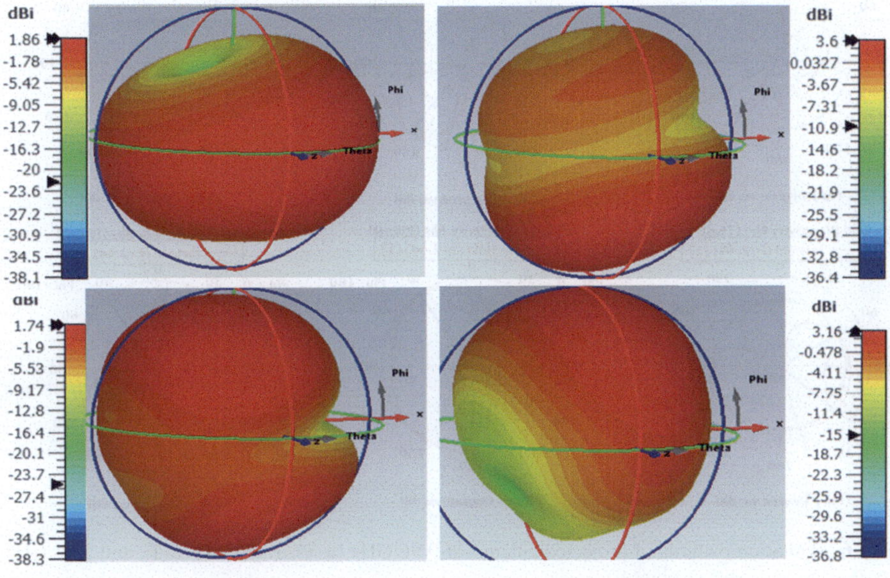

Fig. 5 3D gain plot of proposed antenna at 0.96, 1.91, 2.44, and 3.05 GHz

3.3 Radiation Pattern

Figure 6a–i illustrates the simulated radiation pattern plots of the proposed antenna at the various resonant frequencies (0.96, 1.91, 2.44, and 3.05 GHz). The plot shows the antenna far-field region at phi = 0° (*yz*-plane) and phi = 90° (*xy*-plane). The radiation patterns/gains of the various antenna frequency bands measured at 0.96, 1.91, 2.44, and 3.05 GHz are presented in Fig. 6. The antenna peak values are 1.86, 3.6, 1.74, and 3.16 dBi at 0.96, 1.91, 2.44, and 3.05 GHz, respectively. The variation of the gain is less than 1.44 dB.

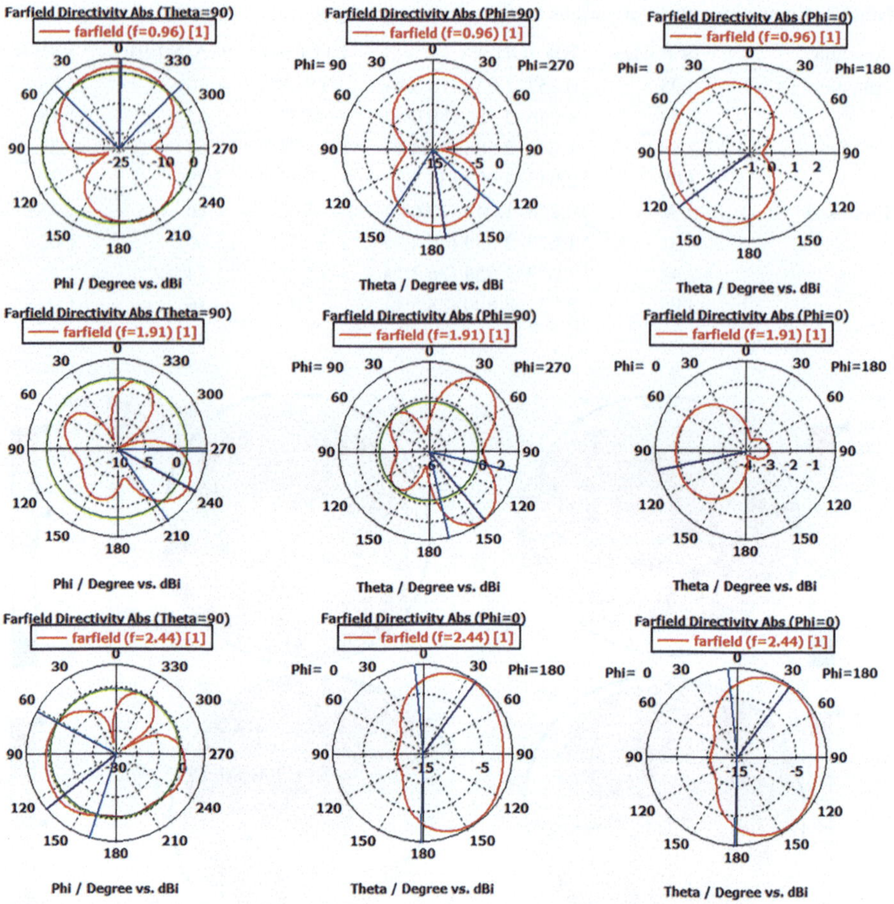

Fig. 6 Radiation patterns of projected antenna at 0.96 GHz (**a–c**), 1.91 GHz (**d–f**), and 2.44 GHz (**g–i**) at *xy*-, *yz*-, and *xz*-planes

3.4 Surface Current Distribution

Figure 7 shows the simulated antennas' electric field current distribution. From the responses, the current is highly concentrated at the lower section of the antenna. This current distribution intensity decreases as it approaches the various patch ends where the fringe effects may occur. In the lighter regions (orange/red), the current has an increasingly larger value. It is observed from the current distribution plots that at specific resonant frequencies, the electric field distribution is stronger at the corresponding monopole antenna. In addition, current distributions are also observed in nearby antennas because of mutual couplings helping to improve the bandwidth and resonance.

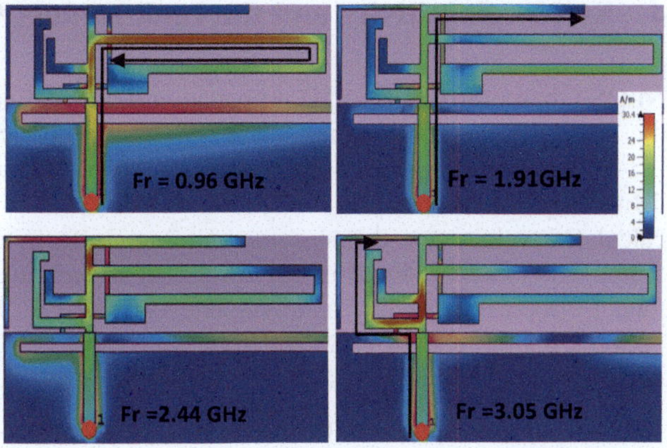

Fig. 7 Surface current distribution of projected antenna at various resonance frequencies

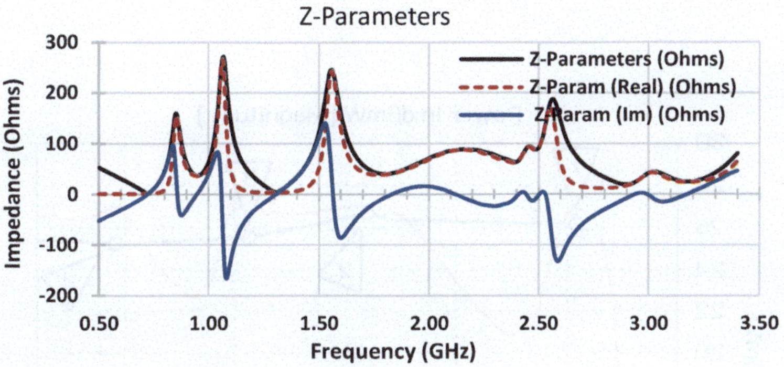

Fig. 8 Input impedance of the proposed antenna system

3.5 *Input Impedance*

The variation of the input impedance and the real and imaginary parts of the proposed antenna are shown in Fig. 8. From the plot, it is observed that the first resonance occurring at around 960 MHz has slowly changing impedance values versus frequency, indicating that it is of the series type. Below the 960 MHz frequency, the antennas' impedance is capacitive (negative imaginary parts), and an inductive impedance occurs for frequencies above 960 MHz (positive imaginary parts). The second, third, and fourth resonant frequencies (1.91, 2.44, and 3.05 GHz) also have slowly varying impedances with inductive and capacitive impedances at frequencies below and above the resonant frequencies, respectively.

4 System Sensitivity and the Link Budget Calculation

NLOS urban macro (UMa) 5G propagation model defined in [15] is considered for the link budget calculations. This model is applicable to frequency bands 0.5–100 GHz. Figure 9 shows the power received and transmitted (radiated) in dBmW by the antenna at various frequencies. Table 3 shows the link budget calculated values for the system path loss, effective isotropic radiated power (EIRP), the receive signal power level at the receiving antenna (sensitivity), and total isotropic sensitivity. A waveguide loss of 0.3 dB, a transmission distance between a base station (BS) and user equipment (UE) of 500 m, and a mobile antenna height of 2 m are assumed.

(i) *The path loss*, PL, of a RF link is given by the following:

$$PL_{(UMA,NLOS)} = 13.54 + 39.08^{*} \log_{10}(d) + 20^{*} \log_{10}(fc) - 0.6^{*}(h_{UT} - 1.5) \quad (1)$$

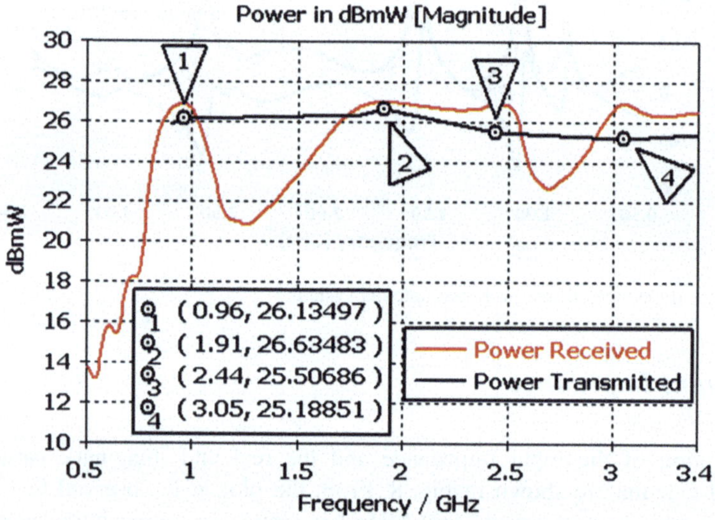

Fig. 9 Received and transmitted antenna power for the sub-6 GHz frequencies

Table 3 Link budget analysis for the proposed antenna for a 500 m link distance

Freq (GHz)	λ (m)	Trx power (dBm)	Tx Ant gain (dBi)	PL (dB)	EIRP (dBm)	Sensitivity (dBm)	TIS (dBm)
0.96	0.312	26.15	1.86	118.39	27.71	−88.82	−88.12
1.91	0.157	26.64	3.60	124.36	29.94	−90.82	−90.47
2.44	0.123	25.51	1.74	126.49	26.95	−97.80	−96.35
3.05	0.098	25.18	3.16	128.42	28.04	−97.22	−95.53

From (1), we have the path loss for a 500 m distance at 0.96 GHz, PL (0.96), given by the following:

$$\begin{aligned} PL_{(\mathrm{UMA,NLOS,0.96\ GHz})} &= 13.54 + 39.08^{*} \log_{10}(d) + 20^{*} \log_{10}(fc) - 0.6^{*}(h_{\mathrm{UT}} - 1.5) \\ &= 13.54 + 39.08^{*} \log_{10}(500) + 20^{*} \log_{10}(0.96) – 0.6^{*}(2 - 1.5) \\ &= 118.39\ \mathrm{dB} \end{aligned}$$

(ii) *Effective isotropic radiated power* (EIRP):

$$\mathrm{EIRP}_{\mathrm{dB}} = P_{\mathrm{Tx\ (dBm)}} + G_{\mathrm{Tx(dBi)}} – L_{\mathrm{wg(dB)}} \tag{2}$$

$$\mathrm{EIRP}_{(0.96)\mathrm{dB}} = 26.83 + 1.86 - 0.3 = 28.39\ \mathrm{dBm}$$

(iii) *Receiver sensitivity* (minimum receive power signals):

$$\mathrm{RSL}_{\mathrm{dBm}} = \mathrm{EIRP}_{\mathrm{dB}} - \mathrm{PL} + G_{\mathrm{Rx(dBi)}} \tag{3}$$

$$\mathrm{RSL}_{\mathrm{dBm}(0.96)} = 28.39 - 118.39 + 1.86 = -88.14\ \mathrm{dBm}$$

(iv) *Total isotropic sensitivity:*

$$\mathrm{TIS}_{[\mathrm{dBm}]} = \mathrm{Receiver\ Sensitivity}^{[\mathrm{dBm}]} - \mathrm{Antenna\ Efficiency}_{[\mathrm{dB}]} \tag{4}$$

$$\mathrm{TIS}_{[0.96\mathrm{GHz}]} = -88.82 - (10^{*} \log 10^{*}(85.14/100)) = -88.12\ \mathrm{dBm}$$

Mobile radio communication path loss is affected by various factors. Such factors include the antenna height of the base station and the height of the mobile antenna. Path loss models that incorporate these various factors have been developed such as 5G urban macro propagation model. The proximal antenna front-end deployment at a maximum of 0.5 m was chosen to ensure the lowest possible cable loss (with a maximum standard deviation of 0.2 and mean TL loss of 0.5 dB) introduction during each transmission regime. As the implementation is aimed at achieving energy- and spectrum-efficient communication, a wide range/coverage of multi-radio solutions [4, 5] for 5G/5G+ is possible with this novel design.

The presented novel antenna can enable reliable multiband RF energy harvesting and utilization and satellite-cellular convergence connectivity applications [10, 11] with a hybrid RF power divider combiner. The reported multi-broadband planar monopole antenna can utilize a low-carbon adaptive broadband beamforming algorithm [12–14] to deterministically adjust the directional response of the array to reduce noise, null interferences, and enhance the gain and recognition of the desired RF 5G signal. This holds a promising intrinsic enhancement of the sensitivity of the 5G, 6G, and future adaptive mobile multi-broadband handsets [9].

5 Conclusion

In this paper, a simple compact, planar, and low-profile multiband antenna for a sub-6 GHz IoT wireless communication system has been designed and presented. The simulated antenna performance metrics satisfy the requirements of the IoT devices operating at GSM850/900/1800/1900, UMTS/LTE2300/2500, WLAN/Bluetooth, and sub-6 GHz 5G frequency bands (3.4–3.8 GHz). It uses a simple and straightforward approach without the need for lumped elements. Operational bandwidths of 19.6% (0.854–1.043 GHz), 46.4% (1.673–2.559 GHz), 36.3% (1.673–2.559 GHz), 28.5% (2.932–3.800 GHz), and 25.4% (2.932–3.800 GHz) are achieved. The proposed antenna also achieves gains in the associated frequencies of 1.86, 3.6, 1.74, and 3.16 dBi, respectively. Good VSWR was also achieved with the antenna having a VSWR <2 in all resonance frequency bands, i.e., 1.43, 1.02, 1.22, and 1.09. The proposed antenna can satisfy a modern handheld communication system device in the sub-6 GHz FR1 bands. It also demonstrates good radiation characteristics, optimal gain, and high radiation efficiency that meet small satellites (CubeSats and nanosatellites) and smart IoT sensor connectivity requirements.

References

1. Wang, Y.P.E., Lin, X., Adhikary, A., Grovlen, A., Sui, Y., Blankenship, Y., Bergman, J., Razaghi, H.S.: A primer on 3GPP narrowband internet of things. IEEE Commun. Mag. **55**(3), 117–123 (2017)
2. Sim, C., Cheng, P., Lee, C.: Low-profile multiband planar hybrid antenna design for mobile handset applications. Microw. Opt. Technol. Lett. **53**(4), 910–915 (2011). https://doi.org/10.1002/mop.25889
3. Ekpo, S.C.: Parametric system engineering analysis of capability-based small satellite missions. IEEE Syst. J. **13**(3), 3546–3555 (2019). https://doi.org/10.1109/JSYST.2019.2919526
4. Ekpo, S., George, D.: A deterministic multifunctional architecture for highly adaptive small satellites. Int. J. Satell. Commun. Policy Manag. **1**(2–3), 174–194 (2012). https://doi.org/10.1504/IJSCPM.2012.049543
5. Ekpo, S.C., George, D.: A system engineering analysis of highly adaptive small satellites. IEEE Syst. J. **7**(4), 642–648 (2013). https://doi.org/10.1109/JSYST.2012.2198138
6. Ekpo, S.C., George, D.: Reconfigurable cooperative intelligent control design for space missions. Recent Patents Sp. Technol. **2**(1), 2–11 (2012). https://doi.org/10.2174/1877611611202010002
7. Jalali Khalilabadi, A.: Planar multi-broadband antenna for LTE/5G/GPS/GSM/UMTS and WLAN/WiMAX wireless applications. Wirel. Pers. Commun. **118**(4), 2611–2620 (2021). https://doi.org/10.1007/s11277-021-08145-4
8. Yang, L., Cui, Y., Li, R.: A multiband uniplanar antenna for LTE/GSM/UMTS, GPS, and WLAN/WiMAX handsets. Microw. Opt. Technol. Lett. **57**(12), 2761–2765 (2015). https://doi.org/10.1002/mop.29430
9. Uko, M., Ekpo, S.: A 23–28 GHz pHEMT MMIC low-noise amplifier for satellite-cellular convergence applications. Int. Rev. Aerospace Eng. J. **14**(5), 1–10 (2021). https://doi.org/10.15866/irease.v14i5.20361

10. Ekpo, S.C., Adebisi, B., George, D., Kharel, R., Uko, M.: System-level multicriteria modelling of payload operational times for communication satellite missions in LEO. Recent Prog. Space Technol. **4**(1), 67–77 (2014). https://doi.org/10.2174/2210687104666140620221119
11. Ekpo, S.C.: Thermal subsystem operational times analysis for ubiquitous small satellites relay in LEO. Int. Rev. Aerosp. Eng. **11**(2), 48–57 (2018). https://doi.org/10.15866/irease.v11i2.13663
12. Enahoro, S., Ekpo, S., Uko, M.C., Ansari, U., Altaf, A., Zafar, M.: Adaptive beamforming for mmWave 5G MIMO antennas. In: Proc 21st IEEE Annual Wireless & Microwave Conference, pp. 1–4, Sand Key, Clearwater Beach, Florida, USA (Online/Virtual) (28–29 April 2021). https://doi.org/10.1109/WAMICON47156.2021.9443616
13. Ekpo, S.C., Adebisi, A., Wells, A.: Regulated-element frost Beamformer for vehicular multimedia sound enhancement and noise reduction applications. IEEE Access. **5**, 27254–27262 (2017). https://doi.org/10.1109/ACCESS.2017.2775707
14. Enahoro, S., Ekpo, S., Gibson, A.: Massive multiple-input multiple- output antenna architecture for multiband 5G adaptive beamforming applications. In: 2022 IEEE22nd Annual Wireless and Microwave Technology Conference (WAMICON), pp. 1–4 (2022). https://doi.org/10.1109/WAMICON53991.2022.9786100
15. 3GPP TR 38.901 version 15.0.0 Release 15

Novel Techniques for 5G/6G Power Amplifier Operation Constraints Mitigation for Satellite-Cellular Applications

Arslan Altaf, Sunday Cookey Ekpo, Muhammad Ijaz, Andy Gibson, and Peter H. Aaen

1 Introduction

The imminent arrival of the fifth generation of wireless communication standard (5G) brings with it the challenge of meeting the skyrocketing demand for mobile data traffic [1, 2]. Unlike its predecessor, 5G represents a complete transformation of wireless technology infrastructure, rather than just an incremental improvement. It promises significantly faster data rates for enhanced mobile broadband (eMBB) as depicted in Fig. 1, along with the support for a multitude of applications such as Internet of Things (IoT), Industry 4.0, virtual reality, UHD video streaming, and satellite communications [3–9].

To achieve the ambitious goals of 5G, an "ultradense" deployment is required [1], necessitating the use of many Multiple Input Multiple Output (MIMO) antennas operating across multiple bands and standards. These MIMO antennas must deliver higher output power while consuming less power, all while maintaining signal integrity. The power amplifier (PA), as the final output stage of the transmitter front-end module (FEM), plays a critical role in the performance of MIMO antennas [3].

Extensive research and development efforts are being conducted worldwide to enhance the performance metrics of PAs for 5G applications [10]. This paper builds upon previous work [10] that achieved impressive PA performance using gallium nitride (GaN) technology but was limited to a narrow frequency range. Narrowband PAs are unsuitable for 5G [11], thus necessitating the exploration of wideband PA

A. Altaf (✉) · S. C. Ekpo · M. Ijaz · A. Gibson
Department of Engineering, Manchester Metropolitan University, Manchester, UK
e-mail: arslan.altaf@mmu.ac.uk

P. H. Aaen
Department of Electrical Engineering, Colorado School of Mines, Golden, CO, USA

S. C. Ekpo (ed.), *The Second International Adaptive and Sustainable Science, Engineering and Technology Conference*, Signals and Communication Technology,
https://doi.org/10.1007/978-3-031-53935-0_12

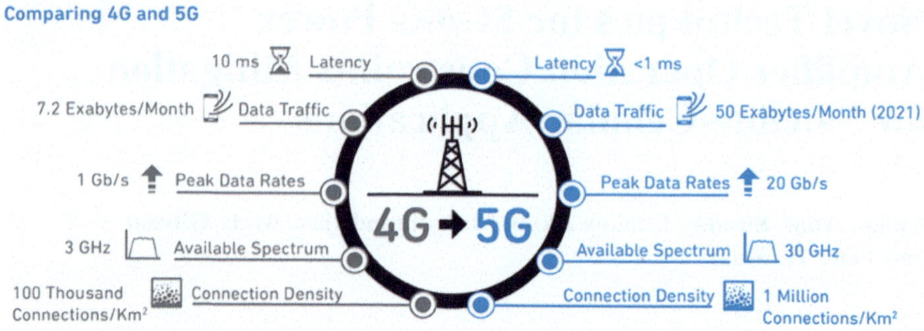

Fig. 1 Comparison of 5G expectations with current 4G wireless. (Source: Qorvo [13])

designs. Some previous designs focused on achieving ultrawideband [11, 12] operation but lacked performance due to the use of SiGe-BiCMOS technology instead of GaN.

The objective of this paper is to address the constraints faced by PAs in the context of 5G. A systematic trade-off between bandwidth and performance is achieved through the design and tuning of matching networks (MNs). MNs plays a crucial role in the overall performance of PAs [10]. The proposed single-stage PA design utilizes the WIN NP25–00 0.25 μm GaN/SiC high electron mobility transistor (HEMT) technology, aligning with the new frequency bands n77, n78, and n79 outline for use in 5G mobile wireless communication (3GPP).

This paper is structured as follows: Sect. 2 provides an overview of available PA design technologies and their applicability to 5G. Section 3 explores the methods for achieving multiband and multi-standard PA operation. Section 4 focuses on improving the analysis of device output characteristics (DC-IV) through time-domain analysis. Section 5 presents the proposed PA design specification, topology, and its results. Section 6 offers a comprehensive analysis of the proposed design. Section 7 concludes the paper.

2 Device Technology for 5G

In the realm of 5G power amplifier (PA) design, semiconductor technology plays a critical role. Among the options, type III-V semiconductors, particularly GaN, offer superior performance and higher-frequency capabilities for 5G PA operation. Key selection criteria for a given application include power-added efficiency, output power, cost, size, and robustness [3]. For 5G, frequencies are divided into sub-6 GHz (FR1) and mmWave ranges beyond 24 GHz [2, 10, 12, 14, 15]. GaN stands out as the preferred choice, providing higher effective isotropic radiated power (EIRP), fewer array elements, and smaller die area compared to other III–V semiconductors [12].

Table 1 Active device technology comparison. Source: [1]

Parameter	Type of device technology			
	CMOS	SiGE/SOI	GaAs	GaN
EIRP (dBm)	35	45	55	65–75
Array elements	–	1024	384	128
Die area mm^2	–	2084	1402	544

GaN-based MIMO antennas reduce cell area by 67% and die area by 61% [12], while offering enhanced breakdown voltage capabilities. GaN surpasses silicon germanium (SiGe) in power consumption, delivering a 40% reduction in energy consumption at higher frequencies. With increasing power demands in 5G, GaN excels in power density in the cmWave and mmWave ranges, outperforming CMOS, SiGe, and GaAs technologies. GaN's wide bandwidth capabilities make it ideal for 5G deployment [10]. Although silicon-based semiconductors offer advantages like monolithic integration and functionalities such as digital pre-distortion (DPD), GaN's unmatched potential mitigates these constraints. Respecting a 5G UE PA design for FWA UE applications, GaN analysis highlights its unparalleled potential. By harnessing GaN's unique characteristics, designers can meet the demanding requirements of 5G, enabling high-performance and efficient PAs for 5G UE devices (Table 1).

3 Multiband PA Constraints and Mitigation

In the pursuit of multiband/multi-standard PA operation, three methods have been identified. The first optimizes the matching network (MN) within a narrow band, outperforming wideband PAs [10]. However, it introduces constraints for 5G, including higher costs and increased antenna size. The second option integrates reconfigurable components like MEMS switches or varactors into the MNs, primarily beneficial for dynamic load modulation in Doherty PAs (DPAs).

Our proposed approach combines cascading LC sections for wideband operation, enabling concurrent PA operation across multiple bands using a single unit. With MMIC implementation and suitable active devices, a compact form factor can be achieved. Challenges highlighted in [3], [10–12], and [16] emphasize the significance of MN design. Optimizing MNs with lower Q-levels facilitates wideband operation, requiring optimal impedances at these levels for power, efficiency, and linearity. Our PA design leverages wideband MNs, extracting optimal impedances through iterative load pull in ADS. Tuned and optimized matching networks strike an ideal balance between bandwidth and performance. Figure 3 showcases the success of our approach, validating its efficacy.

4 DC-IV Analysis

Traditionally, power amplifier (PA) evaluation has focused on frequency domain analysis [17]. However, to achieve a comprehensive understanding of PA design, delving into the time domain is crucial. Time-domain analysis allows designers to observe voltage and current waveforms, enabling optimal bias conditions. Equations (1) and (2) estimate power and efficiency, independent of frequency impact. In our PA design, we extract DC-IV characteristics using these estimation methods and load line analysis, Table 2. Efficiency depends on bias current, while power relates to device size. Choosing class AB operation (100 mA/mm), we strike a balance between efficiency and linearity. Output power is 29.4 dBm (0.25 mm device) and 35.9 dBm (1.25 mm device), making the latter suitable for high-power FWA UEs. Our proposed PA adopts 100 mA/mm bias and a 1.25 mm device size, delivering desired power and efficiency while accepting some nonlinearity. For a comprehensive understanding of power amplifier classes, refer to [18]. The insights gained from this analysis pave the way for more robust and efficient power amplifier solutions in the context of 5G UE applications.

$$Pavg = \frac{1}{8} \cdot Vpp \cdot Ipp \cdot \cos(\varnothing) \tag{1}$$

where $\varnothing$ = phi = phase offset between I–V waveforms; Vpp = peak-to-peak voltage in time domain; and Ipp = peak-to-peak current in time domain.

Table 2 DC-IV characteristics of a 2 × 125 μm NP25–00 GaN HEMT at drain voltage of 28

DC-IV (output characteristics) at drain voltage of 28 V				
Model size: 2 × 125 μm (0.25 mm)				
Knee voltage [V]	Gate voltage [V]	Quiescent current *Iq* [mA/mm]	Estimated pout [dBm]	Estimated efficiency (%)
2 V	−2.65	104	29.39	61.00
	−2.85	48	29.39	68.45
	−2.9	36	29.39	70.60
Model size: 10 × 125 μm (1.25 mm)				
Knee voltage [V]	VGS [V]	*Iq* [mA/mm]	Estimated pout [dBm]	Estimated efficiency (%)
2 V	−2.5	100	35.87	61.31
	−2.7	55.2	35.87	66.46
	−2.8	34	35.87	70.60

$$\eta = \frac{p_{\text{gen}}}{p_{\text{gen}} + P_{\text{diss}}} \tag{2}$$

where P_{diss} and P_{gen} are given by (3) and (4), respectively.

$$P_{\text{diss}} = \frac{1}{T} \int_0^T v(t) \cdot i(t)\, dt \tag{3}$$

$$P_{\text{gen}} = \frac{1}{T} \int_0^T [v(t) - V_{dc}] \cdot [i(t) - I_{dc}]\, dt \tag{4}$$

5 Power Amplifier Circuitry

Figure 2 showcases the proposed circuit design topology for the power amplifier (PA), which is implemented using ideal lumped components. To ensure stability and performance, the input gate terminal incorporates a stability network composed of a parallel RC network and a shunt RL network connected to the gate voltage source. The shunt inductor serves a dual purpose, acting as both an RF choke and a vital component of the stability network. By incorporating a shunt RL network, the power amplifier is stabilized at lower frequencies, bolstering its overall performance. The input MN (IMN) comprises of a lumped LC network and an LL network with a design aim of obtaining low Q. Lumped elements were used to their advantage of achieving a compact form factor. On the output side of the PA (drain), a shunt inductor serves as an RF choke, effectively controlling the flow of RF signals. This is followed by an output matching network (OMN) and a DC block capacitor. The OMN comprises of a single LC network and carefully designed using the ADS smith chart utility.

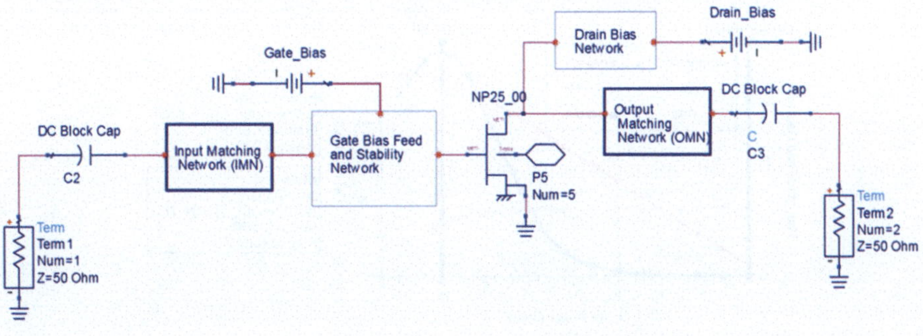

Fig. 2 Design architecture of the proposed PA with IMN/OMN blocks

6 Results and Analysis

Figure 3 showcases the remarkable S-parameter performance of the proposed single-stage power amplifier. Forward transmission reveals an impressively flat gain response with an in-band ripple of less than 3 dB, effectively spanning the 3–5 GHz frequency range. The input return loss demonstrates exceptional attenuation, maintaining levels below -15 dB across the entire 3.1–5.0 GHz range. Notably, the output return loss is slightly poor due to the trade-off for achieving better PAE. Excellent S22 and PAE could be achieved using a two-stage PA approach which helps to alleviate the strain from a single transistor unit. In comparison with existing studies [10, 11], our PA design exhibits expertly matched S-parameters precisely at the operational frequency, elevating its overall useability in 5G application. Stability is a crucial aspect of PA design and is expertly addressed by the proposed PA, as depicted in Fig. 3. The stability factor (k) reaches its lowest value of 3.3 at 2.2 GHz. Importantly, k remains above 1, b1 remains above 0, and mu1 surpasses 1 (up to 400 GHz). This is well beyond the cutoff frequency limit for this active device. These findings ensure that when the PA is manufactured using real components, any oscillatory behavior will be effectively controlled, minimizing heat dissipation and ensuring a stable and reliable operation. The large-signal performance of the proposed PA, simulated at 4 GHz, is depicted in Fig. 3c, highlighting

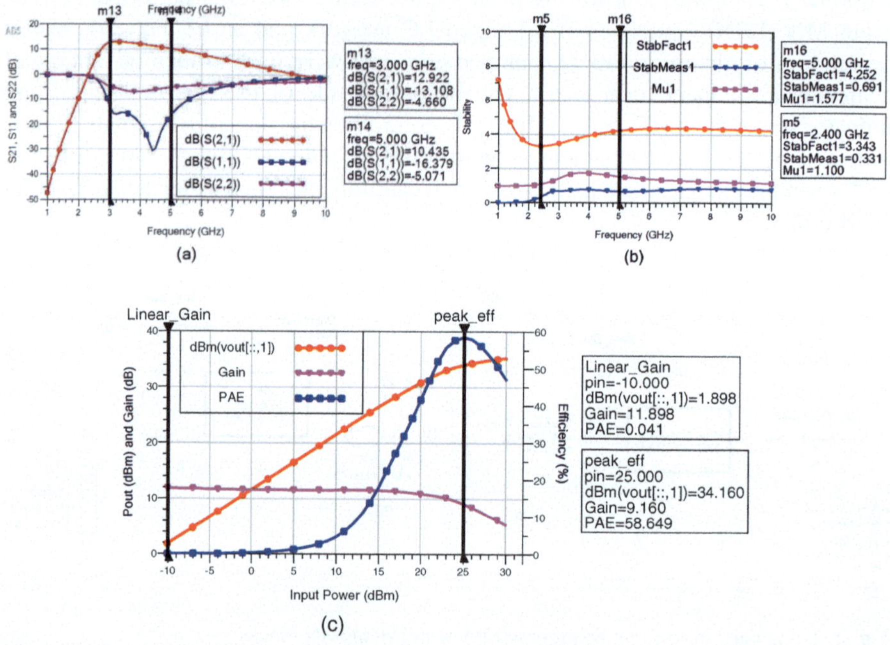

Fig. 3 Simulated (**a**) S-parameter, (**b**) stability, and (**c**) LS-power sweep performance of a single-stage PA

its optimal operating conditions. At an input power (Pin) level of 25 dBm, the PA achieves an impressive output power (Pout) of 34.1 dBm and PAE 58.6%. These performances are obtained at a gain compression of 2.7 dB, which surpasses the performance presented in [10]. This ensures a good signal integrity that would require less aid from linearization methods. However, this was obtained at the cost of a degradation in the efficiency bandwidth and S22. This emphasizes the need to incorporate a multistage PA design that can extend the efficiency bandwidth and enable a better S22 performance while maintaining high linearity and sufficient power.

7 Conclusion

In conclusion, this conference paper presented a novel wideband single-stage class AB power amplifier design for 5G fixed wireless access user equipment. The design addressed the challenges of 5G multiband/multi-standard PA operation, achieving exceptional performance with high efficiency and stability. The study highlighted the trade-offs between power-added efficiency and S22, emphasizing the need for an optimal balance. The research contributes to the advancement of front-end modules for next-generation 5G/6G satellite-cellular convergence use cases and/or applications [19–23], providing insights for optimizing power amplifier performance metrics. For further work, it is recommended to explore the use of multistage PA designs to mitigate the degraded S22 and gain caused by the stability and high PAE trade-offs. Additionally, incorporating harmonic optimization techniques and voltage/current waveform engineering can improve baseline performance before implementing performance enhancement techniques.

Acknowledgments This work was supported by the Industry Engagement Fund Project 214959.

References

1. Higham, E.: Semiconductor trends in Sub-6 GHz 5G networks. Microwave. **63**(6), 22–34 (2020)
2. Anritsu: Ensuring The Connected 5G Experience (2016)
3. Lie, D.Y.C., Mayeda, J.C., Li, Y., Lopez, J.: A review of 5G power amplifier design at cm-wave and mm-wave frequencies. Wirel. Commun. Mob. Comput. (2018). https://doi.org/10.1155/2018/6793814
4. Ekpo, S.C.: Parametric system engineering analysis of capability-based small satellite missions. IEEE Syst. J. **13**(3), 3546–3555 (2019). https://doi.org/10.1109/JSYST.2019.2919526
5. Ekpo, S.C.: Thermal subsystem operational times analysis for ubiquitous small satellites relay in LEO. Int. Rev. Aerosp. Eng. (IREASE). **11**(2), 48 (2018). https://doi.org/10.15866/irease.v11i2.13663

6. Udeshi, S., Uko, M., Zafar, M., Altaf, A., Adebisi, B., Ekpo, S.: Integrated space-enabled hybrid 5G-V2X communications link modeling. In: 37th International Communications Satellite Systems Conference (ICSSC-2019). Institution of Engineering and Technology (2019). https://doi.org/10.1049/cp.2019.1255
7. Uko, M., Zafar, M., Altaf, A., Udeshi, S., Ekpo, S., Adebisi, B.: K/Ka-band transceiver sensitivity modeling and link characterization for integrated 5G-LEO communication applications. In: 37th International Communications Satellite Systems Conference (ICSSC-2019). Institution of Engineering and Technology (2019). https://doi.org/10.1049/cp.2019.1256
8. Altaf, A., Uko, M.C., Enahoro, S., Zafar, M., Ekpo, S.: IET Seminar on Active and Passive RF Devices. IET, Cardiff (2019) Accessed 8 June 2023
9. Zafar, M., Ekpo, S., George, J., Sheedy, P., Uko, M., Gibson, A.: Hybrid power divider and combiner for passive RFID tag wireless energy harvesting. IEEE Access. **10**, 502–515 (2022). https://doi.org/10.1109/ACCESS.2021.3138070
10. Chen, Z., Li, K., Jin, K., Huang, C., Li, P., Geng, L.: A 2.6GHz class-AB GaN power amplifier with maximum output power of 56W achieving 70% power added efficiency. In: 2018 1st Workshop on Wide Bandgap Power Devices and Applications in Asia (WiPDA Asia), pp. 328–331. IEEE (2018). https://doi.org/10.1109/WiPDAAsia.2018.8734543
11. Moon, J.-S., et al.: Wideband linear distributed GaN HEMT MMIC power amplifier with a record OIP3/Pdc. In: 2016 IEEE Topical Conference on Power Amplifiers for Wireless and Radio Applications (PAWR), pp. 5–7. IEEE (2016). https://doi.org/10.1109/PAWR.2016.7440127
12. Xie, C.: Characterization of broadband Monolithic Gallium Nitride distributed power amplifier using thermal imaging technique. In: 2011 IEEE Topical Conference on Power Amplifiers for Wireless and Radio Applications, pp. 77–80 (2011). https://doi.org/10.1109/PAWR.2011.5725379
13. Miller, L.: 5G RF for Dummies: Qorvo, vol. 1, Spec edn. Wiley (2017)
14. 3GPP: 5G NR User Equipment (UE) Radio Transmission and Reception; Part 2: Range 2 Standalone (3GPP TS 38.101–2 Version 17.5.0 Release 17) (2022)
15. 3GPP: NR User Equipment (UE) Radio Transmission and Reception; Part 1: Frequency range 1 Standalone (3GPP TS 38.101–1 Version 17.5.0 Release 17) (2022)
16. Li, S., Fritsche, D., Carta, C., Ellinger, F.: Design and characterization of a 12–40 GHz power amplifier in SiGe technology. In: 2018 IEEE Topical Conference on RF/Microwave Power Amplifiers for Radio and Wireless Applications (PAWR), pp. 23–25. IEEE (2018). https://doi.org/10.1109/PAWR.2018.8310057
17. Cripps, S.C.: RF power amplifiers for wireless communications. Second. Artech House. (2006) [Online]. Available: http://gaasmantech.org/Digests/2003/2003PDF/2-2.pdf
18. Grebennikov, A.: Radio Frequency and Microwave Power Amplifiers. Volume 1: Principles, Device Modeling and Matching Networks. The Institution of Engineering and Technology (2019)
19. Lau, I., Ekpo, S.C., Zafar, M., Ijaz, M., Gibson, A.: Hybrid mmWave-Li-fi 5G architecture for reconfigurable variable latency and data rate communications. IEEE Access. **11**, 42850–42861 (2023). https://doi.org/10.1109/ACCESS.2023.3270777
20. Ekpo, S., Adebisi, B., Wells, A.: Regulated-element frost beamformer for vehicular multimedia sound enhancement and noise reduction applications. IEEE Access J. **5**, 27254–27262 (2017). https://doi.org/10.1109/ACCESS.2017.2775707
21. Saha, D., Ghosh, S., Ekpo, S., Zafar, M., Gibson, A.: Gap-coupled resonator loaded ultra-wide band filtenna for frequency-notching applications. In: 3rd European Conference on Communication Systems, pp. 1–5, Vienna (2023)
22. Sowande, O., Idachaba, F., Ekpo, S., Faruk, N., Uko, M., Ogunmodimu, O.: Sub- 6 GHz 5G spectrum for satellite-cellular convergence broadband internet access in Nigeria. Int. Rev. Aerosp. Eng. J. **15**(2), 85–96 (2022)
23. Uko, M., Ekpo, S.: A 23–28 GHz pHEMT MMIC low-noise amplifier for satellite-cellular convergence applications. Int. Rev. Aerosp. Eng. J. **14**(5), 1–10 (2021)

Deep Learning-Enabled Smart Wearables to Assist People with Autism Spectrum Disorder in Dynamic Environments

Przemyslaw Danowski, Sunday Cookey Ekpo, Fanuel Elias, Muhammad Ijaz, Umar Raza, and Raul Ochoa Carbrero

1 Introduction

Autism is an uncurable spectrum condition believed to affect more than one per one hundred people in the United Kingdom (Autism.org.uk). It has different variations and intensities. Many (mild) types aren't socially recognisable at all. As a result, it cannot be treated all at once, but each symptom, and even different symptoms severity, must be looked at separately. It makes it more time-consuming and resource demanding, but this problem cannot be overlooked. This report addresses autism by its medical term, autism spectrum disorder (ASD).

Besides the issues it causes, there are, on the other hand, strengths it creates. People with ASD can focus on a single thing very hard and keep doing it no matter what. They need to enjoy doing it. It increases their chances of being successful in it. Some people, such as Albert Einstein, have contributed to science significantly and 'would qualify as having autism spectrum disorder today' (behavioral-innovations.com, 2021).

The problem appears when they cannot do what they desire and their syndrome's downside triggers. They lose interest in activities they are proficient in due to external, independent factors such as bad weather. And that's the better scenario, compared to losing control and being aggressive to self or others. They need to control their surroundings to avoid unexpected results.

One of the main struggles people with autism spectrum disorder face is difficulty recognising people's emotions. If they interpret them wrong, they may take the wrong action and will not understand why others react the way they do. A systematic literature review on 'The Impact of Technology on People with Autism Spectrum

P. Danowski (✉) · S. C. Ekpo · F. Elias · M. Ijaz · U. Raza · R. O. Carbrero
Manchester Metropolitan University, Manchester, UK
e-mail: 19064895@stu.mmu.ac.uk

S. C. Ekpo (ed.), *The Second International Adaptive and Sustainable Science, Engineering and Technology Conference*, Signals and Communication Technology,
https://doi.org/10.1007/978-3-031-53935-0_13

Disorder' [1] is available online and has helped complete this research. Relevant studies completed and their extent have been identified to ensure the same work is not repeated. Wojciechowski et al. used Estimote Beacon sensors for object recognition [2]. For this project, similar action has been done with a programmable OAK-D camera wearable device that will detect faces and analyse facial expressions to help the user understand the current situation.

This project intends to provide equal interpersonal skills potential for everybody, regardless of their mental state. It aims to improve the ability of effective socialising and communication for people with ASD. They often struggle to read other people's intentions and emotions or even their feelings. Therefore, it addresses the issues of interpersonal communication as well as self-understanding. Nobody still comes up with a complete solution to this problem; however, some relevant studies have been conducted with further work to be done. Vidakis et al. show that computer-based tools enhance young children with ASD's body language understanding through interaction with natural user interface (NUI) games [3]. Still, it is limited by their ability to follow instructions linked with age and does not cover facial expressions. This work aims to fill that gap by prototyping a smart wearable that will help people with ASD identify different factors and their meanings and guide them in taking the right actions, leading to the best, intended results. This project has been completed using an OAK-D lite camera and a personal laptop. User testing results of the algorithm are presented.

2 Literature/Background/Context Review

Autism spectrum disorder is a neurodevelopmental disorder known for causing issues with social communication, repetitive behaviour, and limited interests. It's common for people with ASD to struggle to interpret emotional expressions, often causing them to be socially isolated and have a hard time building relationships. Recent research has investigated the use of technology to aid such people. This literature review summarises current research on this topic and narrows it down to using emotion recognition for the same purpose.

Emotion recognition technology requires an algorithm, software, and hardware to run it on. It can also involve sensors and machine learning methods. It has been used in different environments, such as education and healthcare, to find the best fit for people with ASD. Other studies have shown that people with ASD struggle with recognising emotional expressions, which is connected to a lack of social communication skills [4, 5].

The literature review revealed that autistic people's social skills can be significantly improved with the help of emotion recognition technology. Studies have shown that people with autism spectrum disorder can enhance their ability to recognise emotional expressions while using emotion recognition technology [6]. However, more results are needed to determine the long-lasting benefits of this technology as the effectiveness of using emotion recognition positively impacting social skills is still unclear.

The use of emotion recognition technology to help autistic people interpret the environment better and improve their social skills is an exciting new area of research. However, several challenges need to be addressed. For example, the accuracy of emotion recognition technology is not as high as some would expect, and it is challenging to capture complex emotional expressions and define starting conditions as they differ for various individuals. On top of that, this solution might not be suitable for all people with ASD and likely won't work for those with extreme communication difficulties [7].

In conclusion, emotion recognition technology has the potential to improve social skills in people with ASD by helping them better interpret their surroundings and recognise the emotional expressions of people around them. While some challenges still need to be addressed in further research, using this technology is another step towards having equal abilities in social interactions for people with ASD and everybody else. This project will bring it closer to the goal by gathering more statistical data and exploring new potential use cases for similar applications.

3 Method

After first looking into the topic of emotion recognition/prediction algorithms online, the hardware required was a host and a camera that would connect to it.

The host device would provide power to the camera, execute the code, and display results, while the camera would give input images and do most of the processing. Specific devices used for this project are as follows:

- Lenovo IdeaPad L340 Gaming.
- OAK-D lite camera.

Two options for the host were a Raspberry Pi or a personal laptop. Since the laptop was a more powerful device in this case, it would be a safer decision. It was Lenovo IdeaPad L340 Gaming, and its significant parameters were 8 GB DDR4 RAM memory, Intel Core i5-9300H @2.4 GHz, NVIDIA GFORCE GTX 1650 graphics card, and 2× USB 3.1 ports. It was run on Windows 11. It is likely that many similar devices would also be sufficient but would risk facing new issues not described in this work. Another essential component is the OAK-D lite camera with the parameters shown in Table 1.

This device was chosen because it was available at the university immediately. It is built on top of RVC2 (Robotic Vision Core 2), which provides it with 4 TOPS of processing power, allows it to run any AI model, and uses the 'computer vision' feature that does a variety of operations such as ImageManip, resize, and crop input data (Luxonis.com).

Getting familiar with how this camera operates is beneficial by going through OpenCV tutorials on OAK-D devices and DepthAI, available on learnopencv.com. OpenCV is an open-source software library especially useful for machine learning (opencv.org). DepthAI is a spatial AI platform that 'allows robots and computers to perceive the world like a human can' (luxonis.com documentation).

Table 1 OAK-D lite parameters (Luxonis.com)

Camera specs	Colour camera	Stereo pair
Sensor	IMX214 (PY014 AF, PY114 FF)	OV7251 (PY013)
DFOV/HFOV/VFOV	81°/69°/54°	86°/73°/58°
Resolution	13MP (4208 × 3120)	480P (640 × 480)
Focus	AF: 8 cm–∞ OR FF: 50 cm–∞	Fixed-focus
Max frame rate	35 FPS	120 FPS
F-number	2.2 ± 5%	2.0 ± 5%
Lens size	1/3.1 in	1/7 in
Effective focal length	3.37 mm	1.3 mm
Distortion	<1%	<1.5%

Fig. 1 Steps of emotion recognition

The algorithm used in this project was found on GitHub as 'gen2-emotion-recognition'. It is written in Python; therefore, python3 must be installed. After downloading it, the command 'python3 -m pip install -r requirements.txt' will be needed to run the program. The path would be set to the newly downloaded GitHub code location.

A type-C to type-A USB cable will connect the OAK-D lite camera to the host. A 3.0+ USB is a default setting for this device; combining it with older versions will result in not detecting the camera while running the script. It is still possible but requires an extra line of code to force USB 2.0 communication, and the output quality will be affected by latency and lower FPS count. Moreover, the host device should not be in battery saver mode as it also increases the latency of received output.

The program is a two-stage NN (neural network) model with a machine learning algorithm using CNN (convolutional neural networks). It stands out by acting similarly to a human brain and being most efficient with deep learning algorithms. It consists of three layers: convolutional, pooling, and fully connected. The complexity increases from layer one onwards, but the required computational power decreases similarly.

As you can see in Fig. 1 below, this algorithm first performs object detection on the input image and then extracts and selects relevant features, and in the end, the result is classified.

For this project, a hybrid approach has been chosen where some of the database is labelled initially, but most of it is categorised by the neural network described above. This choice has been made to keep the accuracy relatively high while considering available hardware. On the other hand, a purely deep-learning approach would likely give more accurate results, but it would also demand more powerful hardware.

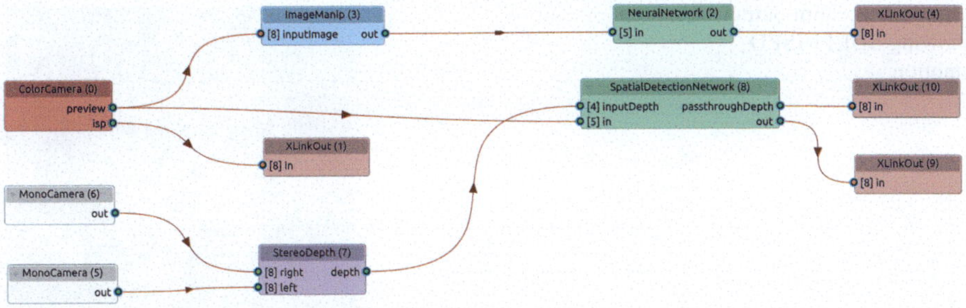

Fig. 2 A two-stage neural network pipeline graph

While running, the algorithm takes the input image from the colour camera sent to the host, which will be downscaled to prepare for first-stage NN, object, or face detection. After processing in this node, if a face is detected, it will go through emotion recognition NN, which uses the AffectNet dataset. The result will be displayed on the host in a new window. If detected, it will say at the bottom of the face what emotion has been estimated from five emotions available: happy, sad, angry, surprised, or neutral. DepthAI is equipped with a pipeline graph function, seen in Fig. 2, for a better understanding of the process.

For debugging purposes, a second window may be set to open on the host, showing the result of the first stage of the algorithm. Key Q has been assigned to stop the algorithm and close the window(s).

4 Results and Discussion

This project aimed to prototype smart wearables using an OAK-D lite camera to detect and analyse facial expressions to define emotions to help people with ASD communicate more effectively. Due to the camera heating up quickly and the inability to run on its battery, designing a smart wearable wasn't possible. This means the original idea could have been completed with a stronger, similar device with a rechargeable battery instead of relying on a cabled connection with the host. This wasn't possible to achieve in this project due to time and budget limitations. However, the algorithm has been run on the device, and its accuracy was examined. Below, you can see the algorithm's output showing different emotions.

From Figs. 3, 4, 5, 6, and 7, the algorithm works for five emotions: surprised, sad, neutral, angry, and happy, Figs. 3, 4, 5, 6, and 7, respectively.

Figure 8, however, represents the accuracy of different emotions predicted by this algorithm. The greatest accuracy prediction is for 'happy' emotion, and for 'sad', 'neutral', and 'surprise' emotions, it is slightly lower. Also, the last emotion, 'angry', appears to be the hardest to predict with accuracy below 50%. Some volunteers' faces naturally looked more happy, sad, or angry. When they tried recreating other facial emotions, the algorithm often showed the same emotion or switched between

Fig. 3 Algorithm output showing SURPRISED emotion

Fig. 4 Algorithm output showing SAD emotion

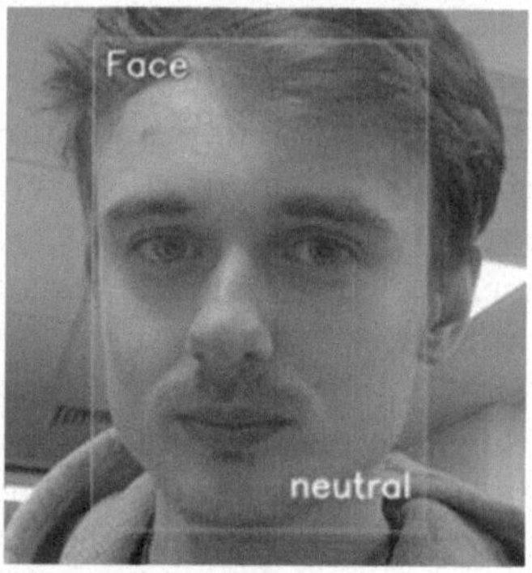

Fig. 5 Algorithm output showing NEUTRAL emotion

them. This suggests that some calibration could be implemented to adjust the recognition scale from the starting point of 'neutral'. For example, if a person naturally has a happy face, the algorithm will recognise their regular expression as 'happy'. However, with calibration, the algorithm would remember that person's face and know that they are naturally 'happier' than average facial neutral expression, therefore adjusting the scale and taking it as a neutral expression and, from

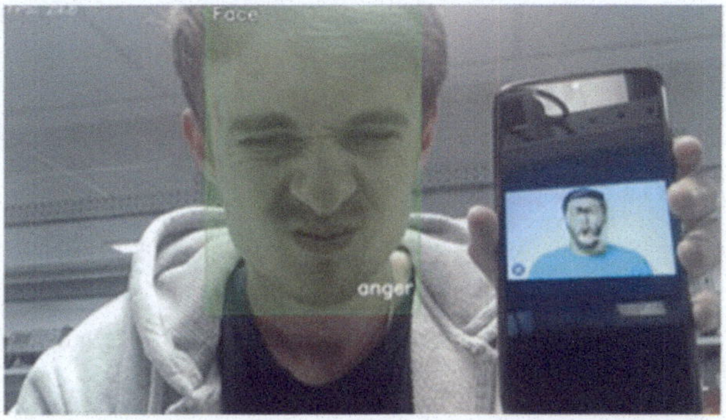

Fig. 6 Algorithm output showing ANGRY emotion

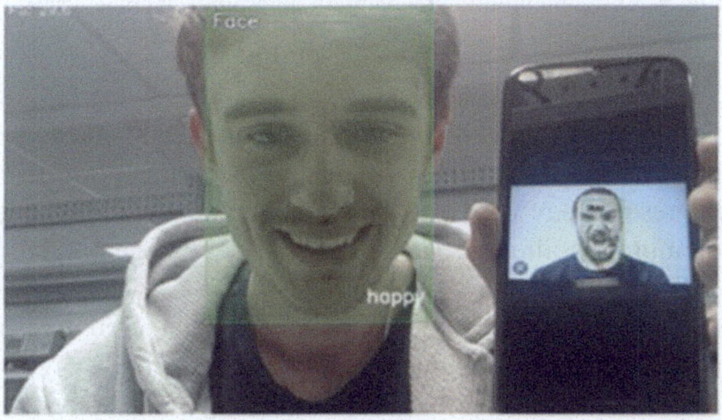

Fig. 7 Algorithm output showing HAPPY emotion

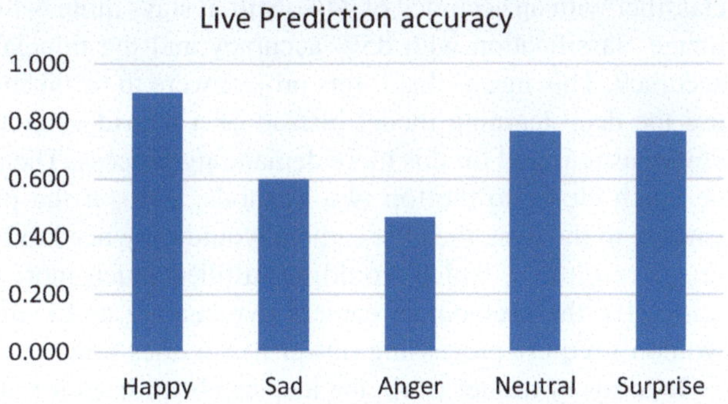

Fig. 8 Accuracy live results

there, change all other emotions as well to avoid repetition and increase accuracy. Also, volunteers were asked to recreate different emotions on the spot, which wasn't easy for some. A better way would be to monitor them in a controlled room where they would, for example, watch something evoking their certain emotions. It would be a more natural and reliable way of testing the algorithm. On the other hand, it would require greater arrangements and a time frame dedicated to data gathering alone.

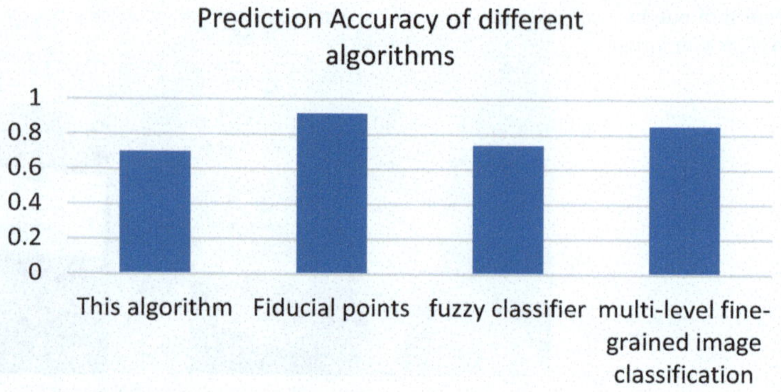

Fig. 9 Different emotion recognition algorithms average accuracy

The data was gathered from 30 volunteers who were engineering students, with most participants also being male. Since the algorithm wasn't tested on ASD users, the results may vary from real accuracy on the targeted group and require retesting it on more specified research groups. The average prediction accuracy turned out to be 70%.

It was intended to compare results gathered live and non-live, but very few participants were willing to support the second way; therefore, there isn't enough data to complete this task. However, Fig. 9 shows average accuracy compared with other emotion-detecting algorithms. This algorithm compares closely with a fuzzy classifier with an accuracy of 74%. Still, it stays quite behind multi-level fine-grained image classification with 85% accuracy and the fiducial points method with 92% accuracy. This means that if this project were to be taken further, it would be best to use the deep learning model instead of a hybrid version and have a stronger host device as required for this more demanding process. Then, the accuracy would likely be much closer to the top two methods. Also, if the processing could be moved entirely to the host, then the camera would only need to be wireless and provide HQ image resolution, which would be fulfilled much more easily. It could potentially also solve the problem of camera overheating and increasing latency with it as it wouldn't require processing power like it does now.

In terms of further work, the idea can be utilised for other use cases, for example, monitoring patients' average moods in waiting rooms. The emotion recognition algorithm could also be linked and correlated with other sensors, for example, a heart rate sensor. When people feel strong emotions, whether positive or negative, their heart rate increases significantly. That would be used to confirm or deny certain users' emotions when their facial expressions didn't fully match the algorithm's criteria.

Another use of the algorithm could also be identifying suspects based on CCTV footage. That would require adapting the algorithm by changing the input source from a live camera image to manually feed it with videos or images. If somebody in a crowd committed a crime, then they likely would show signs of negative or aggressive behaviour and have angry facial expressions. Moreover, the idea could be expanded on to body language.

5 Conclusion

In conclusion, creating smart wearables to help people with ASD interpret the environment better is achievable but would require more time and a greater budget. Data gathered clearly indicates that robotic emotion recognition works, but the algorithm could be improved, resulting in better accuracy with the hardware due to enhancement.

This research work can be taken further by applying a purely deep learning algorithm approach, changing the focus from disabled users to CCTV footage identifying potential wrongdoers and monitoring the average mood of people in areas like hospital waiting rooms.

References

1. Valencia, K.: 2019. Online. Available: https://www.mdpi.com/1424-8220/19/20/4485
2. Wojciechowski, A.: 2016. Online. Available: https://link.springer.com/article/10.1007/s11042-016-3995-9
3. Vidakis, N.: 2014. Online. Available: https://doiserbia.nb.rs/img/doi/1820-0214/2014/1820-02141400039C.pdf
4. Harms, M.: 2010. Online. Available: https://link.springer.com/article/10.1007/s11065-010-9138-6
5. Uljarevic, M.: 2016. Online. Available at: https://journals.sagepub.com/doi/pdf/10.1177/1362361316661855?casa_token=X8R8pNM76eQAAAAA:0YOxknLJFicxBg_a1nkrJQJue2sRskkdSJIGoSKJ7yfTPRjQlINc7IItu2oSTR21PPwrGzyiFg42mg
6. Golan, O.: 2015. Online. Available: https://link.springer.com/article/10.1186/s13229-015-0018-z
7. Lu, X.: 2022. Online. Available: https://www.frontiersin.org/articles/10.3389/fpsyg.2021.818833/full

Downlink Power Budget and Bit Error Analysis for LoRa-Based Sensor Node-to-Satellite Link in the Industrial, Scientific, and Medical Frequency Bands

Anietie Essien, Ubong Ukommi, and Emmanuel Ubom

1 Introduction

Over the years, wireless sensor networks have gained applications in many fields and Internet of Things-based smart systems [1, 2]. Again, sensor nodes are usually resource-constrained, requiring low power with long-range communication capabilities [3, 4]. Accordingly, several low-power transceiver technologies like NB-IoT, LoRa, Zigbee, and Sigfox have been designed for such sensor networks [5, 6]. Among the several options, LoRa transceiver technology has been identified as the most suitable for low-power and long-range applications [7, 8].

Furthermore, across the globe, LoRa transceiver technologies have been implemented in some frequencies like 433 MHz, 779–787 MHz, 863–870 MHz, and 915–928 MHz [9, 10]. While the present implementations of the LoRa transceiver technologies are within 1 GHz, further studies have also examined the application of LoRa transceiver in the various industrial, scientific, and medical (ISM) frequency bands above 1 GHz [11, 12].

In addition, some studies have also examined the application of LoRa-based sensor networks in satellite communication applications [13, 14]. As such, this paper presents the analysis of the downlink power budget and bit error probability for LoRa-based sensor node-to-satellite link in the industrial, scientific, and medical (ISM) frequency bands ranging from 433 MHz to 24 2.4 GHz. The essence of the study is to evaluate the feasibility of wireless communication in such LoRa-based satellite link operated in the higher-frequency bands with their attendant high rain

A. Essien (✉) · U. Ukommi · E. Ubom
Department of Electrical and Electronic Engineering, Akwa Ibom State University, Mkpat Enin L. G. A., Akwa Ibom State, Nigeria
e-mail: ubongukommi@aksu.edu; emmanuelubom@aksu.edu

S. C. Ekpo (ed.), *The Second International Adaptive and Sustainable Science, Engineering and Technology Conference*, Signals and Communication Technology,
https://doi.org/10.1007/978-3-031-53935-0_14

fading and path loss. The study presents the requisite mathematical analysis and numerical examples for the discussion of the link performance as function of frequency and rain attenuation.

2 Methodology

Long range (LoRa) transceiver technology pioneered by Semtech is suitable for low data rate, low power consumption, and long-range applications. It uses chirp spread spectrum (CSS) modulation scheme and is operated at different frequency bands in different regions across the globe. Generally, LoRa transceiver and modulation scheme have many parameters, and the configuration of those parameters depends on the region of operation. In the European region, LoRa can transmit with up to 20 dBm power [15], and it can operate with sensitivity as low as −139 dBm at bandwidth of 125 KHz with data rate ranging from 250 bps at spreading factor of 12 to data rate of 5.5 kbps at spreading factor of 7 [16]. Also, LoRa uses cyclic error coding which is a form of forward error detection and correction scheme with cyclic coding rate (CR) ranging from 4/5 to 4/8. In addition, the bit error performance of LoRa transceiver depends on the configurations which include the values for parameters like the spreading factor, the coding rate, the bandwidth, the receiver sensitivity, and the operating link margin. In any case, the minimum required carrier to noise ratio for LoRa ranges from −7.0 dBm for spreading factor of 7 to −20 dBm for spreading factor of 12.

The focus in this paper is to conduct downlink power budget and bit error analysis for a LoRa transceiver-based sensor node-to-satellite communication link, especially considering the impact of rain attenuation, as shown in Fig. 1. The case study

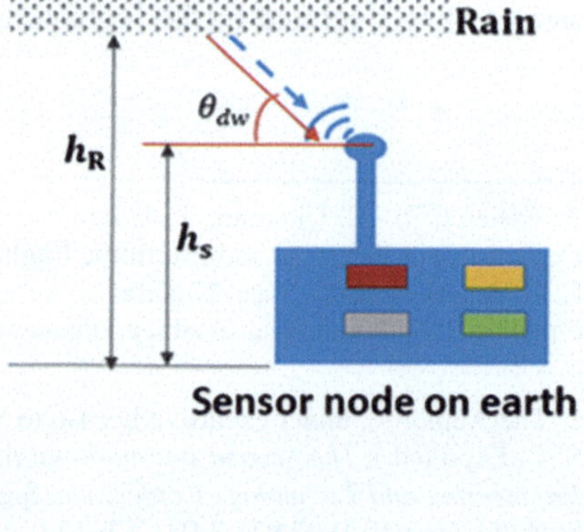

Fig. 1 Schematic diagram of a typical satellite to earth station (sensor node) downlink with rain attenuation

satellite is EGS (AJISAI) geodetic satellite with NORAD ID of 16,908 and orbital height of 1480.33 km. Also, the earth station sensor node is located at longitude of 7.761340 and latitude of 4.625907. In addition, 10-day predicted visible pass elevation angles of the case study satellite with respect to the earth station were obtained using N2YO satellite tracking web tool [17], and the elevation angles obtained range from 13° to 90°. The elevation angles are used in the computation of the rain fading (rain attenuation) [18, 19] and the slant range of the satellite link. Specifically, the rain attenuation is calculated using the same ITU Recommendation P.618-13 rain attenuation model steps and equations presented in the work by [20, 21]. The case study earth station site is in the ITU rain zone N with rainfall intensity of 95 mm/h exceeded at 0.01% of an average year.

2.1 Calculation of the Link Margin

The link margin is calculated starting from the slant range, d_{dwRn}, of the satellite link given as follows:

$$R_{\mathrm{SdwRn}} = H_{\mathrm{sdwRn}} + R_E \tag{1}$$

$$d_{\mathrm{dwRn}} = \sqrt{\left[(R_{\mathrm{SdwRn}})^2 + {R_E}^2 - 2(R_E)(R_{\mathrm{SdwRn}})\left(\sin(\varphi_{\mathrm{dwRn}})\right)\right]} \tag{2}$$

$$\varphi_{dwRn} = \theta_{\mathrm{eldwRn}} + \sin^{-1}\left\{\left(\frac{R_E}{R_{\mathrm{SdwRn}}}\right)(\cos(\theta_{\mathrm{eldwRn}}))\right\} \tag{3}$$

where H_{sdwRn} is the orbital height in km, $R_E = 6378$ km is the earth radius, and θ_{eldwRn} is the elevation angle in degree.

Then, the free space path loss, L_{dwRnfspl}, is given as follows:

$$L_{\mathrm{dwRnfspl}} = 32.45 + 20\,(\log(f_{\mathrm{dwRn}})) + 20\,(\log(d_{\mathrm{dwRn}})) \tag{4}$$

The receiver figure of merit ($G_{\mathrm{dwRnrx}}/T_{\mathrm{dwRnrx}}$) is give as follows:

$$\frac{G_{\mathrm{dwRnrx}}}{T_{\mathrm{dwRnrx}}} = G_{\mathrm{dwRnrx}} - 10\log_{10}(T_{\mathrm{dwRnrx}}) \tag{5}$$

$$T_{\mathrm{dwRnrx}} = T_{\mathrm{antdwRn}} + T_{\mathrm{recvdwRn}} \tag{6}$$

$$T_{\mathrm{antdwRn}} = T_m\left(1 - 10^{-\left(A_{\mathrm{pdwRn}}/10\right)}\right) + (T_C)\left(10^{-\left(A_{\mathrm{pdwRn}}/10\right)}\right) \tag{7}$$

where T_{dwRnrx} is the system noise temperature, T_{antdwRn} is the antenna noise temperature, T_{recvdwRn} is the noise temperature associated with the receiver components, T_C is the total cosmic temperature, T_{m} is the absorptive medium temperature, and all the temperature parameters are in Kelvin. Also, A_{pdwRn} denotes the downlink rain attenuation and G_{dwRnrx} denotes the downlink receiver antenna gain.

The carrier to noise ratio, $C/N|_{dwRn}$ is given as follows:

$$C/N|_{\mathrm{dwRn}} = P_{\mathrm{dwRntx(dB)}} + G_{\mathrm{dwRntx(dB)}} + \mathrm{G}_{\mathrm{dwRnrx}} - L_{\mathrm{dwRnfspl}} - L_{\mathrm{dwRnothers}} - A_{\mathrm{pdwRn}} - N_{\mathrm{dwRn}} \tag{8}$$

$$N_{\mathrm{dwRn}} = T_{\mathrm{dwRnrx}} + 10\mathrm{Log}_{10}(\mathrm{BW}_{\mathrm{dwRn}}) + 10\mathrm{Log}_{10}\left(1.3807 * 10^{-23}\right) + 30 \tag{9}$$

where N_{dwRn} is the noise power in dBm, G_{dwRnrx} is the transmitter antenna gain in dBm, $P_{\mathrm{dwRntx(dB)}}$ is the transmitter power, $L_{\mathrm{dwRnothers}}$ is other losses in the system apart from rain fade and free space path loss, and $\mathrm{BW}_{\mathrm{dwRn}}$ is the downlink noise bandwidth in Hz. The minimum required $C/N|_{\mathrm{dwRnrqd}}$ is given as follows:

$$C/N|_{\mathrm{dwRnrqd}} = S_{\mathrm{dwsens}} + 174 - 10\mathrm{Log}_{10}(\mathrm{BW}_{\mathrm{dwRn}}) - \mathrm{NF}_{\mathrm{dwRn}} \tag{10}$$

where S_{dwsens} is the receiver sensitivity in dBm. The operating link margin $\mathrm{OLM}_{\mathrm{dwRn}}$ is given as follows:

$$\mathrm{OLM}_{\mathrm{dwRn}} = C/N|_{\mathrm{dwRn}} - C/N|_{\mathrm{dwRnrqd}} \tag{11}$$

2.2 *Calculation of the Bit Error Probability*

The bit error probability ($\mathrm{BEP}_{\mathrm{dwRn}}$) is determined from the energy per bit to noise density ratio, ${}^{E_b}\!/_{N_O}|_{\mathrm{dwRndB}}$ as follows:

$$\frac{C}{N}\Big|_{\mathrm{dwRn}} = {}^{E_b}\!/_{N_O}\big|_{\mathrm{dwRndB}} + 10\log_{10}(R_{s\mathrm{dwRn}}) + 10\log_{10}(\mathrm{SF}_{\mathrm{dwRn}}) + 10\log_{10}(\mathrm{CR}_{\mathrm{dwRn}})\ 10\log_{10}(\mathrm{BW}_{\mathrm{dwRn}}) \tag{12}$$

$$\begin{cases} \mathrm{CR}_{\mathrm{dwRn}} = \dfrac{4}{4+n} \quad \text{where } n = 1, 2, 3, \text{or } 4 \\ R_{s\mathrm{dwRn}} = \dfrac{2^{\mathrm{SF}_{\mathrm{dwRn}}}}{\mathrm{BW}_{\mathrm{dwRn}}} \end{cases} \tag{13}$$

where $R_{s\mathrm{dwRn}}$ is the symbol rate, $\mathrm{SF}_{\mathrm{dwRn}}$ is the spreading factor, and $\mathrm{CR}_{\mathrm{dwRn}}$ is the coding rate. Then:

$${}^{E_b}\!/_{N_O}\big|_{\mathrm{dwRndB}} = C/N|_{\mathrm{dwRn}} - 10\log_{10}(R_{s\mathrm{dwRn}}) - 10\log_{10}(\mathrm{SF}_{\mathrm{dwRn}}) - 10\log_{10}(\mathrm{CR}_{\mathrm{dwRn}})\ 10\log_{10}(\mathrm{BW}_{\mathrm{dwRn}}) \tag{14}$$

where the value of is in dB. Then:

$$E_b/N_O\big|_{dwRn} = 10^{\frac{\left(E_b/N_O\big|_{dwRndB}\right)}{10}} \tag{15}$$

$$BER_{dwRn} = \begin{cases} Q\left(\left(\frac{\log_{12}(SF_{dwRn})}{\sqrt{2}}\right)\left(E_b/N_O\big|_{dwRn}\right)\right) \\ \frac{1}{2}\left[1 - \operatorname{erf}\left(\left(\frac{\log_{12}(SF_{dwRn})}{\sqrt{2}}\right)\left(E_b/N_O\big|_{dwRn}\right)\right)\right] \end{cases} \tag{16}$$

Notably, the BER_{dwRn} is determined based on the chirp spread spectrum (CSS) modulation scheme.

3 The Simulation Setup and Results

Sample numerical computations based on the case study satellite and earth station sensor node were carried out for the following five frequencies in the ISM frequency bands: 0.45492 GHz, 0.9233 GHz, 2.433 GHz, 5.821 GHz, and 24.021 GHz. The details of the parameters used for the simulation are presented in Table 1.

Table 1 The details of the parameters used for the simulation

S/N	Parameter description	Symbol used in the study	Value used for the simulation
1.	Satellite orbital altitude	H_{sdwRn}	1480.33 km
2.	Earth station longitude		7.761340
3.	Earth station latitude		4.625907
4.	Earth station height	Ho	0 km
5.	Satellite link elevation angle	θ_{eldwRn}	13°
6.	Rain height	HR	4.8 km
7.	Rainfall intensity at $P = 1.0\%$	$R_{1.0}$	5 mm/h
8.	Rainfall intensity at $P = 0.01\%$	$R_{0.01}$	95 mm/h
9.	Rainfall intensity at $P = 0.003\%$	$R_{0.03}$	140 mm/h
10.	Spreading factor	SF_{dwRn}	12
11.	Receiver sensitivity	S_{dwsens}	−129 dBm
12.	Bandwidth	BW_{dwRn}	500 KHz
13.	Transmitter antenna gain	$G_{dwRntx(dB)}$	10 dBm
14.	Receiver antenna gain	G_{dwRnrx}	2 dBm
15.	Coding rate	CR_{dwRn}	0.8
16.	Transmitter power	$P_{dwRntx(dB)}$	20 dBm
17.	Other losses	$L_{dwRnothers}$	3 dBm
18.	Total cosmic temperature	T_C	280 K
19.	The absorptive medium temperature	T_m	10 K
20.	The noise temperature	$T_{recvdwRn}$	290 K

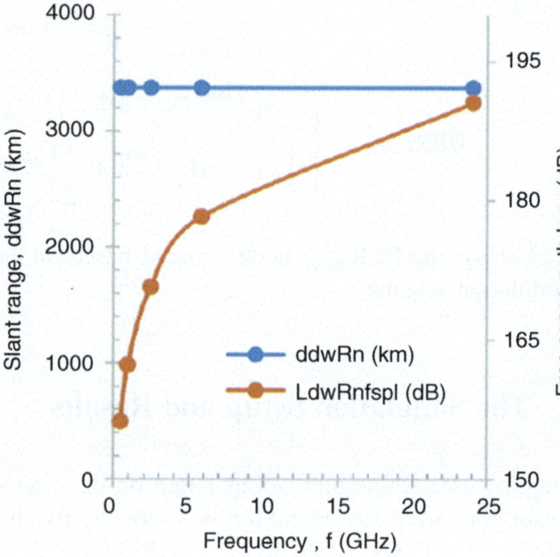

Fig. 2 The graph of the slant range and free space path loss versus frequency

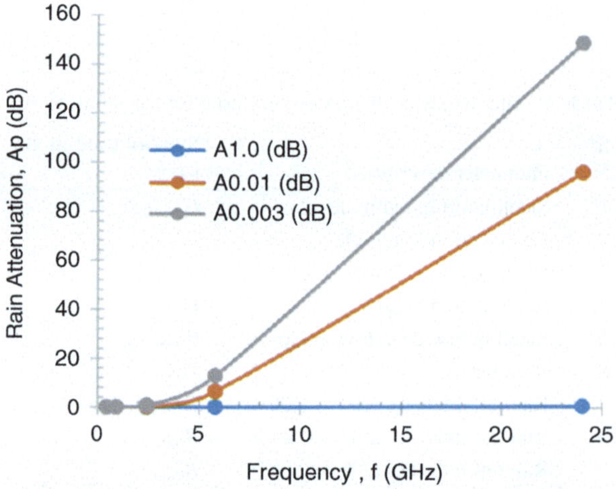

Fig. 3 The graph of the rain attenuation versus frequency

The results of the slant range and free space path loss versus frequency are shown in Fig. 2. The results in Fig. 2 show that the slant range does not depend on frequency; however, the free space loss associated with the slant range increases with increase in frequency. Specifically, while the slant range remained constant at 3374.8 km, the free space path loss increased from a value of 156.2 dB at frequency of 0.45492 GHz to a value of 190.6 dB at frequency of 24.021 GHz.

The results of the rain attenuation versus frequency and percentage (p) exceeded in an average year are shown in Fig. 3. The results in Fig. 3 show that for any given value of p the rain attenuation increases with increase in frequency. As such, in the graph, for $p = 0.01\%$, the rain attenuation increased from a value of 0.00534 dB at

Fig. 4 The graph of the downlink carrier to noise ratio, C/N versus frequency

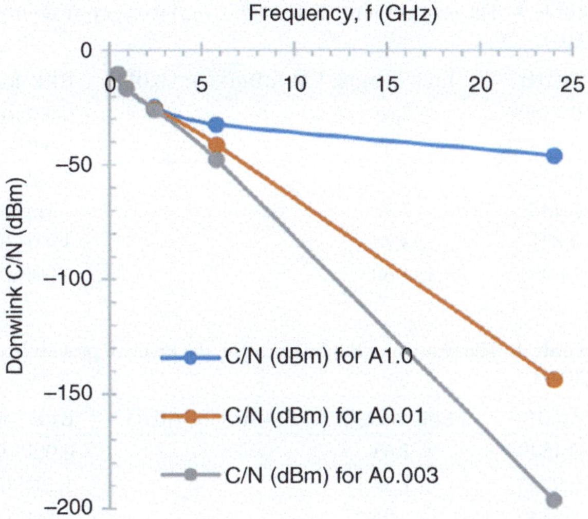

Table 2 The results for the link margin, the bit error probability, and link status versus frequency at A1.0

f (GHz)	Link margin, LM (dBm) for A1.0	BER for A1.0	Link status for A1.0
0.45492	7.65	0.00E+00	Feasible
0.9233	1.51	5.51E-08	Feasible
2.433	−6.91	2.22E-01	Not feasible
5.821	−14.5	4.47E-01	Not feasible
24.021	−26.99	4.98E-01	Not feasible

frequency of 0.45492 GHz to a value of 96.09238 dB at frequency of 24.021 GHz. On the other hand, for any given frequency, the rain attenuation increases with decrease in the percentage (p) exceeded in an average year. Again, for $f = 24.021$ GHz, the rain attenuation increased from a value of 0.668716 dB at $p = 1.0\%$ to a value of 148.8963 dB at $p = 0.003\%$.

The results of the downlink carrier to noise ratio, C/N versus frequency, are shown in Fig. 4. The results in Fig. 4 show that for any given value of p the C/N decreases with increase in frequency. As such, in the graph, for $p = 0.01\%$, C/N decreased from a value of −10.36 dB at frequency of 0.45492 GHz to a value of −143.67 dB at frequency of 24.021 GHz. On the other hand, for any given frequency, the C/N decreases with decrease in the percentage (p) exceeded in an average year. Again, for $f = 24.021$ GHz, C/N decreased from a value of −10.36 at $p = 1.0\%$ to a value of −196.47 dB at $p = 0.003\%$.

Link margin, the bit error probability and link status versus frequency are presented in Table 2. The results showed that in all the frequencies and rain attenuation configurations considered, the link margins are positive for the frequency of 0.45492 GHz and 0.9233 GHz. Hence, wireless communication is feasible in the downlink for the two frequencies, 0.45492 GHz and 0.9233 GHz. On the other hand,

Table 3 The results for the link margin, the bit error probability, and link status versus frequency at A0.01

f (GHz)	Link margin, LM (dBm) for A0.01	BER for A0.01	Link status for A0.01
0.45492	7.65	0.00E+00	Feasible
0.45492	7.64	0.00E+00	Feasible
0.9233	1.46	7.58E-08	Feasible
2.433	−7.41	2.48E-01	Not feasible
5.821	−23.24	4.93E-01	Not feasible
24.021	−124.68	5.00E-01	Not feasible

Table 4 The results for the link margin, the bit error probability, and link status versus frequency at A0.03

f (GHz)	Link margin, LM (dBm) for A0.03	BER for A0.03	Link status for A0.03
0.45492	7.63	0.00E+00	Feasible
0.9233	1.38	1.24E-07	Feasible
2.433	−8.05	2.78E-01	Not feasible
5.821	−29.88	4.98E-01	Not feasible
24.021	−177.48	5.00E-01	Not feasible

for all the rain attenuation configurations considered, the link margins are negative, and hence the links are not feasible for the frequencies of 2.433 GHz, 5.821 GHz, and 24.021 GHz. Furthermore, the bit error probability (BEP) of the feasible links is less than 1.0×10^{-06}. However, the BEP of the links that are not feasible is far above 1.0×10^{-06}.

In all, the study has shown that the ISM frequency bands less than 1 GHz are suitable for LoRa transceiver applied in direct satellite communication links. However, the high rain attenuation for higher frequencies and the corresponding increase in free space path loss with frequency will limit attainable orbital altitudes when LoRa transceiver is employed in direct satellite communication link. The link margin, bit error rate, and link status in relation to frequency are shown in Tables 2, 3, and 4 at different percentage exceeded in an average year of A1.0, A0.01, and A0.03.

4 Conclusion

Analysis of satellite downlink power budget and bit error for LoRa transceiver used in direct satellite-to-communication link is presented. The study considered the transceiver operations in five frequencies selected for the industrial, scientific, and medical (ISM) frequency bands. The essence of the study is to evaluate the feasibility of wireless communication in such satellite link using LoRa transceiver in each of the ISM frequencies considered. The study considered the satellite downlink with rain fading.

Presently, LoRa transceiver is operated in the frequencies below 1 GHz. However, some studies have considered the possibility of using LoRa technologies in the higher-frequency bands like the 2.4 GHz. The results obtained for this study showed that the higher frequencies present higher free space path loss and higher rain attenuation which when combined limits the feasibility of wireless communication on the satellite link. However, the results showed that wireless communication is feasible and the LoRa transceiver is operated at frequencies below 1 GHz.

References

1. Rizvi, T., Mishra, R., Chopkar, P.N., Gomkale, A., Upadhye, P., Bhonsle, D.: Internet of Things-based cognitive wireless sensor networks: applications, merits, and demerits. Cognit. Sens. **1**, 21–22 (2022)
2. Nourildean, S., Hassib, M., Mohammed, Y.: Internet of things based wireless sensor network: a review. Indones. J. Electr. Eng. Comput. Sci. **27**(1), 246–261 (2022)
3. Chaudhari, B.S., Zennaro, M., Borkar, S.: LPWAN technologies: emerging application characteristics, requirements, and design considerations. Future Internet. **12**(3), 46–70 (2020)
4. Pereira, F., Correia, R., Pinho, P., Lopes, S.I., Carvalho, N.B.: Challenges in resource-constrained IoT devices: energy and communication as critical success factors for future IoT deployment. Sensors. **20**(22), 6420 (2020)
5. An, Z., Lin, Q., Li, P., Yang, L.: General-purpose deep tracking platform across protocols for the internet of things. In: 18th International Conference on Mobile Systems, Applications, and Services, pp. 94–106. Association for Computing Machinery, New York (2020)
6. Shahroury, F.R.: Design of a low-power CMOS transceiver for semi-passive wireless sensor network application. Integration. **71**(6), 95–104 (2020)
7. Taha, A., Feteiha, M.F., Abdul, W.: Performance evaluation for LoRa transceiver. Int. J. Comput. Sci. Softw. Eng. (IJCSSE). **8**(2), 25–39 (2019)
8. Hossain, M.I., Markendahl, J.I.: Comparison of LPWAN technologies: cost structure and scalability. Wirel. Pers. Commun. **121**(1), 887–903 (2021)
9. Rashid, N.F.A., Abu-Samah, A., Noh, A.M., Azam, N.Z.S., Wahid, N.N., Chiang, C.Q., Alobaidy, H., Abdullah, N.F., Hamid, S.A., Nordin, R.: Development of smart campus applications based on wireless technologies using open-source platforms. Jurnal Teknologi. **84**(3), 173–184 (2022)
10. Eklund, C., Johansson, K.: A Comparison of Energy Usage Between LoRa 433Mhz and LoRa 868MHz. JÖNKÖPING (2021) https://www.essays.se/essay/89c811bd34. Last accessed 2023/05/22
11. Polak, I., Milos, J.: Performance analysis of LoRa in the 2.4 GHz ISM band: coexistence issues with Wi-Fi. Telecommun. Syst. Model. Anal. Des. Manag. **74**(3), 299–309 (2020)
12. Paredes, M., Bertoldo, S., Carosso, L., Lucianaz, C., Marchetta, E., Allegretti, M., Savi, P.: Propagation measurements for a LoRa network in an urban environment. J. Electromag. Waves Appl. **33**(15), 2022–2036 (2019)
13. Ezeh, H.I., Aneke, M., Young, E.G.: Evaluation of direct satellite communication capability of lora transceiver. Int. Multiling. J. Sci. Technol. (IMJST). **7**(8), 1–11 (2022)
14. Temim, M.A.B., Ferré, G., Tajan, R.: A new LoRa-like transceiver suited for LEO satellite communications. Sensors. **22**(5), 1830 (2022)
15. Loubany, A., Lahoud, S., Samhat, A., EHelou, M.E.: Improving energy efficiency in LoRaWAN networks with multiple gateways. Sensors. **23**(11), 5315 (2023)
16. Kufakunesu, R., Hancke, G.P., Abu-Mahfouz, A.M.: A survey on adaptive data rate optimization in LoRaWAN: recent solutions and major challenges. Sensors. **20**(18), 5044 (2020)

17. Maiti, A., Majumdar, S., Shukla, S., Koti, S.R., Gupta, P.K.: An open source web-GIS based precise satellite tracking and visualisation tool using two line element data. Ann. Int. Soc. Photogramm. Remote Sens. (ISPRS), 109–114 (2018)
18. Ukommi, U.S., Ubom, E.A., Ikpaya, I.: Ground station design for satellite communication and space technology development. Am. J. Eng. Res. **10**(8), 12–19 (2021)
19. Ekanem, K., Ubom, E., Ukommi, U.: Analysis of rain attenuation for satellite communication in Akwa Ibom State, Nigeria. In: 18th International Conference and Exhibition on Power and Telecommunication (ICEPT), pp. 25–34. IEEE Xplore, Abeokuta (2022)
20. Alam, J., Hossain, M.S., Ansari, J., Imran, A., Kamrul, M.I.: Estimation of rain attenuation of earth-to-satellite link over Nepal for Ku and Ka bands. In: International Conference on Robotics, Electrical and Signal Processing Techniques (ICREST), pp. 169–173. IEEE Xplore, Khaka (2021)
21. International Telecommunication Union, Recommendation ITU-R P.618-12, Propagation data and prediction methods required for the design of Earth-space telecommunication systems, P Series Radiowave propagation. https://www.itu.int/rec/R-REC-P.618/en. Last accessed 2023/05/07

Evaluation of Energy Consumption and Battery Life Span for LoRa IoT Multisensor Node for Precision Farming Application

Oduoye Israel Olufemi and Ubong Ukommi

1 Introduction

Nowadays, farmers across the globe are increasingly adopting technologies for maximization of farm resource utilization [1, 2]. Such precision farming technologies enable timely application of minimal farm resource with corresponding maximum yield [3, 4]. Notably, precision farming is a series of strategies and tools employed in farming management to optimize productivity [5, 6]. It requires the use of sensor-based data acquisition systems for monitoring, measuring, and transmission of vital farm parameters which are used in farm management decision-making. In practice, the sensor nodes deployed in the farms are part of Internet of Things [7] network equipped with one or more sensors for data capturing along with requisite transceiver that enables the node to connect and communicate to base stations and servers via wireless communication links. Generally, sensor nodes are resource constrained [8], and in remote locations like the farm, the sensor nodes are battery powered. As such, determination of the energy consumption [9] and battery lifetime of the sensor node is essential.

Moreover, LoRa technology has gained popularity in recent years as the most widely adopted transceiver technology for such resource-constrained IoT sensor networks [10, 11]. However, available studies on the energy consumption model for LoRa-based sensor nodes had focused on the single sensor node scenario [12, 13]. In practice, multisensor node is usually deployed whereby multiple sensor units are installed on a single sensor node with one transceiver [14, 15]. In such case, the energy consumption model is expected to factor in the diversity of data capture

O. I. Olufemi · U. Ukommi (✉)
Department of Electrical and Electronic Engineering, Akwa Ibom State University, Mkpat Enin L. G. A., Akwa Ibom State, Nigeria
e-mail: ubongukommi@aksu.edu.ng

S. C. Ekpo (ed.), *The Second International Adaptive and Sustainable Science, Engineering and Technology Conference*, Signals and Communication Technology,
https://doi.org/10.1007/978-3-031-53935-0_15

time requirements associated with the multiple sensors per node. Accordingly, in this paper, a multisensor energy consumption model was derived from some existing single energy models. A case study multisensor node for capturing five different parameters relevant for precision farming was used for sample numerical computations. A Visual Basic for Application program was developed to implement the numerical computations and to demonstrate the applicability of the multisensor energy model presented in this paper.

2 Methodology

This paper considered battery-powered precision farming multisensor node used for monitoring five key farm parameters, namely, temperature, relative humidity, CO_2, soil moisture, and light intensity. Generally, as a battery-powered sensor node captures and transmits data, it draws certain amount of current from the battery. According to [13], sensor node energy consumption model is required to determine (i) the average current drawn per cycle, (ii) the battery life span, and (iii) the energy consumed per cycle. Accordingly, notable energy consumption model for LoRa-based sensor node with single sensor unit has been presented in [12] and another model in [13]. In the two models presented in [12, 13], the sensor node operated in cycle that consists of data collection, data transmission, sleep, and wake-up phases. These four major phases are broken into ten phases in the model presented in [12], whereas in [13] about 11 phases are adopted. In any case, in each of the two models presented in [12, 13], there is a single sensor unit, whereas in the precision farming multisensor node presented in this paper, there are five sensor units, one sensor for each of the five farm parameters considered.

Accordingly, in this paper, the model presented in [13] is adapted by adding the data capture phases for the five different sensor units contained in the sensor node, as shown in Table 1. Next, the approach used for the determination of the sensor node average current per cycle, battery life span, and energy consumption per cycle is presented. Then, information on the case study multisensor node and numerical computation input parameters are presented along with the results.

2.1 Determination of the Sensor Node Average Current per Cycle, Battery Life Span, and Energy Consumption per Cycle

The multisensor energy model has 15 states, as listed in Table 1. The sleep state is the 15th state and the turnoff sequence is the 14th state. In order to determine the average current drawn per cycle, the values of the current, $I_{(j)}$, drawn in each of the sensor node state j listed in Table 1 and the duration, $t_{(j)}$, of the state are required.

Table 1 Summary of the comparison of the proposed multisensor node energy consumption model states and those of the two cited single sensor node energy models

State index, j	State description	Energy model for single sensor node presented in [12]	Energy model for single sensor node presented in [13]	Proposed energy model for multisensor node
1	Wake up	✓	✓	✓
2	Measure atmospheric temperature	×	✓	✓
3	Measure relative humidity	×	×	✓
4	Measure light intensity	×	×	✓
5	Measure soil moisture	×	×	✓
6	Measure CO_2	×	×	✓
7	Process data	×	✓	✓
8	Wake-up transceiver or radio preparation	×	✓	✓
9	Data transmission	✓	✓	✓
10	Wait first window or receive delay 1	✓	✓	✓
11	1st receive window or data reception 1	✓	✓	✓
12	Wait second window or receive delay 2	✓	✓	✓
13	2nd receive window or data reception 2	✓	✓	✓
14	Radio off	✓	×	×
15	Post-processing	✓	×	×
16	Sleep preparation or turnoff sequence	✓	✓	✓
17	Sleep	✓	✓	✓
Total number of sensor node states		10	11	15

The values of the current drawn by the sensor node components in each of the sensor states are obtained from the data sheet of the sensor node components. Also, the values of $t_{(j)}$ are presented in [13] for all the states except for the states 9, 11,13, and 15 which are computed using analytical models given in [13] as follows:

$$t_{11} = (N_{\mathrm{dsym}})\left(\frac{2^{\mathrm{SF}}}{\mathrm{BW}}\right) \tag{1}$$

$$t_{12} = 2000 - t_{11} \tag{2}$$

where SF is the spreading factor, BW is bandwidth in KHz, and N_{dsym} is 8 symbols for SF = 11 or 12, while N_{dsym} is 12 symbols for SF = 7, 8, 9, or 10. The details of the expressions for the determination of $t_{(9)}$ which is the packet time on air are presented in [13], and in this paper, $t_{(9)}$ is computed using an online LoRa packet time on air calculator tool [16]. The duration of the sleep state, $t_{(15)}$, is given as follows [13]:

$$t_{(15)} = t_{\mathrm{cycle}} - \sum_{j=1}^{j=14} \left(t_{(j)}\right) \tag{3}$$

Notably, while the last state (15th state) is the sleep state, the active state of the sensor node consists of state 1 to 14. Hence, according to [13], the average current, I_{avg}, drawn by the multisensor node with 15 states per cycle and with cycle time denoted as t_{cycle} is given as follows:

$$I_{\mathrm{avg}} = \frac{\sum_{j=1}^{j=15} \left[\left(I_{(j)}\right)\left(t_{(j)}\right)\right]}{\sum_{j=1}^{j=15} \left(t_{(j)}\right)} \tag{4}$$

where $t_{(j)}$ is the duration of sensor node state j and $I_{(j)}$ is the current drawn by the sensor node in state j. The sensor node battery life span, t_{Lh}, in hours is given as follows:

$$t_{Lh} = \frac{C_{\mathrm{bat}}}{I_{\mathrm{avg}}} \tag{5}$$

where C_{bat} is the battery capacity in mAh and I_{avg} is in mA. The energy consumed by the sensor node in each cycle, E_{cycle}, is given as follows:

$$E_{\mathrm{cycle}} = V\left(I_{\mathrm{avg}}\right)\left(t_{\mathrm{cycle}}\right) \tag{6}$$

where V the sensor node voltage in volts, I_{avg}, is in mA, t_{cycle} is in ms, and E_{cycle} is in μJ.

Table 2 The case study multisensor node parameters and the size of data generated by the sensors per cycle

Sensed parameter	Measurement range and resolution	Number of bits per measurement	Data capture repeat time (s)	Total number of bits captured per cycle
Temperature	−55 to 125 °C	13	67.5	104
	Res.: 0.035 °C			
Relative humidity	0%–100%	15	67.5	120
	Res.: 0.005%			
CO_2	0 to 25 m/s	14	67.5	112
	Res.: 0.002 m/s			
Soil moisture	0° to 359°	14	33.75	224
	Res.: 0.035 m/s			
Light	900 to 1100	12	33.75	192
	Res.: 0.05 hPa			
Total data capture in a cycle				752 bits or 94 bytes

2.2 *Case Study Multisensor Node and Simulation Parameters*

The case study multisensor node has five sensors for capturing data on the following farm parameters: temperature, relative humidity, CO_2, soil moisture, and light, as shown in Table 2. The LoRa transceiver is required to operate with duty cycle of not more than 1%. As shown in Table 2, the multisensor node generates a total of 94 bytes of data in each cycle with cycle time of 540 s (or 540,000 ms). The spreading factor of 12 with maximum payload size limit of 51 bytes requires two cycles (with pay load size of 47 bytes) to transmit the 94 bytes in each cycle. The packet time on air for 47 bytes payload is 2629.6ms (or 2.6296 s ≈2.7 s) for the SF of 12 which gives a transmitter repeat time of approximately 270 s for a duty cycle of 1%. Hence, the LoRa transmitter will be required to transmit 47 bytes in every 270 s to be able to transmit all the 94 bytes in each sensor cycle time of 540 seconds. The duty cycle for the transmitter is 2.6296/270 = 0.009739259≈0.97%.

The details of the current and timing dataset for the case study multisensor node operated with spreading factor of 12, bandwidth of 125 KHz, transmitter power of 10 dBm, and cycle time of 540,000 ms are presented in Table 3.

3 Results and Discussions

As stated earlier, in the cited model presented in [12, 13], numerical results were obtained by using the current and time dataset (in Table 3) to determine the average current drawn per cycle, battery life time, and energy consumption per cycle based on Eqs. 4, 5, and 6, respectively. The computations of these energy consumption parameters were performed using a program written in Visual Basic for Application (VBA). They can be performed with MATLAB, Python program, or any other tool;

Table 3 The current and timing dataset for the case study multisensor node operated with SF of 12, BW of 125 KHz, transmitter power of 10 dBm, and cycle time of 540,000 ms

State index, *j*	Current draw in state *j* (mA)	Duration of each occurrence of state *j* (ms)	Number of state j occurrence per sensor node cycle time of 540,000 ms	Total state j duration per sensor node cycle (ms)
1	15.021	1	1	1
2	16.520	750	8	6000
3	15.353	400	8	3200
4	15.059	6000	16	96,000
5	15.371	5000	16	80,000
6	15.371	5000	8	40,000
7	15.021	1	2	2
8	16.621	5	2	10
9	60.321	2629.632	2	5259.264
10	15.021	1000	2	2000
11	24.821	262.144	2	524.288
12	15.021	1737.856	2	3475.712
13	24.821	265	2	530
14	15.021	1	1	1
15	14.521	302,996.736	1	302,996.736
Total for the active states 1 to 16				237,003.264
Total per cycle				540,000.000

however, the choice of VBA was due to its simplicity and compatibility with most of the Microsoft Office packages like Excel and Word.

In the computation, the sensors and the transceivers are assumed to be operated at 3.3 V with a battery capacity of 2400 mAh. The results of the energy consumption for the case study multisensor node operated with spreading factor of 12, bandwidth of 125 KHz, transmitter power of 10 dBm, and cycle time of 540,000 ms are presented in Table 4. Based on the results in Table 4, the following values are obtained for the current, battery life span, and total energy per cycle:

$$\text{The average current is } I_{\text{avg}} = \frac{8{,}264{,}138.71}{540{,}000} = 15.30\ \text{mA}$$

$$\text{The battery life span in hours, } t_{Lh} = \frac{2400}{15.30} = 156.82\ \text{h}$$

The total energy consumed per cycle is:

$$E_{\text{cycle}} = 3.3(15.30)(540000) = 27{,}271{,}657.75\ \mu\text{J}$$

Also, the results in Table 4 show that the sensor node active state (consisting of state 1 to 14) took about 43.89% of the cycle time and consumed 46.76 5% of the total energy per cycle, whereas the sleep state (consisting of state 15) took 56.11% of the cycle time and consumed about 53.24% of the total energy.

Table 4 The results of the energy consumption for the case study multisensor node operated with spreading factor of 12, bandwidth of 125 KHz, transmitter power of 10 dBm, and cycle time of 540,000 ms

State index, *j*	State description	Ist(*j*) * tst(*j*)	Energy consumed per cycle in state *j* (μJ)	Percentage of energy consumed per cycle in state *j* (%)	Percentage of the cycle time spent in state *j* (%)
1	Wake up	15.021	49.5693	0.0002	0.0002
2	Measure atmospheric temperature	99,120	327,096	1.1994	1.1111
3	Measure relative humidity	49,129.6	162,127.68	0.5945	0.5926
4	Measure light intensity	1,445,664	4770691.2	17.4932	17.7778
5	Measure soil moisture	1,229,680	4,057,944	14.8797	14.8148
6	Measure CO_2	614,840	2,028,972	7.4399	7.4074
7	Process data	30.042	99.1386	0.0004	0.0004
8	Wake-up transceiver or radio preparation	166.21	548.493	0.0020	0.0019
9	Data transmission	317,244.1	1,046,905.41	3.8388	0.9739
10	Wait first window	30,042	99,138.6	0.3635	0.3704
11	1st receive window or reception 1	13,013.35	42,944.063	0.1575	0.0971
12	Wait second window	52,208.67	172,288.61	0.6317	0.6437
13	2nd receive window or reception 2	13,155.13	43,411.929	0.1592	0.0981
14	Sleep preparation or turnoff sequence	15.021	49.5693	0.0002	0.0002
15	Sleep	4399815.60	14519391.49	53.2399	56.1105
Total for the active states 1 to 14		**3,864,323.11**	**12,752,266.26**	**46.76**	**43.89**
Total per cycle		**8,264,138.71**	**27,271,657.75**	**100.00**	**100.00**

Similar, the numerical computations were performed for the spreading factors of 7, 8, 9, 10, 11, and 12. The results of the average current drawn per cycle shown in Fig. 1 indicate that the average current drawn per cycle increases as the spreading factor (SF) increases from 7 to 12 which is in agreement with the results presented in

Fig. 1 The results of the average current drawn per cycle

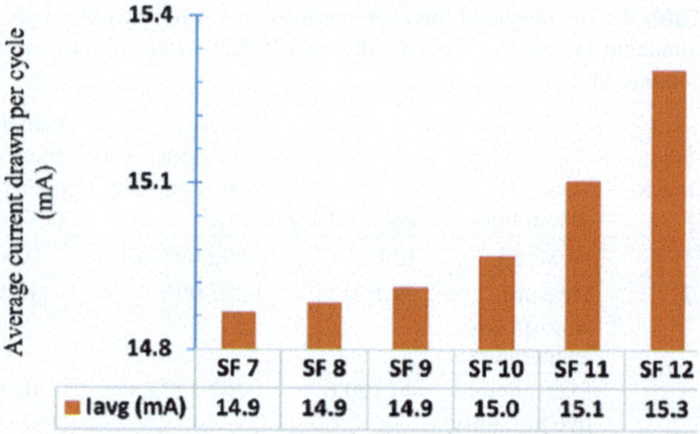

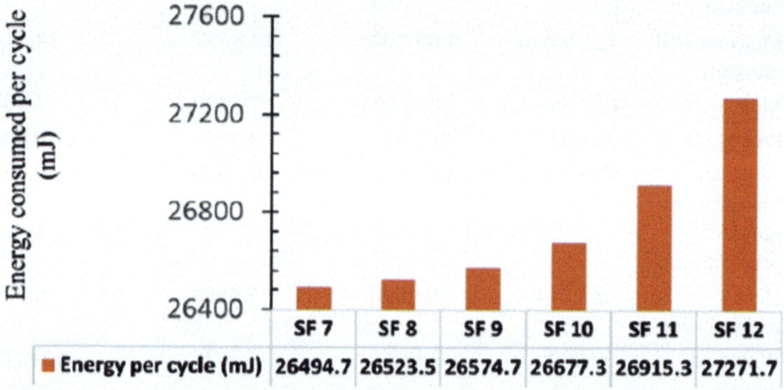

Fig. 2 The results of the energy consumed per cycle

[17]. Similarly, Fig. 2 indicates that the energy consumed per cycle increases as the spreading factor (SF) increases from 7 to 12. These results agree with the results presented in [13, 17–19]. On the other hand, the results of the battery life span versus spreading factor shown in Fig. 3 indicate that the battery life span decreases as the spreading factor (SF) increases from 7 to 12. These results agree with the results presented in [13, 17].

4 Conclusion

Energy consumption model for multisensor node based on LoRa transceiver technology is presented. The study considered battery-powered precision farming multisensor node used for monitoring five key farm parameters, namely, temperature, relative humidity, CO_2, soil moisture, and light intensity. The model consists of requisite analytical expressions used to determine the average current drawn per cycle, the battery life span, and the energy consumed per cycle by the case study multisensor node. Visual Basic for Application (VBA) was developed and used to

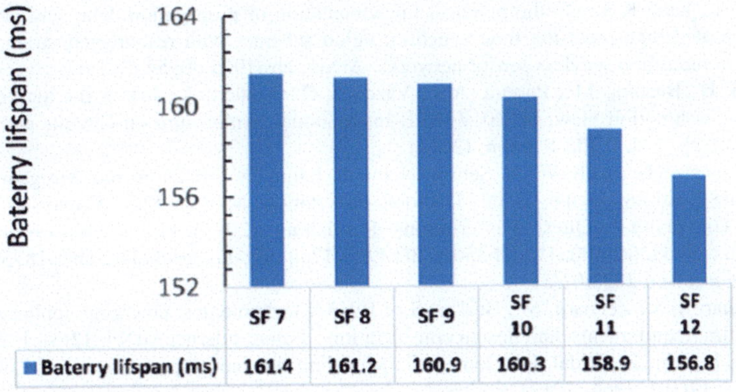

Fig. 3 The results of the battery life span versus spreading factor

carry out sample numerical computations based on the analytical models and input dataset for the case study multisensor node. The results showed that the energy consumption increases with increase in the spreading factor, while the battery life span decreases with increase in the spreading factor. Conditions of rain fade may have effect on the outcome [20], and further investigation is encouraged.

References

1. Pal, D., Joshi, S.: AI, IoT and robotics in smart farming: current applications and future potentials. In: Proceedings of the 2023 International Conference on Sustainable Computing and Data Communication Systems (ICSCDS), pp. 1096–1101. IEEE Xplore, Tamil Nadu (2023)
2. Javaid, M., Haleem, A., Khan, I.H., Suman, R.: Understanding the potential applications of Artificial Intelligence in agriculture sector. Adv. Agrochem. **2**(1), 15–30 (2023)
3. Sharma, A., Jain, A., Gupta, P., Chowdary, V.: Machine learning applications for precision agriculture: a comprehensive review. IEEE Access. **9**, 4843–4873 (2020)
4. Monteiro, A., Santos, S., Gonçalves, P.: Precision agriculture for crop and livestock farming – brief review. Animals. **11**(8), 2345–2362 (2021)
5. Tran, D.V., Nguyen, N.V.: The concept and implementation of precision farming and rice integrated crop management systems for sustainable production in the twenty-first century. Int. Rice Comm. Newslett. **55**, 91–102 (2006)
6. Aubert, B.A., Schroeder, A., Grimaudo, J.: IT as enabler of sustainable farming: An empirical analysis of farmers' adoption decision of precision agriculture technology. Decis. Support Syst. **54**(1), 510–520 (2012)
7. Poongodi, T., Rathee, A., Indrakumari, R., Suresh, P.: IoT sensing capabilities: sensor deployment and node discovery, wearable sensors, wireless body area network (WBAN), data acquisition. In: Peng, S.L., Pal, S., Huang, L. (eds.) Principles of Internet of Things (IoT) Ecosystem: Insight Paradigm. Intelligent Systems Reference Library, vol. 174, pp. 127–151. Springer, Cham (2020)

8. Huan, X., Kim, K.S.: On the practical implementation of propagation delay and clock skew compensated high-precision time synchronization schemes with resource-constrained sensor nodes in multi-hop wireless sensor networks. ArXiv, abs/1905.00554 (2019)
9. Bellini, B., Becoña, J.P., Pereira, A.S., Vázquez, C., Arnaud, A.: IoT in the agribusiness, a power consumption view. In: 2019 IEEE International Symposium on Circuits and Systems (ISCAS), pp. 1–4. IEEE, Sapporo (2019)
10. Richardson, S.G.: LoRaWAN Sensor Network Jamming Detection and Mitigation Using Machine Learning in the Cloud. Doctoral dissertation. Morgan State University, Morgan State University ProQuest Dissertations Publishing (2023) https://www.proquest.com/openview/8ba826e4400a414d2b350ef73f139ffb/1?pq-origsite=gscholar&cbl=18750&diss=y. Last accessed 2023/07/21
11. Chaudhari, B.S., Zennaro, M., Borkar, S.: LPWAN technologies: emerging application characteristics, requirements, and design considerations. Future Internet MDPI. **12**(3), 1–25 (2020)
12. Casals, L., Mir, B., Vidal, R., Gomez, C.: Modeling the energy performance of LoRaWAN. Sensors. **17**(10), 2364 (2017)
13. Eriksen, R.: Energy consumption of low power wide area network node devices in the industrial, scientific and medical band. (Student thesis, KTH, School of Electrical Engineering and Computer Science). Diva2, 1351642 (2019) https://kth.diva-portal.org/smash/record.jsf?pid=diva2%3A1351642&dswid=4119. Last accessed 2023/07/21
14. Marinov, M.B., Nikolov, D.N., Ganev, B.T., Djamiykov, T.S.: Smart multisensor node for remote elevator condition monitoring. In: 21st International Symposium on Electrical Apparatus & Technologies (SIELA), pp. 1–4. IEEE Xplore, Bourgas (2020)
15. Kazlauskas, M., Navakauskas, D.: Case study of a multisensor patient network and microservices managed by fog computing. In: 2021 IEEE 9th Workshop on Advances in Information, Electronic and Electrical Engineering (AIEEE), pp. 1–4. IEEE Xplore, Riga (2021)
16. Pötsch, A., Hammer, F.: Towards end-to-end latency of LoRaWAN: experimental analysis and IIoT applicability. In: 15th IEEE International Workshop on Factory Communication Systems (WFCS), pp. 1–4. IEEE Xplore, Sundsvall (2019)
17. Fialho, V., Fortes, F.: Battery lifetime estimation for LoRaWAN communications. Int. J. Innov. Technol. Explor. Eng. (IJITEE). **9**(11), 306–310 (2020)
18. Johansson, K., Eklund, C.: A comparison of energy usage between LoRa 433Mhz and LoRa 868MHz. Student thesis, Jonkoping University, JTH, Department of Computing. Diva portal, (2021). https://hj.diva-portal.org/smash/record.jsf?pid=diva2%3A1581948&dswid=4804. Last accessed 2023/07/21
19. Mnguni, S., Mudali, P., Abu-Mahfouz, A.M., Adigun, M.: Performance evaluation of spreading factors in LoRa networks. In: Towards New e-Infrastructure and e-Services for Developing Countries: 12th EAI International Conference, AFRICOMM 2020, Ebène City, Mauritius, LNICST, vol. 361, pp. 203–215. Springer (2021)
20. Ekanem, K., Ubom, E., Ukommi, U.: Analysis of rain attenuation for satellite communication in Akwa Ibom State, Nigeria. In: 18th International Conference and Exhibition on Power and Telecommunication (ICEPT), pp. 25–34. IEEE Xplore, Abeokuta (2022)

Vibration Analysis for Predictive Maintenance and Improved Machine Reliability of Electric Motors in Centrifugal Pumps

Fortune Jameson, Emmanuel Ubom, and Ubong Ukommi

1 Introduction

Vibration analysis can be used as a predictive maintenance procedure as well as a support for machinery maintenance decisions [1]. By measuring and analyzing the machine's vibration, the nature and severity of defects can be determined and actions taken to prevent machine breakdown. The practice requires the measurement and analysis of vibration in rotating machines, utilizing different vibration sensors such as accelerometers, velocity transducers, or displacement probes. These can be mounted on a case of a machine using a permanent bolt or potable magnet to hold them in place to measure the vibration, relative to a gravitational pull level. The measured vibration signal can then be processed to provide a measured output velocity in inches/second or millimeters/second.

Vibration analysis can be defined as a process that monitors vibration levels and investigates the patterns in vibration signals within a component, machinery, or structure to detect abnormal vibration events and to evaluate the overall condition of the test object. It is mostly conducted both on the time waveforms of the vibration signal directly and on the frequency spectrum, which is usually obtained by applying Fourier transform on the time waveform. The time domain analysis may reveal when and how severe an abnormal vibration event has occurred. This is achieved by extracting and studying parameters such as the root mean square (RMS), standard deviation, peak amplitude, and skewness from the acquired data and is capable of evaluating the overall condition of the machine being monitored [2]. It is one of the most widely used predictive maintenance techniques; with very few exceptions, can

F. Jameson · E. Ubom (✉) · U. Ukommi
Department of Electrical and Electronic Engineering, Akwa Ibom State University, Mkpat Enin, Akwa Ibom State, Nigeria
e-mail: emmanuelubom@aksu.edu.ng

S. C. Ekpo (ed.), *The Second International Adaptive and Sustainable Science, Engineering and Technology Conference*, Signals and Communication Technology,
https://doi.org/10.1007/978-3-031-53935-0_16

be applied to a wide variety of rotating equipment; and can be used for detection of performance deviation, diagnosis of impending failure, and validation of maintenance actions. This work focused on the vibration analysis of centrifugal pumps as it has rampant application in industries with special attention given to the electric motors. Centrifugal pumps need to be maintained so that their life span is long and the performance is at optimum. Such routine maintenance must be planned in accordance with certain parameters that will enable the quickest detection of discrepancies in the operational behavior of the pump, one of which is the vibration of the moving parts of the electric motors in the pumps.

Other than the electric motor, pump vibrations can be caused by the following:

1. Misalignment of the pump and driver shaft.
2. Imbalance such as unbalanced impeller or shaft, bent shaft, and damaged impeller vanes.
3. Nonrigid foundation or baseplate that the pump is attached to may lack proper internal locking.
4. Related improper installation lapses.

Loose bolts on the feet or soft foot conditions also result in vibration, improper grouting, and inadequate tie-down bolts [3], of great interest in this work is the vibrations caused by the malfunctioning of electric motor such as burnt armature or stator coils, cracked motor casing, or an unequal air gap between the motor armature and stator [4]. To obtain a full-spectrum vibration signature in four axes (vertical, horizontal, vertical, and axial) on both ends of the motor, sensors are attached to the housing of the motor as casing-mounted transducers [5]. Sensors are devices that convert mechanical signals to electrical signals [6]. They are transducers that can be directly mounted on the machine structure to measure the vibration and produce an electrical output that can be processed as a data acquisition source for vibration analysis. Sensors like accelerometers are used to quantify the magnitude of vibration known as vibration amplitude which reports how smooth or rough the machine is running.

2 Methodology

2.1 Data Collection

The research deployed Fluke 810 vibration meter, which was attached to multiple points of the bearing housing of the pump motor to obtain vibration data in the vertical, horizontal, axial, and radial directions. The mean of the collated data after multiple repeated measurements was used to analyze the vibration of the system in time domain. Figure 1 depicts the approach deployed in the work.

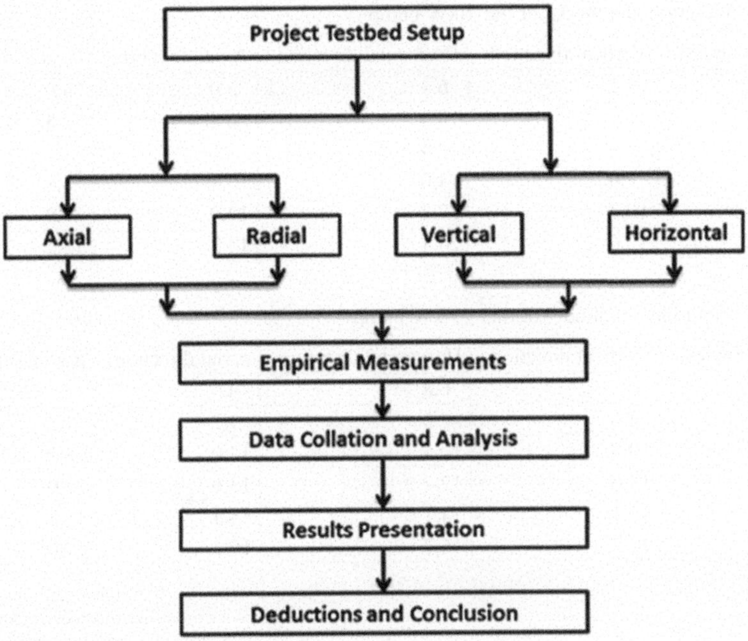

Fig. 1 General approach deployed in the analysis

2.2 *Data Analysis*

The data was analyzed in the time domain using some statistical parameters stated in [7]. The time domain analysis provides all the real values of the signal or function at various separate instances in the case of discrete-time or continuous-time functions. Such parameters include the following:

Standard Deviation Standard deviation (SD) generally represents the variation of the data from the mean. A low standard deviation means the data point is very close to the mean, whereas high standard deviation means the data points are spread over a wide range of values [8]. It is calculated using Eq. (1):

$$\mathrm{SD} = \sqrt{\frac{1}{N}\sum_{i=0}^{N}[X(i) - X]} \tag{1}$$

Root Means Square (RMS) The RMS value is usually used for tracking of the overall noise value and indicates if the amount of energy in the vibration signal is proportional to its RMS value. It is mostly used in detecting out of balance in random signal [9]. The RMS value of the discrete-time signal is given by Eq. (2):

Table 1 Vibration amplitude for the 4-kW pump

Time (minutes)	Vertical direction	Horizontal direction	Axial direction	Radial direction
2	4.2	6.4	5.0	4.0
4	6.0	8.3	6.8	5.8
6	7.8	12.4	8.6	7.2
8	9.0	14.0	10.4	8.8
10	10.2	17.2	12.0	9.8
12	12.6	20.4	14.2	12.2

Table 2 Vibration amplitude for the 5.5-kW pump

Time (minutes)	Vertical direction	Horizontal direction	Axial direction	Radial direction
2	6.2	6.8	8.0	5.6
4	8.0	8.2	9.2	7.3
6	9.8	10.5	12.0	8.2
8	11.6	12.2	13.6	10.0
10	13.1	14.1	14.4	11.8
12	15.3	16.7	16.3	13.0

$$\Psi_x = \sqrt{\frac{1}{n}\sum_{i=0}^{n-1}[X(i)]^2} \tag{2}$$

Crest Factor The crest factor can be defined as the ratio of the peak value to the RMS value; it yields the information about the spikiness of the measured signal. It is a pure number without any dimensions. Crest factor indicates the existence of sharp peaks in signal, and its value is very sensitive toward signal properties and is given by Eq. (3):

$$\text{Crest Factor (CF)} = \frac{\text{Peak Value}}{\text{RMS value}} \tag{3}$$

Peak Value This is the highest value of the vibration intensity and is defined as the maximum vibration speed attained by the machine in a given time period of time [10]. It can be used to indicate the highest speed attained by the machine.

Tables 1 and 2 are the mean of the vibration amplitude in (m/s^2) measured at normal operations for the 4-kW and 5.5-kW pumps, respectively.

2.3 *Measurement Under Faulty Condition*

Another measurement was carried out with the replacement of the electric motor bearing with a confirmed faulty part for the 4 kW pump. The faulty bearing produced

Table 3 Compared vibration measurements of 4-kW pump using a good bearing and a bad bearing

Time (minutes)	Good vertical	Bad vertical	Good horizontal	Bad horizontal	Good axial	Bad axial	Good radial	Bad radial
2	4.2	24.2	10.0	26.4	8.4	25.1	6.8	24.0
4	6.0	26.0	4.2	28.3	10.6	25.8	8.4	25.6
6	7.8	27.2	13.9	31.2	12.0	27.3	10.0	27.3
8	9.0	29.0	15.3	32.4	13.8	28.8	12.1	28.7
10	10.2	31.2	17.2	34.0	15.2	29.4	13.8	29.9
12	12.6	32.6	18.9	37.2	16.8	32.1	15.2	31.4

noise and uneven resistance when the shaft was manually slowly rotated. There was also evidence of lubrication leakages as well as physically identified wear on the exposed surface. The test with the physically confirmed faulty bearing was to ascertain the changes in the vibration data, using the same accelerometer and time spacing, and all conditions remaining the same with the healthy measured conditions. Table 3 presents the differences in the measured data. Table 3 faulty bearing measured vibration data was compared to the ISO 10816-3 vibration severity chart [11], and it showed an escalation in the vibration level above the acceptable limits.

3 Results

3.1 Empirical Measurements of Vibration Rates

The first step was to obtain a full-spectrum vibration signature in four axes. Measurements were carried out as described in Sect. 2.1 with the accelerometer attached vertically, horizontally, axially, and radially. The result presented in Table 1 is shown in Fig. 2 for the 4-kW pump.

The plot shows a higher value of vibration when measured in the horizontal position. This indicates a possible imbalance as high horizontal readings are commonly associated with unbalance in the rotating elements [12]; a high horizontal vibration indicates an electronic or mechanical imbalance, faulty gears, failed bearings, loose foundations, wear and tear, or misalignment. Maintenance action is therefore necessary to check for causes of imbalances in this motor. The vertical, axial, and radial measurements were within acceptable limits and therefore exclude faults like coupling misalignments and bent shafts. Figure 3 presents the vibration measurement report carried out on the 5.5-kW pump.

Measurements carried out on 7.37 HP pump show a relatively stable and fine system as the vibration accelerations measured at the four positions were within acceptable levels as shown in Fig. 3. The axial measurement was at a warning level signifying possible misalignment issue.

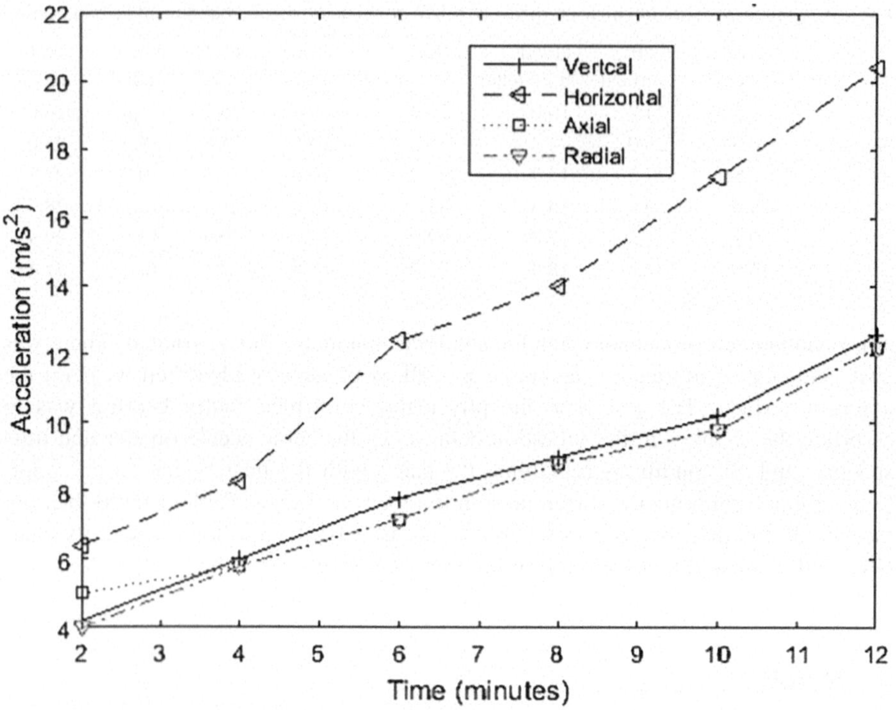

Fig. 2 Vibration measurement result for the 4-kW pump

3.2 Analysis of the Vibration Data

Analysis using the RMS values of the measured data shown in Fig. 4 confirms high horizontal vibrations for the 4-kW pump. The RMS value is generally the most useful because it is directly related to the energy content of the vibration profile and thus the destructive capability of the vibration and takes into account the time history of the waveform.

The crest factor (CF) is the peak-to-RMS ratio mostly used to determine deterioration of bearings by relative comparison. From Figs. 5 and 6, it could be confirmed that there is an existing deterioration in the bearing condition of the 4-kW pump. The CF gives a quick idea of how much impacting is occurring in a waveform. Impacting is often associated with roller bearing wear, cavitation, and gear tooth wear. Figure 5 shows the CF of all the measured positions for the two pumps, while Fig. 6 took the averaged CF. A CF of 1.44 indicates the existence of cavitation, wear and tear, or bearing roller fault and must be monitored to avoid the failure of the pump, although pump is still working perfectly.

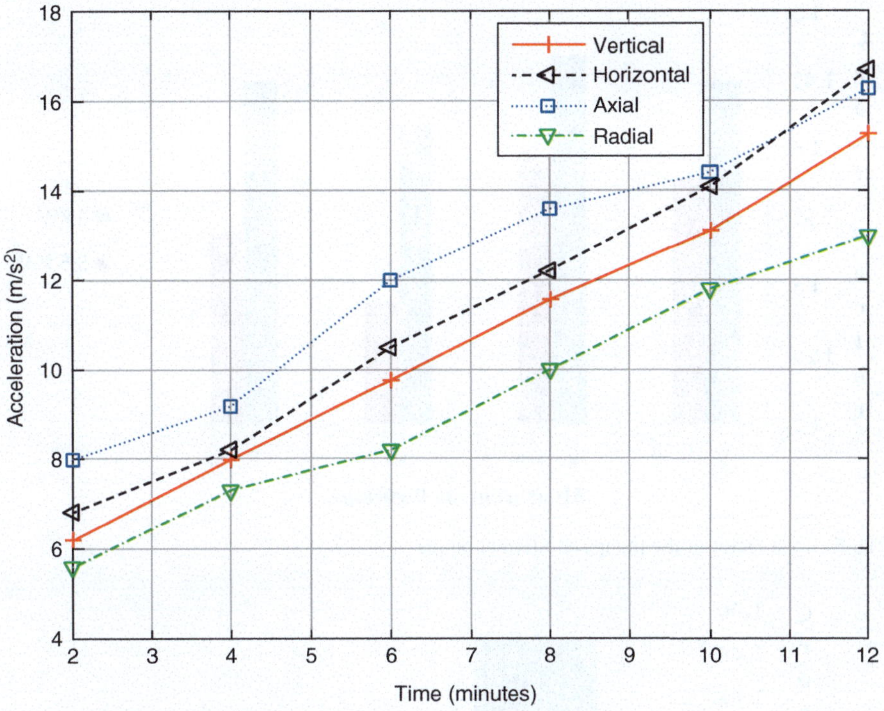

Fig. 3 Measurement results for 5.5-kW pump

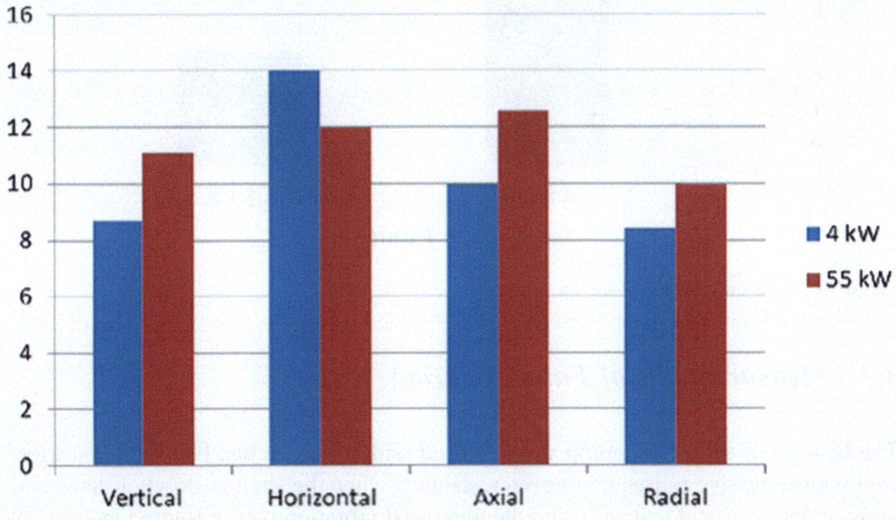

Fig. 4 RMS values of measurement positions

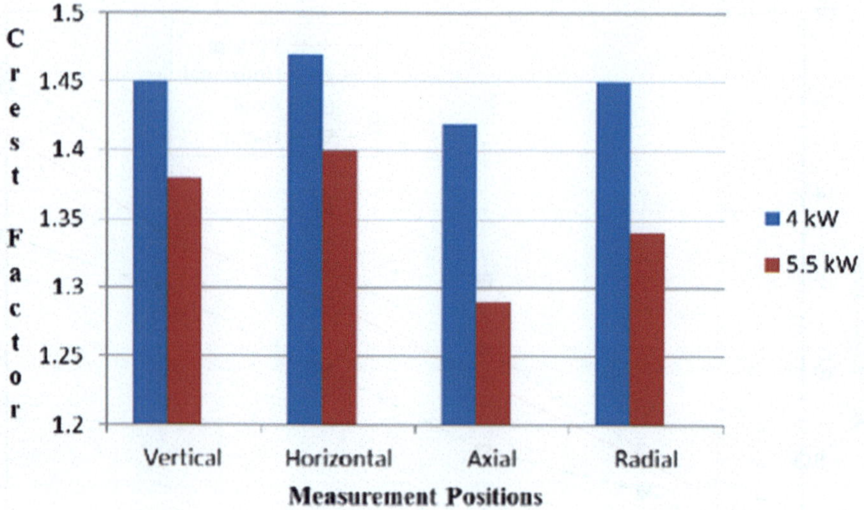

Fig. 5 Crest factor of the pumps at different points

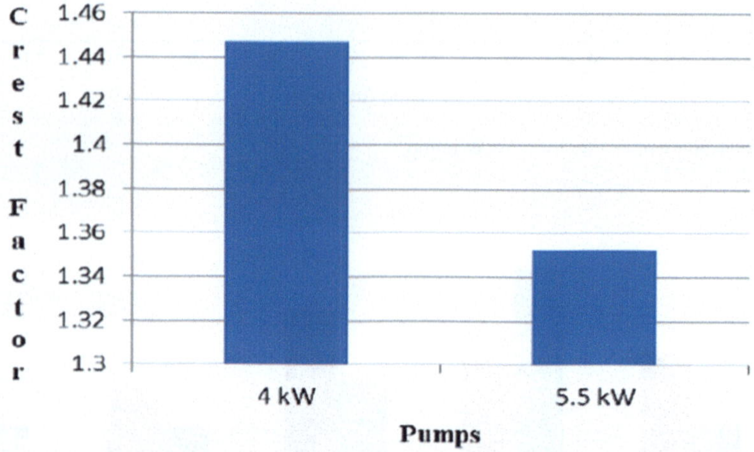

Fig. 6 Average crest factor of the pumps

3.3 Measurements of Faulty Bearing

The bearing of the 4-kW pump was replaced with a known bad bearing (characterized with excessive noise and uneven resistance when the shaft is slowly rotated and with evidence of seal leakage), and the measured vibration was presented in Table 3. Figure 7 depicts the faulty bearing scenario, and all the positions recorded unacceptable vibration levels with the horizontal vibration showing a possible bearing fault compared to the good bearing records. Faulty bearing shaft and imbalance or wear and tear on the bearing casing are possible suspects.

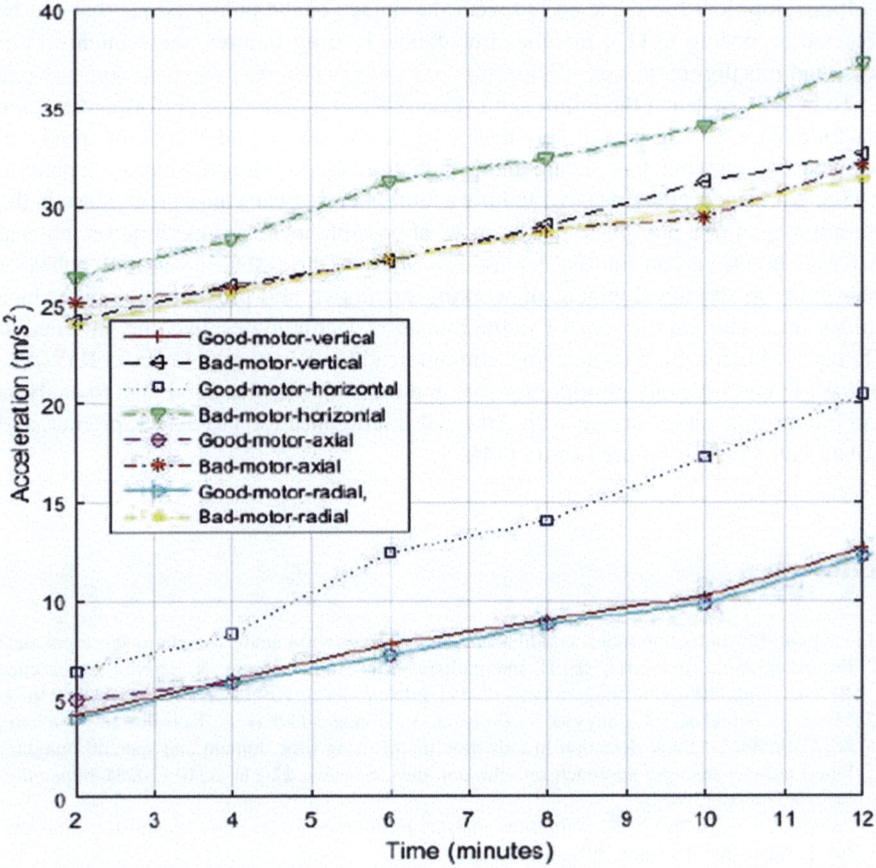

Fig. 7 Comparison of vibration with faulty bearing

4 Conclusion

The saying that prevention is always better than cure can best be likened to the use of vibration analysis to predict the initialization of failures and prevent system downtimes since developing problems can be predicted so that repair happens before disaster strikes.

The work carried out measurements of vibration using an accelerometer in two working centrifugal pump's electric motors in vertical, horizontal, axial, and radial positions. The captured data were analyzed using time domain approaches and deductions made. From the deduced reports, it is convenient to say that vibration measurements can be used to detect failure modes in centrifugal pump motors. Early detection will help the maintenance engineer to plan maintenance actions within the fault development life cycle and prevent the outage of the pump. Issues that can be detected according to [13] includes imbalance, bearing failures, mechanical looseness, and misalignment.

To be able to detect the failures, it is necessary to monitor the performance of the machine from when it is relatively new over a long time to understand the trend and be able to detect when there is/are undue deviation(s) from its usual vibration pattern.

The work succeeded in showing how a faulty bearing can cause an increase in the vibration level of a pump and discusses what possible increases in vibration taken at different directions can signify. A long-term study of the system's vibration behavior may guide in the development of working predictive and preventive maintenance guides and with the advent of communication technologies like the Internet of Things (IoT), machine-to-machine communication, Bluetooth, LoRa/LoRaWAN, and Wi-Fi can integrate multiple sensors and transmit the measured data to analysis and monitoring server driven with Artificial Intelligence (AI) to detect, predict, and isolate faults before failure occurs [14].

References

1. Tsypkin, M.: Induction motor condition monitoring: vibration analysis technique – a practical implementation. In: 2011 IEEE International Electric Machines & Drives Conference (IEMDC), pp. 406–411, Niagara Falls (2011). https://doi.org/10.1109/IEMDC.2011.5994629
2. Misra, S., Satish, K.V.C., Sayyad, S., Bongale, A., Dongare-Jadhav, P., Kotecha, K., Abraham, A., Gabralla, L.: Fault detection in induction motor using time domain and spectral imaging-based transfer learning approach on vibration data. Sensors. **22**(21), 8210 (2022). https://doi.org/10.3390/s22218210
3. Aherwar, A., Khalid, M.S.: Vibration analysis techniques for gearbox diagnostic: a review. Int. J. Adv. Eng. Technol. **3**(2), 4–12 (2012)
4. Albraik, A., Althobiani, F., Gu, F., Ball, A.: Diagnosis of centrifugal pump faults using vibration methods. In: 25th International Congress on Condition Monitoring and Diagnostic Engineering (COMADEM 2012) Journal of Physics: Conference Series, Volume 364, pp. 1–12, Huddersfield (2012)
5. Gopinath, S.: Study on electric motor mass unbalance based on vibration monitoring analysis technique. In: 2010 International Conference on Mechanical and Electrical Technology, pp. 539–542. IEEE Xplore, Singapore (2010). https://doi.org/10.1109/ICMET.2010.5598417
6. Majumder, B.D., Roy, J.K., Padhee, S.: Recent advances in multifunctional sensing technology on a perspective of multi-sensor system: a review. IEEE Sensors J. **19**(4), 1204–1214 (2019)
7. Arunkumar, K.M., Manjunath, T.C.: A brief review/survey of vibration signal analysis in time domain. SSRG Int. J. Electron. Commun. Eng. (SSRG – IJECE). **3**(3), 12–15 (2016)
8. Hur, J., Shifat, T.A.: Motor vibration analysis for the fault diagnosis in non-stationary operating conditions. Int. J. Integr. Eng. **12**(3), 151–160 (2020). https://doi.org/10.30880/ijie.2020.12.03.019
9. de Sá Só Martins, D.H.C., Viana, D.P., De Lima, A.A., et al.: Diagnostic and severity analysis of combined failures composed by imbalance and misalignment in rotating machines. Int. J. Adv. Manuf. Technol. **114**(9), 3077–3092 (2021)
10. Jain, P., Bhosle, S.: Analysis of vibration signals caused by ball bearing defects using time-domain statistical indicators. Int. J. Adv. Technol. Eng. Explor. **9**(90), 700–715 (2022)

11. Güçlü, S., Ünsal, A., Ebeoğlu, M.A.: Vibration analysis of induction motors with unbalanced loads. In: 10th International Conference on Electrical and Electronics Engineering (ELECO), pp. 365–369. IEEE, Bursa (2017)
12. Deore, K.S., Khandekar, M.A.: Design machine condition monitoring system for ISO 10816-3 Standard using Fuzzy Logic. Int. J. Eng. Res. Technol. (IJERT). **4**(1), 726–729 (2015)
13. Aguayo-Tapia, S., Avalos-Almazan, G., Rangel-Magdaleno, J., Ramirez-Cortes, J.: Physical variable measurement techniques for fault detection in electric motors. Energies. **16**(12), 4780 (2023). https://doi.org/10.3390/en16124780
14. Zhao, Z., Wu, J., Li, T., et al.: Challenges and opportunities of AI-enabled monitoring, diagnosis & prognosis: a review. Chin. J. Mech. Eng. **34**(1), 1–29 (2021). https://doi.org/10.1186/s10033-021-00570-7

Preparation of Eco-friendly and Sustainable Adsorbent from *Gmelina arborea* Pulping Black Liquor and Preliminary Study on Its Potentials in the Treatment of Dye Effluents

Aniekan E. Akpakpan, Itoro E. Udo, Okon E. Okon, and Nyeneime W. Akpanudo

1 Introduction

Water is an essential requirement for life and it is used for various household and industrial activities. It is one of the critical natural resources that is highly exploited. Presently, the world is currently facing water crises due to unrestricted and excessive exploitation of water. Anthropogenic activities, especially from textile and allied industries such as pulp mills, leather, cosmetics, printing, food, and plastics, are posing threats to the water bodies and the health of humans because untreated effluents are discharged into water bodies which contain toxic compounds, and these effluents deteriorate the water quality, thereby affecting aquatic animals. Also, the water is unsafe for human consumption [1, 2]. Toxic textile effluents alter the physical and chemical properties of the soil, deteriorating water bodies, and also cause harm to the flora and fauna in the environment since the textile effluent consist of chromophore and some polar groups which make them nonbiodegradable [3, 4]. Harmful dyes cause death to the soil microorganisms, which in turns affect the agricultural productivity [5, 6]. The wastewater released by these industries should be treated using an ecofriendly, cost-effective, and sustainable method.

A. E. Akpakpan (✉) · N. W. Akpanudo
Department of Chemistry, Akwa Ibom State University, Ikot Akpaden, Nigeria
e-mail: aniekanakpakpan@aksu.edu.ng

I. E. Udo
Department of Chemistry, University of Uyo, Uyo, Akwa Ibom State, Nigeria

O. E. Okon
Department of Chemistry and Biochemistry, Akwa Ibom State Polytechnic, Ikot Osurua, Nigeria

S. C. Ekpo (ed.), *The Second International Adaptive and Sustainable Science, Engineering and Technology Conference*, Signals and Communication Technology,
https://doi.org/10.1007/978-3-031-53935-0_17

Lignin which is an amorphous biopolymer, a waste material released from pulp and paper industry, contains various functional groups such as methoxy, phenolic, carboxyl, aromatics, alcohol, and aldehyde [7]. These functional groups can be modified to make them more reactive and hence improve its potentials as adsorbent in the treatment of textile effluents. Chemical oxidation of lignin results in the formation of functional groups such as aldehydes, aromatic carbonyl, and carboxylic groups on the surface of lignin moieties [8, 9]. Oxidation increases the oxygen contents and hence the electron density of the lignin which may increase lignin potentials toward adsorption [10–12].

The aim of this work is to utilize lignin which is the waste material released during wood pulp production in preparation of a cost-effective and environmentally friendly adsorbent for the treatment of effluents from textile industry.

2 Materials and Methods

2.1 *Sample Collection and Preparation*

The sample *Gmelina arborea* wood was collected from Ikot Obio Inyang in Etinan local government area of Akwa Ibom State, Nigeria. The samples were cut into logs and further processed into smaller chips of about 1 cm by 1 cm to facilitate easy extraction of lignin during pulping process.

2.2 *Extraction of Lignin*

Lignin was extracted by the pulping of *Gmelina arborea* wood chips in 10-L laboratory digester. The sample (400 g) was placed inside the digester. About 4 liters of 20% NaOH solution were added (sample-to-liquor ratio of 1:10). The digester was heated to the operating temperature of 130 °C for 3 h. After pulping, the pulp was separated from black liquor via filtration, and the soda lignin was precipitated from the black liquor by acid precipitation at pH of 2 using 3 M H_2SO_4. The precipitated soda lignin was filtered and washed to remove excess acid. The lignin obtained was dried in the oven at 65 °C for 24 h.

2.3 *Oxidation of Soda Lignin*

The precipitated soda lignin was modified by oxidation with hydrogen peroxide using the method of Quantana et al. [10] and Ogunsile and Anomo [13]. About 20 g of lignin was added to 200 ml of H_2O_2 solution and 2 ml of 2 M HCl was added as catalyst. The mixture was placed in a 500-ml fitted conical flask connected to a reflux condenser and heated to boil. The reaction was terminated after 2 h. Oxidized lignin was filtered and washed with distilled water to remove acids and residual peroxide. The oxidized lignin was ovendried at 65 °C. The same procedure was repeated for another 20 g of lignin in order to have greater quantity of oxidized lignin.

2.4 Determination of Functional Groups

The functional group on soda lignin and oxidized soda lignin was determined using Fourier transform infrared spectroscopy (FTIR) Cary 630. The spectra were recorded at the wave number range of 4000 cm^{-1} to 1000 cm^{-1} using a KBr disc containing 1% finely ground lignin or oxidized lignin. The spectrum recorded over 30 scans with a resolution of 4 cm^{-1}.

2.5 Preparation of Simulated Textile Effluents (Dye Solutions)

The dye solutions of 100 mg/L each of phenol red (anionic dye) and gentian violet (cationic dye) both triphenylmethane dyes were prepared by dissolving appropriate amount of phenol red and gentian violet, respectively, in distilled water and made up to 1000 ml in standard volumetric flasks. The structures of these dyes are given in Fig. 1a, b.

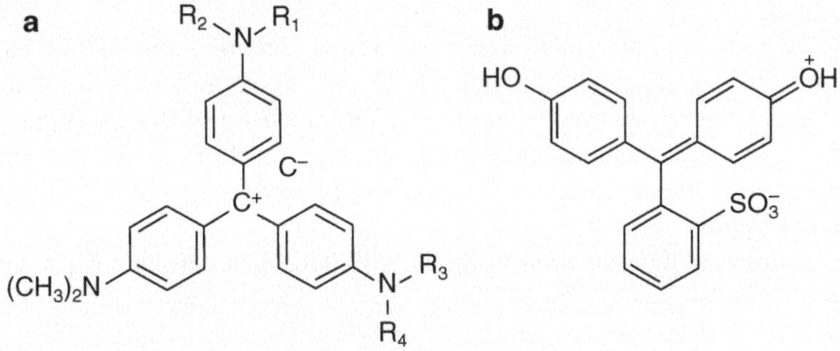

Fig. 1 (**a**) Structure of gentian violet. (**b**) Structure of phenol red

2.6 *Adsorption Experiments*

Batch adsorption experiments were conducted in order to study the potentials of lignin and oxidized lignin as adsorbent for the treatment of the simulated effluents containing phenol red and gentian violet. The effect of adsorbent loading on the adsorption of the dye solutions from the simulated textile effluents was investigated by contacting 100–500 mg of the lignin or oxidized lignin to 50 ml of 100 mg/L of each dye solution in a beaker. The beakers containing the mixture were placed in mechanical shaker and agitated at 150 rpm for 1 h, after which the mixture was filtered and the absorbance taken using UV/visible spectrophotometer. Post-adsorption concentration was calculated using Beer Lambert's law [14]. Adsorption efficiency was calculated using the following:

$$\text{Adsorption efficiency } (\%) = \frac{C - Co}{C} \times 100$$

where

Co = Initial concentration of the dye.
C = Post-adsorption concentration of the dye.

3 Results and Discussion

3.1 *FTIR Characterization of Soda Lignin and Oxidized Soda Lignin*

The results for the FTIR characterization of lignin and oxidized lignin are presented in Figs. 2a, b.

In FTIR spectra of soda lignin, a broad intense absorption band at 3000–3500 cm^{-1} soda lignin corresponds to the stretching vibration of the free aliphatic and aromatic hydroxyl group [15]. The intensity of this absorption band was reduced in oxidized soda lignin, indicating high level of lignin oxidation. This was confirmed by the appearance of the carbonyl absorption band at 1703 cm^{-1} in the oxidized lignin.

The equation for the oxidation of lignin with hydrogen peroxide is presented in Fig. 3.

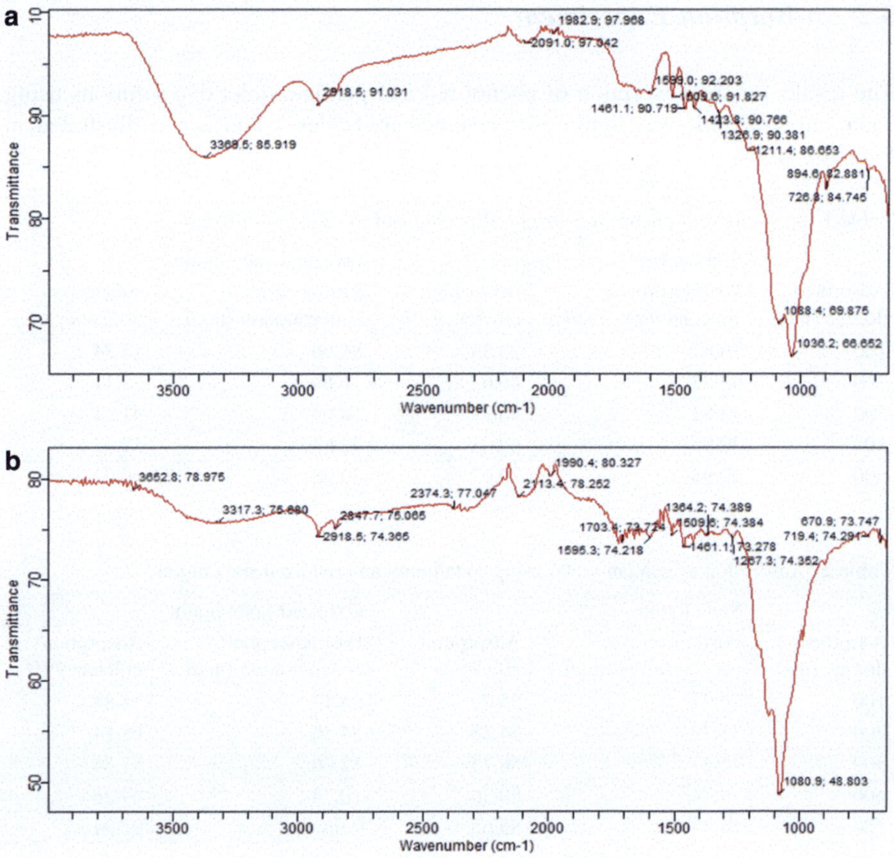

Fig. 2 FTIR spectra. (**a**) FTIR spectra of soda lignin. (**b**) FTIR spectra of oxidized soda lignin

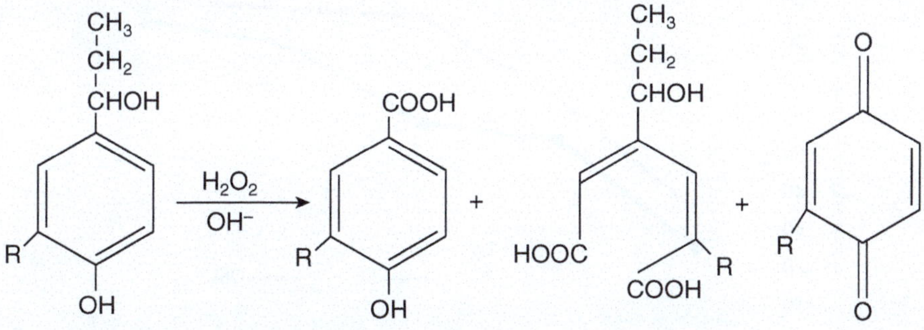

Fig. 3 Oxidation of lignin with hydrogen peroxide [9]

3.2 Adsorption Experiment

The results for the adsorption of phenol red and gentian violet dye solutions using soda and oxidized soda lignin are presented in Tables 1 and 2 and illustrated in Figs. 4 and 5.

Table 1 Adsorption of phenol red using soda lignin and oxidized soda lignin

	Soda lignin		Oxidized soda lignin	
Adsorbent dosage (mg)	Post-adsorption concentration (mg/L)	Adsorption efficiency (%)	Post-adsorption concentration (mg/L)	Adsorption efficiency (%)
100	76.67	23.33	86.66	13.34
200	53.33	46.67	76.89	23.11
300	43.81	56.19	58.09	41.91
400	30.95	69.05	42.68	57.32
500	20.95	79.05	39.98	60.02

Table 2 Adsorption of gentian violet using soda lignin and oxidized soda lignin

	Soda lignin		Oxidized Soda lignin	
Adsorbent dosage (mg)	Post-adsorption concentration (mg/L)	Adsorption efficiency (%)	Post-adsorption concentration (mg/L)	Adsorption efficiency (%)
100	74.91	25.09	65.12	34.88
200	45.72	54.28	34.86	65.14
300	41.27	58.73	12.66	87.34
400	19.74	80.26	10.74	89.26
500	14.35	85.65	7.66	92.34

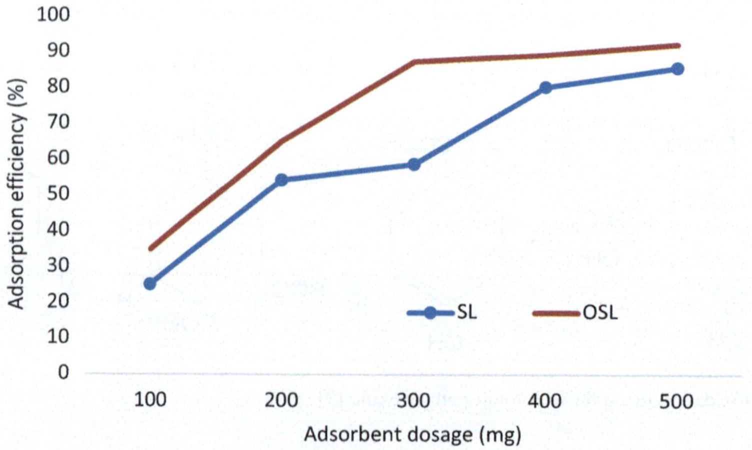

Fig. 4 Adsorption efficiency for the adsorption of gentian violet using soda lignin and oxidized soda lignin

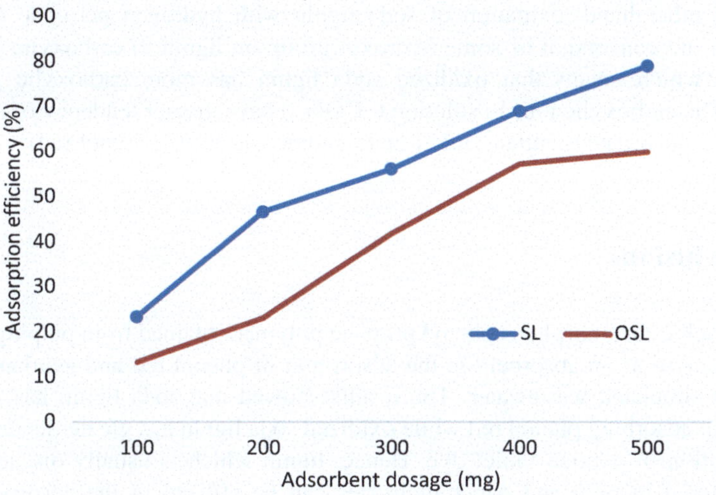

Fig. 5 Adsorption efficiency for the adsorption of phenol red using soda lignin and oxidized soda lignin

The adsorption of phenol red and gentian violet was used for comparative study in order to evaluate the potentials of soda lignin and oxidized soda lignin as adsorbents in the treatment of textile effluents.

Effect of Adsorbent Dosage

The quantity or mass of adsorbent is an important factor in large-scale industrial application of adsorbent in the removal of a desired pollutant. In order to investigate the effect of mass of adsorbent on the adsorption of these dyes, series of adsorption studies were carried out with different masses of adsorbent at fixed initial dye concentration. The results revealed that the higher the amount of adsorbent, the higher the adsorption efficiency (Figs. 4 and 5). This is because the increase in adsorbent mass increases the contact surface area of the adsorbent particles which enable more dye molecules to be adsorbed on adsorption sites and thus adsorption efficiency is increased [16].

Effect of Nature of the Adsorbent and Adsorbate

Soda lignin and oxidized soda lignin were used in the adsorption of phenol red (a typical anionic dye) and gentian violet (a typical cationic dye). The results revealed that soda lignin adsorbed more phenol red dye than gentian violet (Figs. 4 and 5). This is because lignin moiety contains greater percentage of hydroxyl group while phenol red being an acid dye contains negative charge (anionic dye). Hence, the protonated hydroxyl group ($–OH^+$) on the adsorbent surface of soda lignin has greater tendency to adsorb phenol red which has a negative charge, and this resulted in the greater adsorption efficiency of soda lignin.

On the other hand, oxidation of soda lignin with hydrogen peroxide (H_2O_2) resulted in the conversion of some hydroxyl group on lignin to carboxylic groups (Fig. 2b), which means that oxidized soda lignin has more carboxylic groups [17, 18]. The carboxylic ions in solution ($-COO^-$) have greater tendency of removing cationic adsorbate (gentian violet) than anionic adsorbate (phenol red).

4 Conclusions

Lignin which is an amorphous phenyl propane polymer obtained from pulping black liquor was used as an absorbent in the adsorption of phenol red and gentian violet dyes from simulated wastewater. The results showed that soda lignin has greater tendency in adsorbing phenol red while oxidized soda lignin has greater tendency in the adsorption of gentian violet dye. Hence, lignin which is usually discarded as waste material in pulp and paper industries can be utilized in the adsorption of anionic dyes and can also be modified using hydrogen peroxide oxidation and used as adsorbent in the adsorption of cationic dyes from industrial effluents. This will enhance environmental cleaning in both pulp and paper industries and textile/allied industries.

References

1. Gupta, V., Suhas, K.: Application of low-cost adsorbents for dye removal – a review. J. Environ. Manag. **90**, 2313–2343 (2009)
2. Cheng, S., Oatley, D.L., Williams, P.M., Wright, C.J.: Characterization and application of a novel positively charged nanofiltration membrane for the treatment of textile industry waste water. J. Water Res. **46**(1), 33–42 (2012)
3. Liu, T.H., Li, Y.H., Du, Q.J., Sun, J.K., Jiao, Y.Q., Yang, G.M.: Adsorption of methylene blue from aqueous solution by graphene. J. Colloids Surf. **90**, 197–203 (2012)
4. Xu, F., Zhong, X.C., Sun, R.C., Jones, G.L.L.: Lignin distribution and ultra-structure of *Salix psammophila*. Trans. Chin. Pulp. Paper. **20**(1), 6–9 (2005)
5. Ardila-Leal, L.D., Poutou-Pinales, R.A., Pedroza-Rodriguez, A.M., Quevedo-Hidalgo, B.E.: A brief history of colour, the environmental impact of synthetic dyes and removal by using laccases. Molecules. **22–26**(13), 3813
6. Jumasiah, A., Chuah, A.L., Gimbun, J., Idris, A.: Adsorption of basic dye onto palm kernel shell activated carbon: sorption equilibrium and kinetics studies. Desalination. **186**(1–3), 57–64 (2005)
7. Akpakpan, A.E., Inam, E.J., Ita, B.N., Akpabio, U.D.: Physicochemical properties of soda and Kraft lignin extracted from *Gmelina arborea* wood. Int. Res. J. Pure Appl. Chem. **24**(4), 9–19 (2023a)
8. Zhu, H., Qin, C., Zhou, J.R., Wang, S.: Adsorption of heavy metal ions from aqueous single metal solution by aminated epoxy—lignin. Bioresources. **8**, 2257 (2013)
9. He, W., Gao, W., Fatehi, P.: Oxidation of Kraft lignin with hydrogen peroxide and its application as a dispersant for kaolin suspensions. ACS Sustain. Chem. Eng. **5**(11), 10597–10605 (2017)

10. Quintana, G.C., Rocha, G.J., Gonçalves, A.R., Velásquez, J.A.: Evaluation of heavy metal removal by oxidized lignins in acid media from various sources. Bioresources. **3**(4), 1092–1102 (2008)
11. Gellerstedt, G., Agnemo, R.: The reactions of lignin with alkaline hydrogen peroxide. Part III. The oxidation of conjugated carbonyl structures. Acta Chem. Scandinave. **34**, 275–282 (2008)
12. Akpakpan, A.E., Inam, E.J., Itoro, E.U.: Preparation, characterization and application of soda lignin and its ester derivatives as adsorbents in the adsorption of Pb^{2+} and Cd^{2+} from wastewater. J. Mater. Environ. Sci. **14**, 82–96 (2023b)
13. Ogunsile, B.O., Anomo, A.A.: Adsorption of chromium (vi) and cadmium (ii) from aqueous solution by soda lignin obtained from Nypa palm leaves (*Nypa fruitcans*). Ozean J. Appl. Sci. **7**(2), 43–56 (2014)
14. Akpakpan, A.E., Nsi, E.W., Owen, A.E., Ikpe, E.E., Akpan, I.M.: Equilibrium, kinetics and isothem studies on the adsorption of eosin red and malachite green using activated carbon from Hura crepitans seed shells. AASCIT J. Environ. **3**(4), 24–28 (2018)
15. Faix, O., Grunwald, C., Beinhoff, O.: Determination of phenolic hydroxyl group content of milled wood lignins from different botanical origins using selective aminolysis, FTIR, ^{1}H-NMR and UV – spectroscopy. Holzforschung. **46**(5), 425–431 (1992)
16. Nsi, E.W., Akpakpan, A.E., Ekwere, I.O., Udoh, D.P.: Equilibrium and isotherm studies on the adsorption of methylene blue and malachite green by activated carbon prepared from *Irvingia gabonensis* seed shells. Int. J. Ecol. Sci. Environ. Eng. **4**(5), 72–77 (2017)
17. Boeriu, C.G., Bravo, D., Gosselink, R.J.A., Van Dam, J.E.G.: Characterization of structure-dependent functional properties of lignin with infrared spectroscopy. Ind. Crop Prod. **20**, 205–218 (2004)
18. Crestini, C., Crucianelli, M., Orlandi, M., Saladino, R.: Oxidative strategies in lignin chemistry: a new environmental friendly approach for the fictionalization of lignin and lignocellulosic fibers. Catal. Today. **156**, 8–22 (2010)

Scientific Analysis of Engineering Materials Derivable from Smart Electronically Calibrated Nonwoody Agricultural Waste for Production of Pulp

Itoro Esiet Udo, Enefiok Okon Usungurua, Sunday Cookey Ekpo, Perpetual Eze-Idehen, and Joseph Pius Sam

1 Introduction

Straws are agricultural by-products consisting of the dry stalk of cereal plants after the grain and chaff have been removed. It makes up about half of the yield of cereal crops. The fusion of straw can improve soil quality through improved nutrient cycling and soil organic carbon sequestration [1–3]. Straw remains the main organic material easily available for most farmers and serves as an important source of potassium. Reference [1] indicated that at harvest the straw contains 0.57% N, 0.07% P_2O_5, 1.5% K_2O, 0.1% S, and 5% Si by mass. *Luffa cylindrica*, commonly called sponge guard, vegetable sponge, bath sponge, or dish cloth gourd, is a member of Cucurbitaceae family (Fig. 1).

I. E. Udo (✉)
Department of Chemistry, Faculty of Sciences, University of Uyo, Uyo, Nigeria
e-mail: itoroudo@uniuyo.edu.ng

E. O. Usungurua
Department of Mechanical Engineering, School of Engineering, Akwa Ibom State Polytechnic, Ikot Osurua, Ikot Ekpene, Nigeria

S. C. Ekpo
Department of Electrical Electronic, Manchester Metropolitan University, Manchester, UK

P. Eze-Idehen
Department of Chemical Engineering, The University of Manchester, Manchester, UK

J. P. Sam
Department of Chemistry, Faculty of Sciences, University of Uyo, Uyo, Nigeria

Department of Chemistry, University of Uyo, Uyo, Nigeria

S. C. Ekpo (ed.), *The Second International Adaptive and Sustainable Science, Engineering and Technology Conference*, Signals and Communication Technology,
https://doi.org/10.1007/978-3-031-53935-0_18

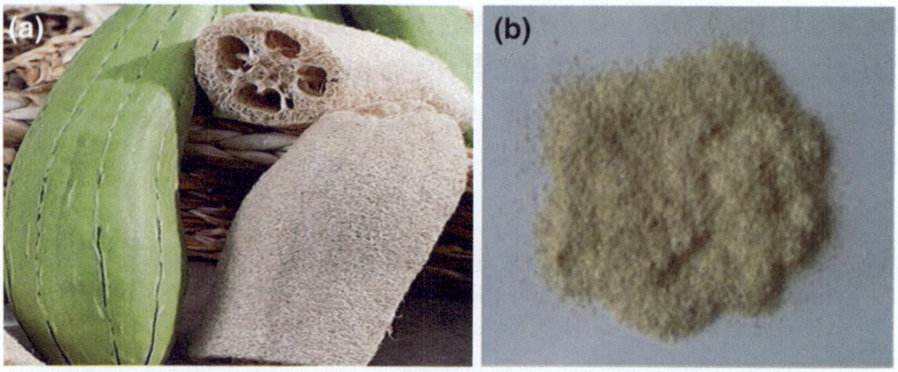

Fig. 1 Depicts the (**a**) dried *Luffa aegyptiaca* and (**b**) grinded *Luffa* fibers

Fig. 2 Depicts dried maize straws [stem (**a**) and grinded (**b**)]

Luffa cylindrica is mainly composed of cellulose, hemicelluloses, and lignin. *Luffa cylindrica* fibers are commonly found in China, Japan, and other countries in Asia and are quite common in the southern parts of the USA. To meet the demand for energy, raw materials, environmental preservation, and economic independence, governments and organizations throughout the world aim to diversify the energy grid and invest in renewable alternative energy sources [4–8]. Energy resources are classified mainly into two, namely, renewable and nonrenewable. The renewable such as wood, water, solar, wind, etc. are better option since the nonrenewable such as kerosene, diesel, gasoline, etc. cannot be replenished and would be exhausted [9]. Conversion of agricultural waste such as *Zea mays* straws and *Luffa cylindrica* fibers into useful feedstock for the production of paper, textiles, etc. becomes imperative. There is no gainsaying that any country that ventures into the use of *Zea mays* straws and *Luffa cylindrica* fibers as raw materials to produce other economic products and biomass energy system is invariably delving into economic and energy diversification (Figs. 2 and 3).

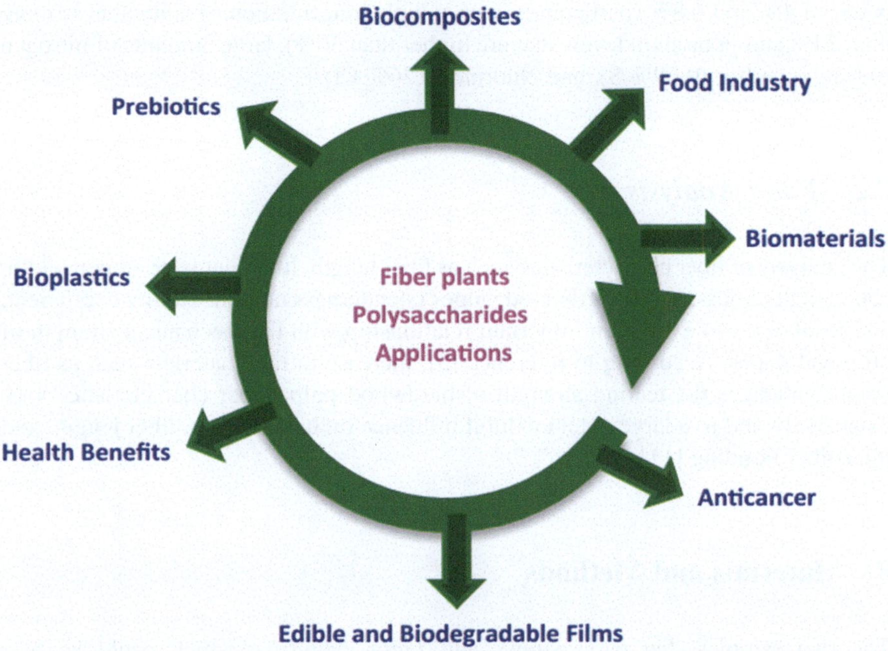

Fig. 3 Depicts the applications of plant fibers and their products

The objectives of this study cover the determination of the physicochemical properties and fiber characteristics of *Zea mays* straws and *Luffa cylindrica* wastes and the development of dissolving pulp from *Zea mays* straws and *Luffa cylindrica* fibers. This makes available a relative cheap and pure source of cellulose as a raw material for the development of other engineering materials and expansion of chemical industries.

2 Scientific Analysis of Chemical Composition

2.1 Ash Content

The ash content was found to be 6.13% which was moderate, and it compared favorably with the range value of 3.00–5.8% reported for those of legumes like cowpea, groundnut, and fluted pumpkin seed [3]. Since ash content gives an idea about the inorganic mineral content, it implies that the seed could provide essential, valuable, and useful minerals needed for good body growth and development and speed up metabolic processes. According to Jenkins [10], when comparing maize straw with other biomass fuels, attention should also be paid to the content of ash between 4% and 6.8% (particularly with a high concentration of silica that is more than 34% and potassium levels that are higher than 30%), large amounts of nitrogen (0.6% N), sulfur (0.09% S), and chlorine (0.36% Cl).

2.2 *Fiber Analysis*

The analysis of fiber characteristics such as fiber length, fiber diameter, lumen width, cell wall thickness, and their derived values (slenderness ratio, flexibility coefficient, and Runkel ratio) exhibit an important relationship with the mechanical strength of pulp and paper. According to reference [5], increase in raw materials such as fiber length enhances the tearing strength of hardwood pulp. Fiber characteristics vary extensively and to a large extent exhibit influence on bulk density, fiber length, and inter-fiber bonding [11].

3 Materials and Methods

The two samples *Zea mays* straws and *Luffa cylindrica* which could be best described as agricultural wastes were used. It was obtained from swamp region of Ikpe Anang River in Ini local government area, Akwa Ibom State, Nigeria. The samples were washed with water and cut into small length of about 2 cm and air-dried for 3 days to reduce moisture before grinding to powder and stored in a container for further analysis.

3.1 *Sample Collection and Characterization*

The agricultural waste materials from home gardens and plantations were collected and characterized using the Association of Pulp and Paper Industries Standards (TAPPI) T 258 °C. Raw material preparation includes selection of leaf folds, mild retting, beating, and washing to free the filaments from other impurities. Other agricultural waste materials were given appropriate treatments before pulping. Locally designed digester and laboratory digesters suitable for pulping each agricultural waste were designed and fabricated (25 cm diameter and 40 cm length). Pulping: Suitable pulping process was adopted after preliminary tests in the laboratory in a specially designed steel drum digester to determine the most suitable pulping liquor and the pulping conditions. Below are other parameters that were determined.

3.2 Solubility in Hot Water

2.0 g of each of the air-dried samples was placed in a 250-ml flask and 100-ml of deionized water was added. The flask and its content were allowed to digest for 3 h at a temperature of 120 °C using the reflux system, after which it was filtered and dried in the oven before weighing.

3.3 Solubility in Cold Water

2.0 g of each of the air-dried samples was placed in three 250-ml beaker. 100 ml of deionized water was added, and it was allowed to stir continuously for 3 h, after which it was filtered and oven-dried before weighing.

3.4 Solubility in 1% NaOH

2.0 g of each of the air-dried samples was macerated and transferred to a 250-ml conical flask. 100 ml of 1% NaOH was added. It was allowed to boil for 3 h in a water bath at the temperature of 120 °C. The samples were filtered, washed thoroughly with hot deionized water, and dried in an oven before weighing. The solubility was then calculated in percentage based on oven-dried sample [4].

Bulk density, fiber length and diameter, fiber lumen width, fiber cell wall thickness, and Runkel ratio were analyzed.

4 Results

4.1 Solubility Test

The results of the soluble contents of *Zea mays* straw and *Luffa cylindrica* fibers are presented in Table 1. These include solubility in cold water, hot water, 1% sodium hydroxide, and bulk density. 1% sodium hydroxide solubility was used to determine the degree of fungus decay that can be taken place in a given wood or nonwood sample. The value obtained in 1% NaOH was 71.5% and 46% for the *Zea mays*

Table 1 Physical parameters of *Zea mays* straw and *Luffa cylindrica* fibers (%)

Characteristic	*Zea mays*	*Luffa*
% solubility in 1% NaOH	71.5	46.0
% solubility in hot water	24.0	26.5
% solubility in cold water	20.0	16.5
Bulk density (gcm^{-3})	0.22	0.12

straw and *Luffa cylindrica* fibers, respectively. *Zea mays* straw recorded 71.5% solubility due to the fact that it softens more easily in 1% NaOH [3]. The value obtained for the wood fibers was very high compared to those obtained for some nonwood raw materials reported by [5]. The value for hot water-soluble were 24% and 26.5% while for cold water-soluble were 20% and 16.5% for the *Zea mays* straw and *Luffa cylindrica* fibers, respectively. The value obtained for the *Zea mays* straw and *Luffa cylindrica* fibers was within the range for those reported for some nonwoods [2].

4.2 Bulk Density

The results of bulk density determined on the samples are presented in Table 3. The value obtained was 0.22 gcm^{-3} for *Zea mays* straw and 0.12 gcm^{-3} for *Luffa cylindrica*. The density value obtained in *Zea mays* straw was lower than that reported by [12]. The bulk density values of *Luffa cylindrica* are within the value of 0.56–0.92 gcm^{-3} reported by [13]. The density of *Luffa* is around 0.82 to 0.92 gcm^{-3}, which is lower than the density of some common natural fibers like sisal (1.26 to 1.45 gcm^{-3}), hemp (1.48 gcm^{-3}), coir (1.25 gcm^{-3}), ramie (1.50 gcm^{-3}), and cotton (1.51 to 1.60 gcm^{-3}) as reported by [12, 14].

4.3 Fiber Characteristics

Variations in fiber characteristics in terms of fiber length, fiber diameter, and fiber width are presented in Table 2. Tables 2 and 3 show the variations in fiber characteristics in *Zea mays* straw and *Luffa cylindrica* fibers. Average fiber length was 0.62 mm and 1.28 mm for the *Zea mays* straw and *Luffa cylindrica* fibers, respectively. *Luffa* has the highest value of 1.28 mm. It is stated that fiber length above 1.60 mm are said to be long and those below 1.60 mm short [15, 16]. Pulping raw

Table 2 Variations in fiber characteristics in *Zea mays* straws

Fiber length (mm)	Fiber diameter (μm)	Fiber width (μm)
1.12585	12.65	6.325
0.3795	12.65	6.325
0.2783	12.65	5.06
0.4048	12.65	6.325
0.7084	11.385	3.795
0.5313	43.01	25.3
0.9361	6.325	2.53
1.012	12.65	6.325
0.5313	6.325	2.53
0.31625	11.385	2.53

Table 3 Variations in fiber characteristics in *Luffa cylindrica* fibers

Fiber length (mm)	Fiber diameter (μm)	Fiber width (μm)
1.265	25.3	12.65
2.13785	25.3	18.975
1.265	25.3	18.975
0.8602	25.3	12.65
0.8602	25.3	12.65
1.25235	37.95	25.3
1.771	25.3	18.975
0.9108	37.95	25.3
1.518	24.035	12.65
0.94875	25.3	18.975

Table 4 Derived values for the characterization of *Zea mays* and *Luffa cylindrica*

Samples	*Zea mays*	*Luffa cylindrica*
FL (mm)	0.62	1.28
FD (μm)	14.17	27.70
FW (μm)	4.43	18.72
LW (mm)	7.09	13.85
CW (mm)	7.09	6.92
RR	1	0.5
FC (%)	6.03	50
SR	0.04	0.05

material with short fiber lengths is less suitable for papermaking than the long fiber [17]; [1]; [17]). In similar observation, [3, 4] reported fiber lengths of less than 1.60 mm in some Nigerian hardwood species. Akpabio and [18] recorded 2.14 and 1.98 mm as fiber length of plantain pseudostem waste and screw pine leaves, respectively. Reference [11] recorded 0.65 mm as fiber length of *Leucaena leucocephala*. Also, similar fiber length ranges of 1.50 to 2.9 mm and 1.46 to 2.91 mm were reported for bamboos in Sudan and Thailand, respectively [14]. Though the fiber length of *Zea mays* straw and *Luffa cylindrica* is short, the samples can be used for preparation of cellulose derivatives. The derived values for the characterization of *Zea mays* straw and *Luffa cylindrica* are presented in Table 2. The average fiber diameter was 14.17 and 27.70 μm for the *Zea mays* straw and *Luffa cylindrica* fibers, respectively. In similar observation, [11] recorded 12.0 μm as fiber diameter for *Nypa fruticans*. It also falls within the range recorded for coniferous and commercial pulp woods [12, 14, 19] (Table 4).

Average fiber width was 4.43 and 18.72 μm for the *Zea mays* straw and *Luffa cylindrica* which was very short compared to those of wood and other common nonwood fibers (approximately 32.0–43.0 μm and 18.0–30.0 μm for softwood and hardwoods, respectively [12, 20] reported 12 μm as fiber width of *Nypa fruticans* frond and petiole. Lumen widths were 7.09 μm and 13.85 μm for the *Zea mays* straw and *Luffa cylindrica* which are within the range of 12 μm and 4.9 μm for *Vitex doniana* sweet as reported by [18]. The cell wall thickness is one of the significant fiber dimensions that determine the choice of a fibrous raw material for pulp and paper production. The values obtained were 3.54 and 6.93 μm for the *Zea mays* straw and *Luffa cylindrica*. Lumen size and cell wall thickness affect the rigidity and strength of papers made from fiber. The value of Runkel ratio obtained for *Zea mays* straw and *Luffa cylindrica* was 1 and 0.5, slenderness ratio was 0.04 and 0.05, while flexibility coefficient was 6.03 and 50, respectively. It is stated that if slenderness ratio of a material is lower than 70, it is invaluable for quality pulp and paper production [1]. High Runkel ratio fibers (i.e., fibers with relatively thick walls, such as those in summerwood) are stiffer and less flexible and form bulkier paper of low bonding area than low Runkel ratio fibers. Fibers with ratio less than or equal to one are very good for papermaking [2, 21]. However, the pulps obtained from *Zea mays* straw and *Luffa cylindrica* will give stiffer, less flexible, and bulkier paper since its Runkel ratio is less than one. Therefore, *Zea mays* straw and *Luffa cylindrica* pulps may be suitable for producing paper boards. The higher the value of the flexibility coefficient, the greater the pliability of the fibers and the well-bonded the sheet formed, although pulp mechanical strength also depends on other processing variables such as pulping conditions, bleaching, beating, etc.

5 Conclusions

The investigation of solubility contents and fiber characteristics of *Zea mays* straw and *Luffa cylindrica* fibers reveals that the species has a fairly long fiber length for papermaking. The derived parameters (lumen width, cell wall thickness, Runkel ratio, and the flexibility coefficients) showed that the pulp could be used to produce paper of different grades. The bulk density is low which is within the acceptable range of some nonwood fibers which makes it easier for transportation. The solubility contents are also within the acceptable range for nonwood pulping raw materials. Therefore, the outcome of this investigation showed that the species is a potential material for pulp and paper production. In view above, thousands of tons of *Zea mays* straws and *Luffa cylindrica* waste allowed to rot away annually in our garden and in commercial farms should be utilized in the production of good cultural papers.

References

1. Akpabio, U.D.: Pulp and Paper Making Raw Materials in Akwa Ibom State of Nigeria. A Display Guide at Raw Material Research and Development Centre, Uyo (2013)
2. Akpakpan, A.E., Akpabio, U.D., Ogunsile, B.O., Eduok, U.M.: Influence of cooking variable on the soda and soda-ethanol pulping of Nypa fruticans petiole. Aust. J. Basic Appl. Sci. **5**, 1202–1212 (2011)
3. Akpakpan, A.E., Ogunsile, B.O., Akpabio, U.D., Eduok, U.M.: Comparative study on the soda-ethanol and soda-butanol pulping of Nypa fruticans petioles. Int. J. Adv. Sci. Res. Technol. **1**, 312–324 (2012)

4. Aniekan, E.A., Ukana, D.A., Obot, I.B.: Evaluation of physicochemical properties and soda pulping of Nypa fruticans frond and petiole. Elixir J. Appl. Chem. **45**, 7664–7668 (2012)
5. Akpabio, U.D., Udiong, D.S., Akpakpan, A.E.: The physicochemical characteristics of plantain (Musa paradisiaca) and banana (Musa sapientum) pseudostem wastes. Adv. Nat. Appl. Sci. **06**, 167–172 (2012a)
6. Akpabio, U.D., Akpakpan, A.E.: Pulp and paper from agricultural waste: plantain pseudostem wastes and screw pine leaves. Int. J. Mod. Chem. **2**, 100–107 (2012)
7. Akpabio, U.D., Effiong, I.E., Akpankpan, A.E.: Preparation of pulp and cellulose acetate from Nypa palm leaves. Int. J. Environ. Bioenergy. **1**, 82–102 (2012b)
8. Akgül, M., Tozluoğlu, A.: Utilizing peanut husk (Arachis hypogaea L.) in the manufacture of medium-density fiberboards. Bioresour. Technol. **99**, 5590–5594 (2008)
9. Akgül, M.: Suitability of stinging nettle (Urtica dioica) stalks for medium density fiberboards production. Compos. Part B. **45**, 925–929 (2013)
10. Jenkins, B.M.: Physical properties of biomass. In: Biomass Handbook, pp. 860–891 (1989)
11. Gary, A.S.: Hand Book of Pulp and Paper Technology, 2nd edn, pp. 355–367. Van No Strand, New York (2013)
12. I. E. Udo, G. U. Obuzor and, M. Horsfall Jnr., "Physicochemical characteristic of exudate of Dacryodes edulis," J. Appl. Sci. Environ. Manag., 20, pp. 209–213, 2016
13. Udo, I.E., Akpabio, U.D., Akpakpan, A.E.: Physicochemical studies on the hydrolysis and saponification products of Dacryodes edulis exudates. Arch. Appl. Sci. Res. **4**, 1741–7147 (2012)
14. I. E. Udo, G. U. Obuzor and, M. Horsfall Jr., "Proximate and phytochemical studies of exudate of Dacryodes edulis," Res. J. Chem. Sci., 6, pp. 1–9, 2016a
15. Hill, A.F.: Economic Botany: A Textbook of Useful Plants and Plant Products, vol. 2, pp. 18–51. Tata McGraw-Hill Publishing Company Limited, New Delhi (1972)
16. Okafor, J.C.: Varietal Delimitation Dacryodes edulis, pp. 102–138, Willey, New Yorks
17. Chen, X., Zhu, H.: Research progress and prospect on producing biogas from crop straws. Trans. Chin. Soc. Agric. Eng. **23**, 279–283 (2007)
18. Ashori, A., Nourbakhsh, A., Karegarfard, A.: Properties of medium density fiberboard based on bagasse. J. Compos. Mater. **43**, 1927–1934 (2009)
19. Gumuskaya, E., Kirci, H., Usta, M.: The effect of various pulping conditions on crystalline structure of cellulose in cotton linters. Polym. Degrad. Stab. **81**(3), 559–564 (2002)
20. Anon, E.A.: Pulp and paper. West Afr. Technol. Revolut. **7**(3), 45–47 (2007)
21. Hurter, A.M.: Utilization of annual plants and agricultural residues for the production of pulp and paper. In: Proceeding of Technical Association of Pulp and Paper Industry (TAPPI) Pulping Conference, pp. 124–129 (2011)

VERD: A Proficient Algorithm for Selecting Cluster Leaders in Wireless Sensor Networks

Emmanuel Oluropo Ogungbemi, Kufre Michael Udofia, Philip Michael Asuquo, and Kingsley Monday Udofia

1 Introduction

Sensors detect external events and physical phenomena like pressure, temperature, humidity, etc. and produce output signals [1]. A collection of sensors that communicates through wireless network infrastructure is referred to as a wireless sensor network (WSN). Therefore, a wireless sensor network is a configured network wirelessly formed using a collection of sensors over a geographical area [2, 3]. A wireless sensor network that communicates using Internet infrastructure is known as the Internet of Things [4]. The fields of applications of wireless network sensors include agriculture [5, 6] [7], medicine [8], smart cities [9], military [10], e-health [11], automotive [12, 13], smart parking [14, 15], and so on.

Some peculiar characteristics define the sensor's network operations, efficiency, and reliability. These characteristics include constrained energy, limited processing power, low power consumption, and limited memory [16]. To facilitate efficiency, the conservation of energy consumption in WSN is essential because all forms of communication, information gathering, and processes cease when the sensor exhausts its energy [17]. Therefore, the energy of the sensors needs to be extended as much as possible. However, optimum energy utilization is a major challenge in WSN's implementation because information exchange consumes a significant amount of energy [18–20].

E. O. Ogungbemi (✉) · K. M. Udofia · K. M. Udofia
Department of Electrical and Electronics, University of Uyo, Uyo, Nigeria
e-mail: ogungbemioluropo@uniuyo.edu.ng; kufremudofia@uniuyo.edu.ng; udofiakingsley@uniyo.edu.ng

P. M. Asuquo
Department of Computer Engineering, University of Uyo, Uyo, Nigeria
e-mail: philipasuquo@uniuyo.edu.ng

S. C. Ekpo (ed.), *The Second International Adaptive and Sustainable Science, Engineering and Technology Conference*, Signals and Communication Technology,
https://doi.org/10.1007/978-3-031-53935-0_19

Several methods have been proposed to reduce energy consumption in WSN, techniques such as clustering or hierarchical routing, putting nodes to sleep when inactive, and packet size optimization. Some applications consider placing sensor nodes to sleep unsuitable [21]. Clustering is the most effective way of achieving optimal energy utilization [20, 22]. Clustering is building a group of nodes (cluster members) that collects and forwards data to their respective cluster heads or cluster leaders [23]. The cluster heads (CHs) then aggregate and send the data to a sink. The sink acts as a middleware between the network and the end user. In a clustering or hierarchical network, every sensor cluster has at least a cluster head (CH) and cluster member (CM). While the responsibility of the CH is to communicate directly with the base station after the aggregation of the data coming from the CMs, the cluster members send data to the base station only through their respective cluster heads. Examples of proposed hierarchal networks include LEACH and other variances of LEACH [17, 20, 24–26], TEEN [27], PEGASIS [28], HPAR [29], etc.

This paper proposes a proficient cluster leader-selecting algorithm for the optimum performance of wireless sensor networks. The algorithm significantly improves the network's lifetime, energy utilization, and stability. The contributions of this study are highlighted as follows:

(a) Improve sensor network lifetime through an efficient selection of cluster heads.
(b) Enhance the network stability through the implementation of an efficient clustering head selection technique.
(c) Evaluation of the residual energy to showcase efficiency in energy consumption.
(d) Implementation of independent sensors with a clustering algorithm to improve the network performance,

VERD is evaluated with HEEL [21], SEP [30], and PEGASIS [28] algorithms, and it shows a significant performance improvement in the conservation of the network's energy. It extended the WSN's lifetime by 36.65% and improved its stability by 29.83%.

2 Review of Related Works

One of the foremost proposed hierarchical algorithms is LEACH [24]. The algorithm increased the network lifetime through correlated data aggregation by the cluster heads. The cluster heads were rotationally selected in the setup phase using a probability-dependent threshold. If given a number of nodes having equal energy and communication range, all the nodes with a value below a threshold value were designated cluster heads. The authors in [25] considered LEACH defective because the CH selection process relies on probability. Therefore, they proposed NR-LEACH to improve the total lifetime of the network by incorporating a node ranking algorithm into LEACH. The improvement is based on energy distribution among the sensors during the CH election. The CH election is predicated on the weight evaluation of each node. The node weight comprises the nodes' residual energy, received signal strength indicator (RSSI), and the number of connections a node has with other nodes.

The impact of heterogeneous nodes in wireless sensor networks that are hierarchically clustered was proposed by authors in [30]. The network heterogeneity is based on the assumption that networks' cluster heads have additional energy resources. The protocol used only the sensors' residual energy as the weight required for the CH selection.

Akhilesh et al. [16] proposed the distribution of the sensors into groups. The sensors' energy and Euclidean distance from the clusters to the base station were the only parameters for deriving the cluster leader fitness function.

PEGASIS [28] is a near-optimal chain-based protocol that improves on the LEACH algorithm. After a sensor node receives data from a neighbor, it aggregates and passes it on to the next neighbor in a chain-like arrangement. The receive and aggregate process is repeated until it reaches a designated node, which eventually transmits to the base station after a final aggregation. To reduce the average energy spent in each round, the nodes in the network take turns becoming the designated nodes. To increase the performance of PEGASIS, residual energy and neighbor's distant thresholds were set for any sensor to become a leader. However, the network implementing this algorithm is considered unstable, and its lifetime is short.

Amruta and Damodar [31] proposed a cluster head election mechanism and construction using fuzzy logic to improve the sensor's network lifetime. The output generated by fuzzy rules selects the cluster heads. The input variable of the fuzzy rules is residual energy, base station's distance, and node degree of sensor nodes.

The HEEL algorithm, proposed by authors [21], uses parameters such as nodes' energy, hop number, neighbors' links quantity, and neighbor's energy level to determine the cluster head eligibility. This algorithm chooses the node with the maximum fitness value within a cluster as the cluster head. The WSN that implements this algorithm has a short lifetime and is unstable.

In the author's [32] proposal, each sensor has a unique identity and weight. The node weight is the criteria used for the cluster heads' election; it is computed using parameters like the base station's distance, residual energy, visible neighbors' number, memory size, etc. An election message initiated by an initiator node transverses along a path through the network. As the election message transverses the path (from parent to son), each node includes its weight, and the initiator node selects the identity of the node with the highest weight as the leader.

3 Network Model

Our motivation is to develop a proficient algorithm for selecting cluster leaders in WSNs. Based on a hierarchical model, the algorithm aims to reduce energy consumption and increase the sensors' network lifetime and stability. We consider a network comprising 100 static and homogeneous sensor nodes randomly dispersed within a 100 m × 100 m geographical area. Our network architecture is as follows:

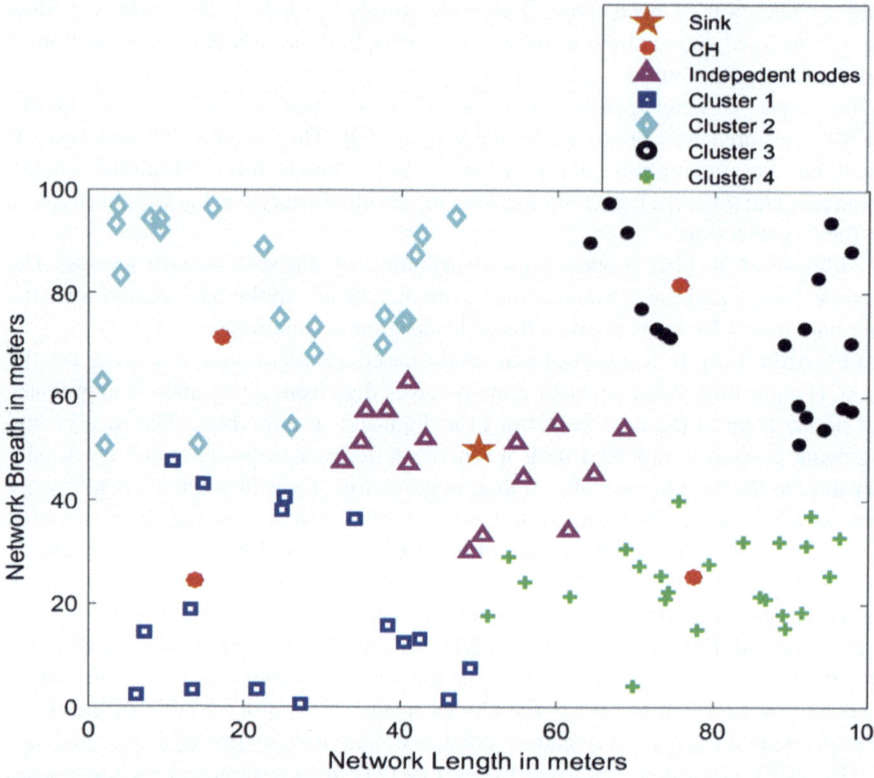

Fig. 1 Sensor network structure

- Individual sensor node ID is unique, has an equivalent transmission range, and can communicate directly to the base station.
- The entire network was segregated into four (4) clusters, each with one cluster head.
- At the beginning of the network, all the sensor nodes have a fixed initial energy of 0.1 J.
- The sink is fixed and located at the network's centroid, i.e., 50 × 50 m^2. The assumption is that the sink has an enormous amount of energy.
- All the sensors within a 15-m radius of the base station are considered independent. This group of sensors is indicated with a magenta color in Fig. 1.

Similar to authors [20, 33], for transmitting n-bit data at an acceptable signal-to-noise ratio (SNR), the energy consumed by any sensor node can be expressed as:

$$E_{TX} = \begin{cases} nE_{\text{trans}} + nE_{fs}d^2 & , \quad d < d_{th} \\ nE_{\text{trans}} + nE_{mp}d^4 & , \quad d \geq d_{th} \end{cases} \tag{1}$$

Receiving n-bits of data can be expressed as:

$$E_{Rx} = nE_{\text{trans}} \tag{2}$$

where nE_{trans} is the energy expended to transmit or receive n-bit of data, E_{fs} is the energy consumed due to free space, E_{mp} is the energy consumed due to multipath, and E_{Rx} is consumed to receive 1 bit of data. The energy consumed depends on the threshold distance dth between the transmitter and the receiver. If the threshold distance dth is greater than distance d, the multipath energy ($E_{mp}d^4$) loss model applies; otherwise, the free space energy loss ($E_{fs}d^2$) model applies. The threshold distance dth is given as:

$$d_{th} = \sqrt{\frac{E_{fs}}{E_{mp}}} \tag{3}$$

Hence, the total energy loss E_{tot} is given as:

$$E_{\text{tot}} = E_{TX} + E_{RX} \tag{4}$$

4 Proposed Scheme

This paper proposes a proficient algorithm for selecting cluster leaders aiming to increase the efficiency of wireless sensor networks. VERD is proposed to optimize the network energy consumption, increase stability, and extend the network lifetime. The clustering-based algorithm used the node's visibility index, the residual energy, the received signal strength indicator (RSSI), and the node's distance to/from the base station as parameters for the CH selection:

Visibility Index: A sensor visibility index is a measure of the node's exposure to its neighbors. It is also a measure of the number of neighbors a sensor can directly send and receive data (i.e., the number of links).

Energy: The sensor node's residual energy is one of the key parameters required for cluster head selection. The CHs are responsible for aggregating data collected from cluster members, and this responsibility increases the CH's energy consumption. A sensor eligible to become a CH must have residual energy greater than a threshold.

Received Signal Strength Indicator: The RSSI is the measure of the signal received from the base station by a sensor. The higher the received signal strength, the better the connection between the node and the sink.

Node Distance: Factors like attenuation and signal spreading affect the node's energy consumption, and these factors are affected by the distance. The most suitable CH node should have the lowest intra-cluster and node-to-sink distances. The nodes' distances were derived by calculating the Euclidean distance, as shown in (5):

$$E.d(a,b) = \sqrt{(y_a - y_b)^2 + (x_a - x_b)^2} \tag{5}$$

The eligibility index is calculated for each sensor within a cluster using (6). The node with the highest eligibility index is considered the most qualified node to be the cluster head:

$$\text{Eligibility Index (i)} = \sum\nolimits_{i=1}^{N} \sum\nolimits_{j=1}^{K} X_i \upbeta_{\text{j}} \tag{6}$$

subject to $\beta_j \geq 0\ \forall j \in \{1, .. a, \ldots, k.\}$

$X \in R^{NXK}$ data matrix. In this case, $k = 4$. X_1, X_2, X_3, and X_4 are visibility index (V), residual energy (E), reciprocal of the distance (D) from the base station, and the received signal strength (R), respectively. β_1, β_2, β_3, and β_4 are the weight assigned to each of the parameters. The weight value depends on the usage and purpose of the application. For instance, in a bad topology, inasmuch as the node's energy is very important, the node's visibility index value is equally very significant.

4.1 *Independent Sensor Distance*

Let y and x be the length and breadth of an entire square-shaped network area. If the area is divided into four equal sub-areas (i.e., cluster area A_c), then A_c is given as:

$$A_c = \frac{x}{2} \times \frac{y}{2} \tag{7}$$

If a sink is located at a point $\left(\frac{x}{2}, \frac{y}{2}\right)$, then the maximum distance D_s to the sink from any point within a cluster is the length of the diagonal of the cluster area, and it is given as:

$$D_s = \sqrt{\left(\frac{x}{2}\right)^2 + \left(\frac{y}{2}\right)^2} \tag{8}$$

If we assume the cluster head is located at the exact centroid of each cluster, then every sensor at a distance less than $\frac{1}{2}\sqrt{\left(\frac{x}{2}\right)^2 + \left(\frac{y}{2}\right)^2}$ from the sink is considered closer to the sink than the cluster head. Therefore, since our network size is 100 m by 100 m, any sensor with a distance less than 35.36 m from/to the sink is considered closer to the sink than to the cluster head. However, since the cluster heads are not usually at the exact centroid of the cluster areas, we have chosen a value of 15 m, which is significantly less than 35.56 m.

5 Performance Evaluation

First, in this session, the environment in which the system was modeled and the values of the parameters used for the simulation were provided. Second, VERD's performance was analyzed and compared to other existing cluster head selection algorithms.

5.1 *Simulation Parameters*

This proposal and other comparison methods have been implemented using a Corei5-4200 CPU 1.6 GHz with 4G of RAM. The VERD algorithm was simulated in MATLAB R2021a using the parameters shown in Table 1.

5.2 *Performance Metrics and Results*

To evaluate the performance of VERD, we also simulated HEEL [21], SEP [30], and PEGASIS [28] algorithms using 100 sensor nodes randomly distributed within a 100 m by 100 m geographical area. Figure 1 shows the network of the 100 sensor nodes with the base station located at the center (i.e., 50 m by 50 m). In VERD, while randomly distributing the sensors within the specified geographical area, 15 of the 100 sensor nodes fell within a designated 15-m radius of the base station; this group of sensors is called independent sensor nodes. The independent sensor nodes are not part of any of the four clusters. Independent sensors send data directly to the sink, thereby reducing the load on the selected cluster heads. Also, since the cluster head selection is based on the trade-offs among the selection parameters, there exists a reasonable probability that the selected CHs will be closer to the cluster centroid than the sink; therefore, it is energy-inefficient for data from sensors already closer to the

Table 1 Network simulation parameters

Parameters	Values
Network size	100 by 100 m^2
Sensor population	100
Data packet size	4000 bytes
Base station location	50 by 50 m^2
Independent node distance to BS	15 m
Initial energy	0.1 J
e_{elec}	50 nJ/bit
e_{mp}	0.0013 pJ/bit/m^4
e_{fs}	10 pJ/bit/m^2
E_{DA}	5 nJ/bit/signal

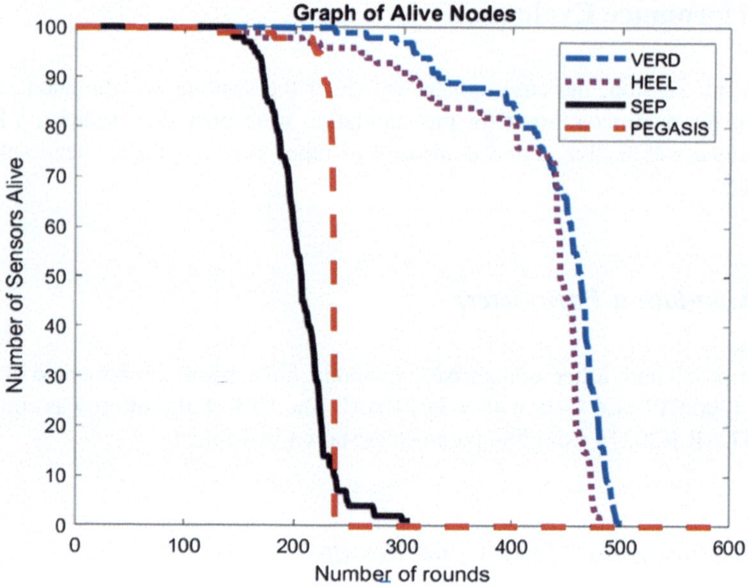

Fig. 2 Number of nodes alive

sink to transverse through the network to the cluster head first before returning to the sink. In our simulation, we assumed β_1, β_2, β_3, and β_4 to be 0.1, 0.7, 0.1, and 0.1, respectively. These values were selected experimentally to obtain optimum performance.

Network Lifetime

The number of active nodes in a wireless sensor network characterizes the network's lifetime. The network lifetime is indicated by the time elapsed between the first round and the round at which the energy of the last node became exhausted (i.e., the round the last node died). We measured the rounds at which the first sensor node died (FND), 50% of the nodes died (half of the nodes died (HND)), and when the last sensor node died (LND) in the network. Figure 2 shows that while the first node in HEEL, SEP, and PEGASIS algorithms died at the 135th, 128th, and 121st rounds, respectively, the network's lifetime using the VERD algorithm was extended till the 238th round. While 50% of the total nodes in HEEL, SEP, and PEGASIS died at the 444th, 221st, and 236th rounds, the lifetime of the nodes in VERD was extended till the 467th round. Finally, while the last node in VERD died at the 492nd round, the last node in HEEL, SEP, and PEGASIS died at the 477th, 369th, and 238th rounds.

Network Lifetime Percentage Evaluation

The percentage performance was evaluated by using (9):

$$\%\text{Performance} = \frac{\text{VERD} - \text{Other Protocol}}{\text{VERD}} \times 100 \qquad (9)$$

Table 2 Network lifetime percentage evaluation

Algorithms	No. of dead nodes	VERD performance	VERD performance %	Average VERD performance %	
HEEL	1	0.44214876	44.21487603	46.969697	FND
SEP		0.47107438	47.10743802		
PEGASIS		0.495867769	49.58677686		
HEEL	50	0.049250535	4.925053533	35.688794	HND
SEP		0.526766595	52.67665953		
PEGASIS		0.494646681	49.46466809		
HEEL	100	0.040241449	4.024144869	27.297116	LND
SEP		0.257545272	25.75452716		
PEGASIS		0.521126761	52.11267606		

Table 3 The round at which the first sensor node, half of the total sensor nodes, and the last sensor node died in VERD, HEEL, SEP, and PEGASIS algorithms

	FND	HND	LND
VERD	238	467	492
HEEL	135	444	477
SEP	128	221	369
PEGASIS	122	236	238

Table 2 shows the performance of the VERD algorithm in comparison to other algorithms. The simulation results show that when 1% of the network's node died (FND), VERD performed averagely better than HEEL, SEP, and PEGASIS algorithms by 46.97%. When 50% of the nodes died (HND), VERD performed averagely better than HEEL, SEP, and PEGASIS algorithms by 35.69%. Lastly, when 100% of the nodes died, VERD performed averagely better than HEEL, SEP, and PEGASIS algorithms by 27.3%. Hence, VERD has extended the network lifetime by an average of 36.65%, which is the value for its overall average performance with respect to the three algorithms used for the comparative analysis. VERD performance is believed to have been augmented by implementing independent sensors.

Network Stability

Network stability is also an important parameter to showcase the performance of a cluster head selection algorithm. It is described as the time from the beginning of the network to the time when the energy of the first node became exhausted (i.e., FND). Table 3 shows that the first node in the VERD algorithm died at the 238th round, the highest round when compared to the HEEL, SEP, and PAGASIS algorithms.

As shown in Table 4, the evaluation established that VERD implementation made the network averagely more stable by 46.1%.

Network Energy Conservation

Lastly, we considered network energy conservation as another parameter for evaluation. Figure 3 shows the total energy available at specific simulation rounds within the wireless sensor network using VERD and other cluster head selection algorithms. As mentioned earlier, the networks comprised 100 sensors, each with an initial energy of 0.1 Joule. Therefore, the total available energy was evaluated to be

Table 4 Network stability percentage evaluation

Algorithms	VERD performance	VERD average performance %
HEEL	0.43277	
SEP	0.46219	46.0784
PEGASIS	0.487395	

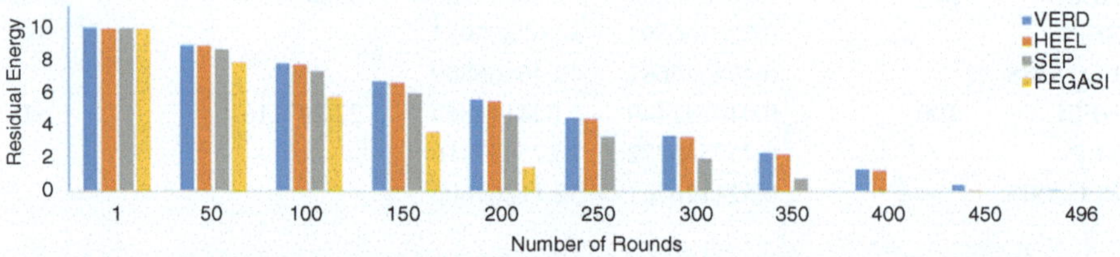

	Number of Rounds										
	1	50	100	150	200	250	300	350	400	450	496
VERD	9.977779	8.911126	7.795031	6.679128	5.571646	4.477660	3.400398	2.340682	1.324281	0.399858	0.086263
HEEL	9.936584	8.874265	7.765063	6.655817	5.544960	4.440620	3.359163	2.310382	1.282980	0.113873	0.000000
SEP	9.934350	8.659733	7.298106	5.961601	4.648096	3.365999	2.010511	0.770194	0.000000	0.000000	0.000000
PEGASI	9.957500	7.913614	5.784600	3.655600	1.528130	0.000000	0.000000	0.000000	0.000000	0.000000	0.000000

Fig. 3 Conservation of energy

10 Joules. The residual energy available in VERD extended to the 492nd round of simulation. The implication is that the sensor network implementing VERD consumes the lowest energy. The energy conservation is due to the incorporation of independent sensors, which alleviate the burden on the cluster heads, resulting in overall energy saving within the network.

6 Conclusion and Future Work

This paper proposed a proficient cluster leader selection algorithm for improving the wireless sensor networks' efficiency. The parameters proposed for selecting the most appropriate cluster heads are the nodes' visibility index, sensor energy, received signal strength, and distance to the base station. The VERD simulation result has showcased a significant improvement in conserving the network's energy. The lifetime of the WSN was extended by 36.65%, and the network stability improved by an average of 29.83%. Incorporating independent sensors in VERD reduces the workload on cluster heads and enhances network energy conservation.

Consequently, VERD has shown a clear improvement over at least three existing algorithms. VERD can be used in various wireless sensor network-related applications such as IoTs, agriculture, E-health, etc. The coefficients of our cluster head selection parameters were obtained experimentally; they can also be obtained through computation intelligent means. Therefore, we will consider using a computational intelligence method to derive our coefficients in future work.

References

1. Ouni, R., Saleem, K.: Framework for sustainable wireless sensor network based environmental monitoring. Sustain. For. **14**(14), 8356 (2022)
2. Raghuvanshi, A.S., Tiwari, A., Tripathi, R., Kishor, N.: Optimal number of cluster in wireless sensor network: an FCM approach. Int. J. Sens. Netw. **12**(1), 16–24 (2012)
3. Thyagarajan, J., Kulanthaivelu, S.: A joint hybrid corona based opportunistic routing design with quasi mobile sink for IoT based wireless sensor network. J. Ambient Intell. Hum. Comput., 1–19 (2020)
4. Johana, A.M., Johan, S.R.-R., Jesus, M.P.: Contrasting internet of things and wireless sensor network from a conceptual overview. In: International Conference on Internet of Things and Green Computing and Communication (2016)
5. Kiani, F., Seyyedabbasi, A.: Wireless sensor network and internet of things in precision agriculture. Int. J. Adv. Comput. Sci. Appl., 99–103 (2018)
6. Kiani, F.: Animal behaviour management by energy efficient wireless sensor networks. Comput. Electron. Agric., 478–484 (2018)
7. Loubna, H., Bouchaib, N.: Application of wireless sensor networks in the field of irrigation: a review. Elsevier Comput. Electron. Agric. **179**, 105782 (2020)
8. Saleh, N., Kassem, A., Haider, A.M.: Energy efficient architecture for wireless sensor networks in healthcare applications. IEEE Access, 6478–6486 (2018)
9. Ahsan, A., Mandar, G., Saadullah, F., Cosimo, L., Kia, D., Hadi, L., Amir, H.: A survey of the role of wireless networks and IoT in disaster management. Geol. Disaster Monit. Based Sens. Netw., 57–66 (2019)
10. Hussain, M.A., Pervez, K., Kyung, S.K.: WSN research activities for military application. In: 11th International Conference on Advance Communication Technology, vol. 1, pp. 271–274. IEEE (2009)
11. Pedro, C., Martinez, J.F., Jesus, R.M., Alexandra, C.: Integration of wearable devices in a wireless sensor network for an e-health application. IEEE Wirel. Commun., 38–49 (2013)
12. Shantanu, B., Avinash, K.A., Om, P., Shailendra, S., Mohit, P., Rishi, K.: Introduction to sensors for aerospace and automotive applications. Sens. Automot. Aerosp. Appl., 1–6 (2018)
13. Philip, A., Haitham, C., Jeremy, M., Chibueze, P.A.O., Ao, L., Waleed, H., Shihan, B., Zhili, S.: Security and privacy in location-based services for vehicular and mobile communications: an overview, challenges, and countermeasures. IEEE Internet Things J., 4778–4802 (2018)
14. Suman, B., Depali, G., Sonal, Reddy, S.R.N.: Smart parking – a wireless sensor networks application using IoT. In: Proceeding of International Conference on Communication, Computing, and Networking. Lecture Note on Networks and Systems, pp. 217–239, Singapore (2019)
15. Hariprasath, M., Yuvaraja, T., Ramya, K., Arun, R.: A novel optimal robotized parking system using advanced wireless sensor network. Hindawi J. Sens., 10 (2021)
16. Akhilesh, P., Rajat, K.S.: EADCR: energy aware distance based cluster head selection and routing protocol for wireless sensor networks. J. Circ. Syst. Comput. (2021)
17. Nellore, N.R., Senthikumar, C.: Energy efficient clustering hierarchy in wireless sensor network using dynamic clustering approach ILEACH-ACHH protocol compared with LEACH-ACHH protocol. In: 2nd International Conference on Innovative Practices in Technology and Management (2022)
18. Pariselvam, S., Manikandan, M., Praveenkumar, P., Vinothvarma, R.: Energy efficient RSSI based low energy node utilization routing in manet. In: Proceeding of International Conference on Systems Computation Automation and Networking, India (2019)
19. Prasanth, A., Pavalarajan, S.: Implementation of efficient intra and inter-zone routing for extending networks consistency in wireless sensor networks. J. Circ. Syst. Comput. (2019)
20. Akhilesh, P., Lackshman, S., Rajat, K.: RCH-LEACH: residual energy based cluster head selection in LEACH for wireless sensor networks. In: International Conference on Electrical and Electronics Engineering (2020)

21. Amir, S., Gulustan, D., Farzad, K.: HEEL: a new clustering method to improve wireless sensor network lifetime. IET Wirel. Sens. Syst., 130–136 (2019)
22. Farzad, K., Amir, S., Sajjad, N.: Improving the performance of hierarchical wireless sensor networks using the metaheuristic algorithm: efficient cluster head selection. Sens. Rev. **41**(4), 368–381 (2021)
23. Ankur, G., Shivam, M., Sandeep, K.: A review of energy- efficient mechanisms for cluster-head selection in WSN for IoT application. In: IOP Conference Series: Material Science and Engineering (2021)
24. Heinzelman, W.B., Anantha, P.C., Hari, B.: An application-specific protocol architecture for wireless microsensor networks. IEEE Trans. Wirel. Commun., 660–670 (2002)
25. Ahmed, A.-B., Ayman, E.S.: A new algorithm for cluster head selection in LEACH protocol for wireless sensor networks. Int. J. Commun. Syst.. Wiley. (2017)
26. Mahmood, D., Javaid, N., Mahmood, S., Qureshi, S., Memon, A.M., Zaman, T.: MODLEACH: A variant of LEACH for WSNs. In: Eighth International Conference on Broadband and Wireless Computing, Communication and Applications, Compiegne (2013)
27. Manjeshwar, A., Agrawal, D.P.: TEEN: a routing protocol for enhanced efficiency in wireless network. In: Proceedings 15th International Parallel and Distributed Processing Symposium, San Francisco (2002)
28. Lindsey, S., Raghavendra, C.S.: PEGASIS: Power efficient gathering in sensor information systems. In: Aerospace Conference Proceeding, USA (2002)
29. Li, Q., Aslam, J., Rus, D.: Hierarchical power aware routing in sensor networks. In: Proceeding of DIMACS Workshop on Pervasive Networking, New Brunswick (2001)
30. Georgios, S., Ibrahim, M., Azer, B.: SEP: A Stable Election Protocol for Clustered Heterogeneous Wireless Sensor Networks. Boston University Computer Science Technical Report, Boston (2004)
31. Amruta, L., Damodar, R.E.: Cluster head selection and cluster construction using fuzzy logic in WSNs. IEEE Xplore. (2020)
32. Haddar, M.A.: Scalable and energy aware K-leaders election algorithm in IoT wireless sensor networks.," Springer. Wirel. Pers. Commun. (2022)
33. Handy, M.J., Haase, M., Timmermann, D.: Low energy adaptive clustering hierarchy with deterministic cluster-head selection. In: 4th International Workshop on Mobile and Wireless Communications Network, Sweden (2002)

Metasurface-Backed Polarization-Diversified Microstrip Patch Antenna for Biomedical Multi-communication Applications

Swarnadipto Ghosh, Dipankar Saha, Ayona Chakraborty, Samik Chakraborty, Sunday Cookey Ekpo, and Fanuel Elias

1 Introduction

The need for biomedical sensors to send information about patients' conditions remotely is of utmost importance nowadays [1]. The use of various invasive and noninvasive biomedical devices is on the rise to make diagnosis and treatment easier. The transceiving of the information is done with the help of biomedically suitable or biocompatible antennas. Antenna miniaturization in this case is vital owing to its use for applications in a human body. Metamaterials with their unique properties can be used for lesser real estate occupation [2–4]. Khan et al. [5] have investigated the challenges of body-centric wideband antenna designs using metamaterials [6–9] and partial ground plane. The use of partial ground plane invariably gives rise to pattern distortion. Reference [10] explains a multiband spiral antenna design with the requisite study of biocompatible insulation and choice of layers to improve radiation efficiency. In an implantable case, the choice of layers is fixed inside the human body formation. Energy bandgap and spring ring resonator have been introduced by Wajid et al. [11] in a wearable dual-band antenna wherein the effect of the cotton substrate material on the gain of the antenna has not been explicitly probed. In [12], 5G biomedical applications have been shown. The coplanar waveguide feed has been

S. Ghosh (✉) · D. Saha
Indian Institute of Space Science and Technology, Thiruvananthapuram, India

A. Chakraborty · S. Chakraborty
Jadavpur University, Kolkata, India

S. C. Ekpo · F. Elias
Manchester Metropolitan University, Manchester, UK

S. C. Ekpo (ed.), *The Second International Adaptive and Sustainable Science, Engineering and Technology Conference*, Signals and Communication Technology,
https://doi.org/10.1007/978-3-031-53935-0_20

introduced to aid for compactness which may give rise to design complications. Metasurface-backed circularly polarized (CP) antenna has been reported. It can be observed from the effective subwavelength meta-unit cell patch is relatively larger.

This work presents a dual-stacked 4X4 array of subwavelength meta-unit cell surface-backed MPA. It consists of CP in all higher-order hybrid modes having dual-polarized nature in the fundamental mode of resonance. Sections 2 and 3 explain the design and the analysis of the dual-stacked antenna, respectively. Section 4 details the biosensing approach to the proposed antenna. Section 5 states the closed-form mathematical relationship to estimate the resonance frequency. Section 5.2 concludes the paper.

2 Design Stages of Proposed Antenna

This article proposes a dual-layer microstrip patch antenna (MPA) architecture by having a subwavelength meta-unit cell-backed slotted MPA. The dual-layer architecture has been achieved by stacking two layers of Teflon dielectric substrate having permittivity ($\epsilon_r = 2.4$) and loss tangent ($\tan\delta = 0.0009$). A slotted MP has been printed on the upper dielectric layer where the 4X4 array of subwavelength meta-unit cell structure has been printed on the lower dielectric layer following a ground plane on the other side of the lower dielectric substrate.

From Fig. 1, two dielectric substrates have been stacked one by one, and the rectangular slotted MP has been excited through a coaxial feed connected through the ground layer of the lower substrate up to the surface layer of the slotted MP. The manufacturing point shows that the outer radius of the coaxial connector is $r_{co\ -\ ax(\text{out})}$= 1.2 mm where the inner radius is $r_{co\ -\ ax(\text{in})} = 0.6$ mm. The detailed layered architecture has been mentioned throughout Fig. 2. It can be mentioned from Fig. 2 the overall dimension of the substrate is ($L_g \times W_g$)=(*60 mm* × *60 mm*) where

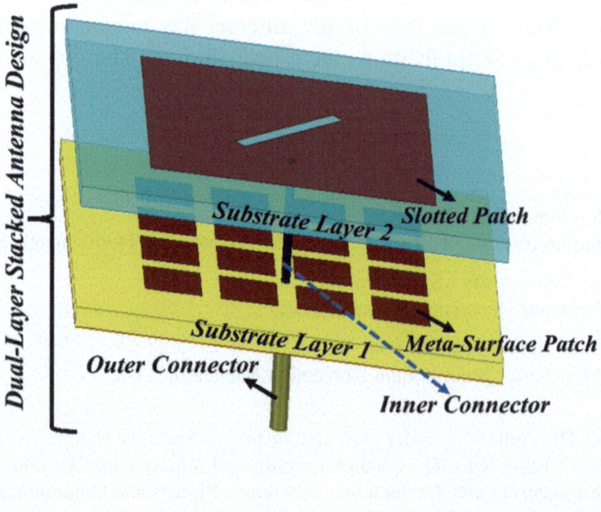

Fig. 1 Schematic of dual-layer architecture of the proposed antenna excited through coaxial connector

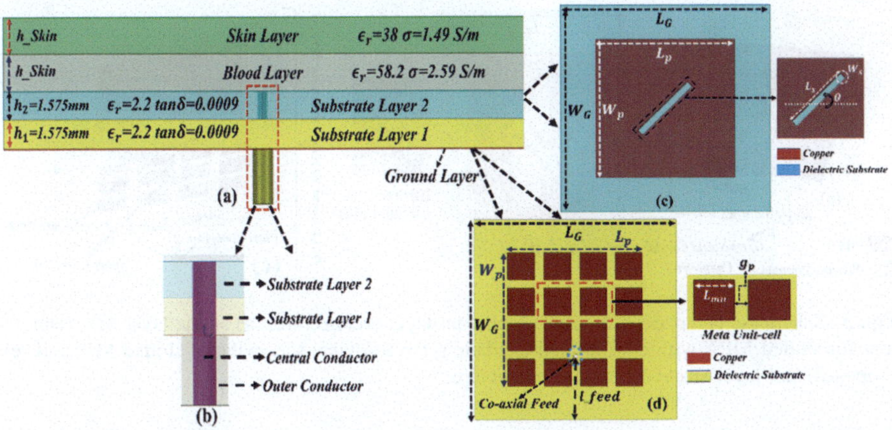

Fig. 2 Schematic dissection of dual-layer architecture of the proposed antenna excited through coaxial connector: (**a**) side view of the stacked dielectric structure, (**b**) dissection of coaxial feed connector, (**c**) top view schematic of rectangular slotted microstrip patch (MP) printed on the upper dielectric substrate, and (**d**) top view schematic of the 4X4 array of subwavelength meta-unit cell printed on the top portion of lower dielectric substrate

the MP printed on the upper dielectric substrate has a dimension of ($L_p \times W_p$)= ($20\ mm \times 20\ mm$). The dimension of the magnetic boundary created in the form of an electrical slot upon the printed MP can be expressed as ($L_s \times W_s$)=($12\ mm \times 2\ mm$) and has been rotated by an angle $\theta = 45°$. Each subwavelength meta-unit cell has a dimension of ($L_{mu} \times L_{mu}$) =($8\ mm \times 8\ mm$), chosen based on the operating frequency (f_r) $= 2.4\ GHz$ and mapped to at least $\frac{\lambda}{15}$ mm for realizing the actual physical behavior of the subwavelength structures. The interelement gap is defined as $g_p = 2.66\ mm$. The ground layer, the microstrip patch, and the 4X4 array of subwavelength meta-unit cell structure have been considered as a perfect electric conductor (PEC) material (Fig. 2).

3 Performance Analysis of Proposed Antenna

A comparative analysis (Fig. 3) between the single- and the dual-layer 4X4 subwavelength meta-unit cell array-backed slotted microstrip patch antenna (MPA) has been analyzed.

The single-layer MPA was excited through a coaxial connector (Fig. 3a). To realize the electromagnetic behavior of the subwavelength 4X4 array of meta-unit cell surface, the structure is placed as an intermediate ground plane in the stacked form of two dielectric substrates. The metasurface (Fig. 3c) is behaving like a periodic defected ground structure (*DGS*) having magnetic boundary bounded by electric walls periodically to modulate the effective surface wave through the metasurface boundary by periodic modulation of phase velocity (v_p) and group velocity (v_g) (Fig. 4).

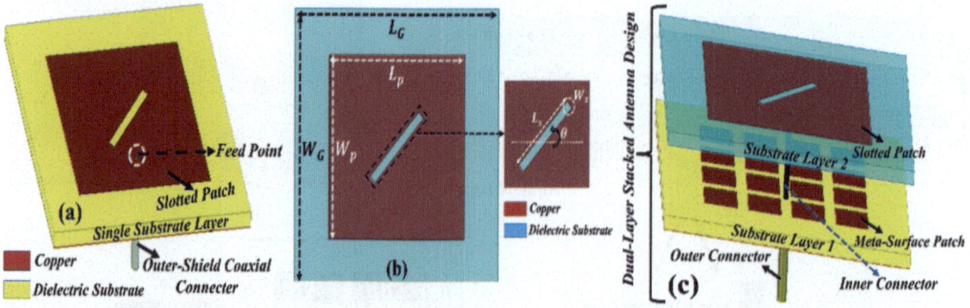

Fig. 3 Schematic comparison of single- and dual-layer architecture: (**a**) single-layer 3D architecture for slotted MPA with coaxial feed excitation, (**b**) top view schematic of slotted MP, and (**c**) proposed dual-layer-stacked antenna architecture.

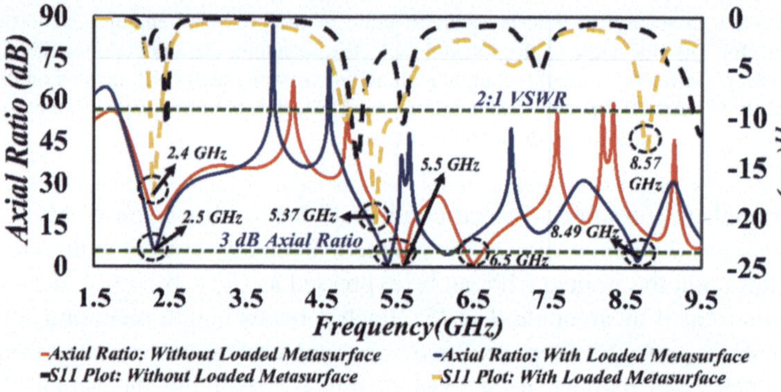

Fig. 4 Simulated $S_{11}(dB)$ and *axial ratio (dB)* plot for both with and without loading of 4X4 array of subwavelength array of meta-unit cell surface

4 Biosensing Application of Proposed Antenna

A closed-form mathematical relationship to express the effect of the various human dielectric loadings on the surface of the dual-layer 4X4 array of subwavelength meta-unit cell surface-backed MPA is presented here. It covers the different loading effects on the variation of the dielectric widths. Two dielectric layers have been stacked upon the proposed dual-layer MPA structure.

This biosensing research involves two dielectrics, viz., (i) human skin [*permittivity* $(\epsilon_r) = 38$, *conductivity* $(\sigma=1.49$ *S/m*)] and (ii) human blood [*permittivity* $(\epsilon_r) = 58.2$, *conductivity* $(\sigma=2.59$ *S/m*)]. The closed-form equation modeled through correlating the two different widths of the proposed dielectric layers can express the overall performance of the proposed antenna at all frequency bands to predict the effect of the dielectric layer thickness. Figure 5 shows the quad-layer architecture.

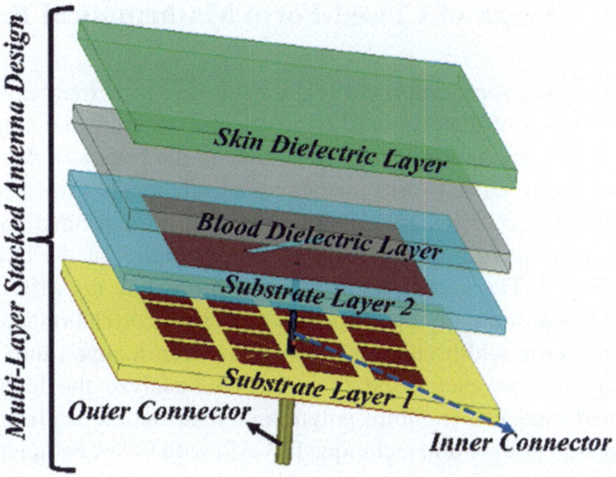

Fig. 5 Schematic of quad-layer architecture of the proposed antenna loaded with skin and blood layer excited through coaxial connector

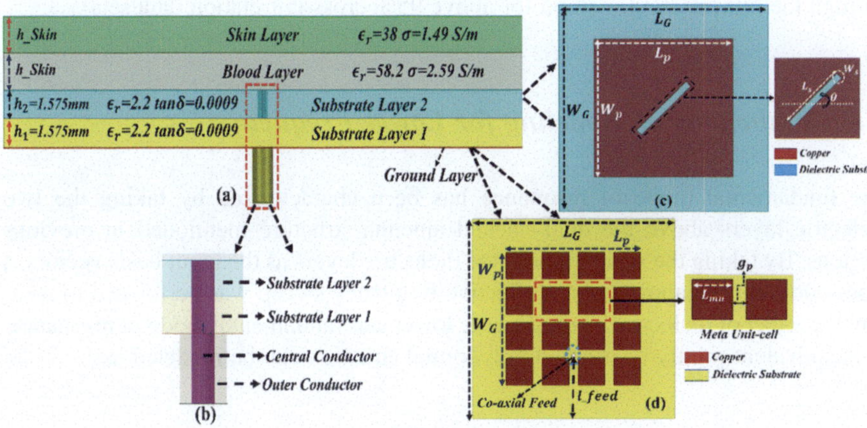

Fig. 6 Schematic dissection of dual-layer architecture of the proposed antenna excited through coaxial connector: (**a**) side view of the stacked dielectric structure, (**b**) dissection of coaxial feed connector, (**c**) top view schematic of rectangular slotted microstrip patch (MP) printed on the upper dielectric substrate, (**d**) top view schematic of the 4X4 array of subwavelength meta-unit cell printed on the top portion of lower dielectric substrate

Two different dielectric layers (viz., human skin and blood) have been stacked on the top of the proposed dual-layer MPA architecture. Figure 6 shows a detailed schematic of the proposed quad-layer structure loaded with the human skin and blood. A closed-form empirical relationship has been modeled to predict the antenna performance at all the three frequency bands based on dielectric widths of the human skin and the blood dielectric layer.

5 Design of Closed-Form Mathematical Relationship

A closed-form empirical relationship for each frequency band based on the width variation of the proposed two dielectric layers is developed to estimate the antenna performance for dielectric loadings on the proposed dual-layer-stacked MPA structure. The closed-form mathematical relationship has been modeled using a jointly distributed or bivariate polynomial distribution function where the two input basis vectors have been taken as dielectric widths of the human skin (h_{skin}) and blood (h_{blood}). The variations of both the blood and the skin dielectric widths have been taken as mutually independent. The cross-correlation matrix of the two independent dielectric width column vectors has been mapped into a particular resonance frequency for each band of interest to analyze the loading effect on the antenna performance. The joint polynomial distribution has been modeled using goodness-of-fit optimization technique [13–15] with 95% coefficient bound. For all the closed-form mathematical relationships, the root mean square error = 0.925 and the adjusted square error = 0.962 (i.e., the prediction tool exhibits the best fitting through the optimization method for above 95% cross-correlation data sets).

5.1 *Mathematical Modeling for Lower Frequency*

The fundamental mode of resonance has been characterized by taking the two dielectric layers above the dual-stacked antenna structure mentioned in previous sections. By taking the width of both the dielectric layers as the input basis vectors, a cross-correlation mapping to the effective frequency sweep was performed to measure the effect of dielectric loading on the lower and fundamental mode of resonance. The equivalent jointly distributed polynomial equation can be modeled as:

$$\mathrm{Freq}_{\mathrm{Low}} = \sum_{m=0}^{N} \sum_{n=0}^{N} P_{mn} x^n y^m \tag{1}$$

In (1), x and y variables have been defined as the height of blood (h_{blood}) and skin (h_{skin}) dielectric layer, respectively. Table 1 shows the respective polynomial coefficient matrices.

Table 1 Polynomial coefficient for lower-frequency estimation

	m				
n	0	1	2	3	4
0	0.1618	−7.464	5.341	−1.174	0.0895
1	7.748	−1.305	−0.2635	0.0235	0
2	−2.347	0.6093	0.0110	0	0
3	0.1858	−0.0677	0	0	0
4	0.0007	0	0	0	0

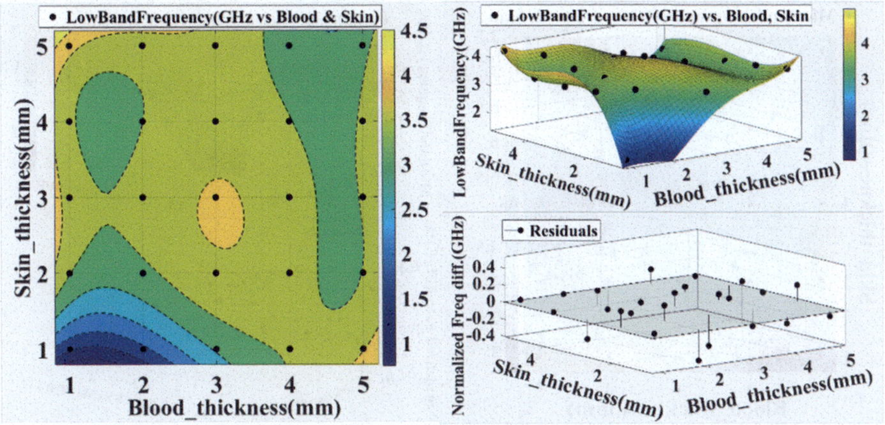

Fig. 7 Contour and residual plot to express the jointly distributed polynomial equation in graphical view

Table 2 Polynomial coefficient for middle-frequency estimation

	m				
n	0	1	2	3	4
0	0.908	−9.47	7.812	−1.705	0.1232
1	13.38	−4.828	−0.2341	0.0596	0
2	−4.062	1.756	−0.0533	0	0
3	0.2493	−0.1437	0	0	0
4	0.01817	0	0	0	0

Figure 7 shows the contour and the residual plots of the mathematical modeling. From Fig. 7, a nonlinear mapping has been obtained from the cross-correlation between the two dielectric layer thickness. The contour plot gives the different cluster separations based on the order of nonlinearity. The residual plot expresses the validation of the polynomial by up to 95% curve-fitting optimization.

5.2 *Mathematical Modeling for Middle Frequency*

The second-order hybrid or quasi-TEM mode of resonance has been characterized by taking the two dielectric layers above the dual-stacked antenna structure. By applying the cross-correlation mapping, the equivalent bivariate polynomial can be expressed as:

In (2), the x and y variables have been defined as the height of blood (h_{blood}), and skin (h_{skin}) dielectric layer. Table 2 contains the respective polynomial coefficient matrices. Figure 8 shows the contour and the residual plots of the mathematical modeling. A nonlinear mapping was performed to estimate the mapping between the cross-correlation vector matrix and the second-order hybrid or quasi-TEM mode of resonance.

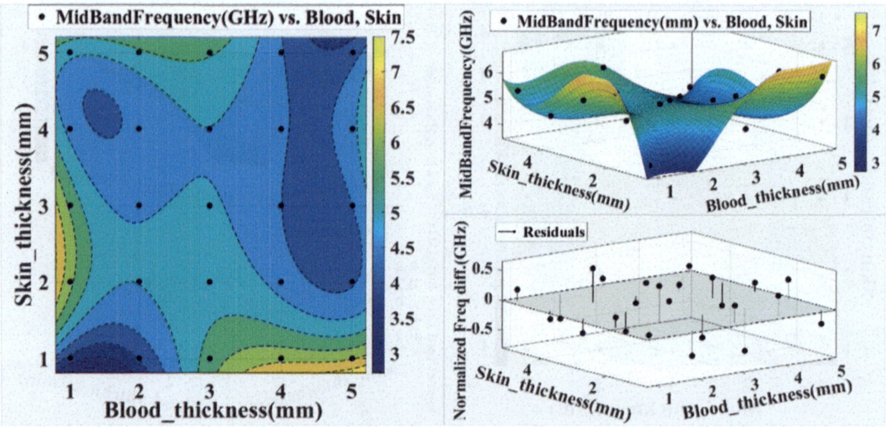

Fig. 8 Second-order contour and residual plots to express the jointly distributed polynomial equation

Table 3 Polynomial coefficient for upper-frequency estimation

	m				
n	0	1	2	3	4
0	1.011	−1.806	3.684	−0.842	0.05508
1	12.33	−6.287	0.3103	0.02983	0
2	−3.762	1.728	−0.0933	0	0
3	0.337	−0.1152	0	0	0
4	−0.00325	0	0	0	0

5.3 *Mathematical Modeling for the Upper Frequency*

By applying the cross-correlation mapping, the equivalent bivariate polynomial can be expressed as:

$$\text{Freq}_{\text{Upper}} = \sum_{m=0}^{N} \sum_{n=0}^{N} P_{mn} x^{n} y^{m} \tag{2}$$

The third-order hybrid or quasi-TEM mode of resonance has been characterized by taking the two dielectric layers above the dual-stacked antenna structure. From (2), the x and y variables have been defined as the height of blood (h_{blood}) and skin (h_{skin}) dielectric layer. Table 3 shows the respective polynomial coefficient matrices.

A nonlinear mapping has been performed to achieve the empirical relationship between the cross-correlated matrix and the frequency points (Fig. 9). By applying the closed-form empirical relationship, a *NXM* matrix can be created to estimate the overall correlation coefficient based on the variation of the dielectric layers implanted on the MPA structure.

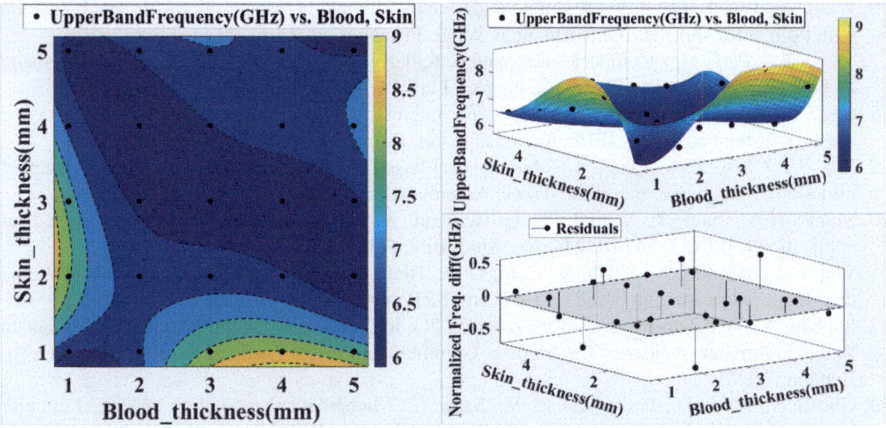

Fig. 9 Third-order contour and residual plots to express the jointly distributed polynomial equation

6 Conclusions

A detailed characterization of a dual-band-stacked 4X4 array of subwavelength meta-unit cell surface-backed microstrip patch antenna architecture is presented. The electromagnetic effect of the periodically defected partial ground structure yields S11 of < -10 dB at 2.4 GHz. The presented antenna structure can be utilized for short-range biomedical communication and implantable sensing applications. On-body and implantable tumor detections are realizable with a highly sensitive adaptive low-noise transceiver [16–22]. With adaptive beamforming, integrated biosensing telemedicine and reconfigurable multimedia biosensor data can be achieved in satellite-cellular convergence connectivity network [23–30].

References

1. Raghavan, S., Rajeshkumar, V.: An overview of metamaterials in biomedical applications. Prog. Electromagn. Res. **25**, 369 (2013)
2. Jamlos, M.A., Othman, N.A., Mustafa, W.A., Marzuki, M.K.: Metamaterial antenna for biomedical application. In: Proceedings of the 11th National Technical Seminar on Unmanned System Technology 2019: NUSYS’19, pp. 519–528. Springer, Singapore (2021)
3. Keltouma, N., Amaria, S., Khadidja, B., Mokhtaria, C., Salima, S., Kada, B., Turkiya, A.: Dual-band half-circular ring implantable antenna with metamaterial SRR for biomedical applications. Telecommun. Radio Eng. **82**(3) (2023)
4. Naik, K.K., Teja, S.C.S., Sailaja, B.V., Sri, P.A.: Design of flexible parasitic element patch antenna for biomedical application. Prog. EM Res. **M 94**, 143–153 (2020)
5. Khan, U.R., Sheikh, J.A., Bashir, S., Ahmed, S.: Metamaterial inspired wideband on-body antenna design for bio-medical applications. Mater. Today Proc. **80**, 1772–1776 (2023)

6. Wang, Z., Liu, J., Long, Y.: A simple wide-bandwidth and high gain microstrip patch antenna with both sides shorted. IEEE Antennas Wirel. Propag. Lett. **18**(6), 1144–1148 (2019)
7. Ta, S.X., Park, I.: Compact wideband circularly polarized patch antenna array using metasurface. IEEE Antennas Wirel. Propag. Lett. **16**, 1932–1936 (2017)
8. Cao, Y., et al.: Broadband and high-gain microstrip patch antenna loaded with parasitic mushroom-type structure. IEEE Antennas Wirel. Propag. Lett. **18**(7), 1405–1409 (2019)
9. Liu, W., Chen, Z.N., Qing, X.: Metamaterial-based low-profile broadband aperture-coupled grid-slotted patch antenna. IEEE Trans. Antennas Propag. **63**(7), 3325–3329 (2015)
10. Malik, N.A., Sant, P., Ajmal, T., Ur-Rehman, M.: Implantable antennas for bio-medical applications. IEEE J. EM RF Microw. Med. Biol. **5**(1), 84–96 (2020)
11. Wajid, A., Ahmad, A., Ullah, S., Choi, D.-y., Islam, F.U.: Performance analysis of wearable dual-band patch antenna based on EBG and SRR surfaces. Sensors. **22**(14), 5208 (2022)
12. Sabban, A.: Wideband wearable antennas for 5G, IoT, and medical applications. In: Advanced Radio Frequency Antennas for Modern Communication and Medical Systems. IntechOpen, London (2020)
13. Ghosh, S., Saha, D., Prahannathan, V., Saha, C.: Mathematical modelling of slotted circular meta-atom for holographic meta-surface applications. In: 2023 International Conference on Microwave, Optical, and Communication Engineering (ICMOCE), vol. 2023, pp. 1–4, Bhubaneswar. https://doi.org/10.1109/ICMOCE57812.2023.10167286
14. Chakraborty, A., Ghosh, S., Roy, S.K., Chakraborty, S., Chatterjee, S., Gupta, B.: Loaded multi loop hexagram shaped printed antennas for multifrequency communication. In: 2022 Microwave Mediterranean Symposium (MMS), vol. 2022, pp. 1–6, Pizzo Calabro. https://doi.org/10.1109/MMS55062.2022.9825536
15. Saha, D., Ghosh, S., Chakrabarti, M., Chattopadhyay, A.: Prism-shaped nonuniformly distributed slotted SIW array with highly reduced SLL for mm-wave ground surveillance application. In: 2023 First International Conference on Microwave, Antenna and Communication (MAC), vol. 2023, pp. 1–5, Prayagraj. https://doi.org/10.1109/MAC58191.2023.10177066
16. Ekpo, S., George, D.: A deterministic multifunctional architecture design for highly adaptive small satellites. Int. J. Satell. Commun. Policy Manag. **1**(2/3), 174–194 (2012)
17. Ekpo, S., George, D.: A system engineering consideration for future-generations small satellites design. In: Proceedings of First IEEE European Satellite Telecoms Conference, pp. 1–4, Rome (2012)
18. Ekpo, S.C.: Thermal subsystem operational times analysis for ubiquitous small satellites relay in LEO. Int. Rev. Aerosp. Eng. J. **11**(2), 48–57 (2018)
19. Ekpo, S., Kettle, D.: Mm-wave LNAs design for adaptive small satellite applications. In: Proceedings of Joint 5th ESA Workshop on mmWave and 31st ESA Antenna Workshop, pp. 843–847, Noordwijk (2009)
20. Ekpo, S., George, D.: Reconfigurable cooperative intelligent control design for space missions. Recent Patents Space Technol. **2**(1), 2–11 (2012)
21. Zafar, M., Ekpo, S., George, J., Sheedy, P., Uko, M., Gibson, A.: Hybrid power divider and combiner for passive RFID tag wireless energy harvesting. IEEE Access J. **10**, 502–515 (2022)
22. Enahoro, S., Ekpo, S., Charles Uko, M., Ansari, U., Altaf, A., Zafar, M.: Adaptive beamforming for mmWave 5G MIMO antennas. In: Proceedings of 21st IEEE Annual Wireless & Microwave Conference, pp. 1–4. Sand Key, Clearwater Beach (Online/Virtual) (2021)
23. Ansari, U., Ekpo, S., Charles Uko, M., Altaf, A., Zafar, M., Enahoro, S., Okpalugo, O.: 5G-enabled mobile operating hospital and emergency care service. In: Proceedings of 21st IEEE Annual Wireless & Microwave Conference, pp. 1–6. Sand Key, Clearwater Beach (Online/Virtual) (2021)
24. Enahoro, S., Ekpo, S., Gibson, A.: Massive multiple-input multiple-output antenna architecture for multiband 5G adaptive beamforming applications. In: Proceeding of 22nd IEEE Annual Wireless & Microwave Conference, vol. 27, pp. 1–4, Marriott Sand Key, Clearwater Beach, (Online/Virtual) (2022)

25. Ekpo, S., George, D., Adebisi, B.: A multicriteria optimisation design of SPSE for adaptive LEO satellites missions using the PSI method. In: AIAA Space 2013 Conference & Exposition, pp. 1–19, San Diego (2013)
26. Ekpo, S., Adebisi, B., Wells, A.: Regulated-element frost beamformer for vehicular multimedia sound enhancement and noise reduction applications. IEEE Access J. **5**, 27254–27262 (2017)
27. Ekpo, S.C., Adebisi, B., George, D., Kharel, R., Uko, M.: A system-level multicriteria modelling of payload operational times for communication satellite missions in LEO. Recent Prog. Space Tech. **4**(1), 67–77 (2014)
28. Uko, M., Ekpo, S.: A 23–28 GHz pHEMT MMIC low-noise amplifier for satellite-cellular convergence applications. Int. Rev. Aerosp. Eng. J. **14**(5), 1–10 (2021)
29. Uko, M., Ekpo, S.: 8–12 GHz pHEMT MMIC low-noise amplifier for 5G and fiber-integrated satellite applications. Int. Rev. Aerosp. Eng. J. **13**(3), 99–107 (2020)
30. George, J., Uko, M.C., Ekpo, S.C., Ijaz, M., Kharel, R., Wang, Q., Ji, H.: Design of a multiband RF slotted-antenna for biosensing applications. In: Proceedings of 12th IEEE/IET International Symposium on Communication Systems Networks and Digital Signal Processing Conference, pp. 1–6, Porto (Online/Virtual) (2020)

Experimental Analysis of Useful Manufacturing Materials Embedded in Agricultural Wastes for Paper Industry Applications

Enefiok Okon Usungurua, Itoro Esiet Udo, Daniel Edet Essang, Sunday Cookey Ekpo, and Perpetual Eze-Idehen

1 Introduction

Paper is an indispensable part of everyday life. Beyond its use as the basic material for written and printed communication, paper in its various forms is used for painting, graphics, signage, design, packaging, decorating, and cleaning and is at work in thousands of constructions and industrial and manufacturing applications. However, the continued use of wood as the main material for conventional pulp and paper production has tremendously increased global deforestation and its attendant problems and has limited the availability of biomass feedstock for clean energy generation. Therefore, for sustainable environment, reduced cost of production, and improved economic versatility, developing more manufacturing (raw) materials embedded in agricultural waste products especially from non-wood wastes to produce pulp for papermaking and other industrial applications becomes imminent.

E. O. Usungurua (✉)
Department of Mechanical Engineering, School of Engineering, Akwa Ibom State Polytechnic, Ikot Osurua, Ikot Ekpene, Nigeria

I. E. Udo · D. E. Essang
Departmen of Chemistry, Faculty of Sciences, University of Uyo, Uyo, Nigeria
e-mail: itoroudo@uniuyo.edu.ng

S. C. Ekpo
Department of Electrical Electronic, Manchester Metropolitan University, Manchester, UK
e-mail: S.Ekpo@mmu.ac.uk

P. Eze-Idehen
Department of Chemical Engineering, The University of Manchester, Manchester, UK
e-mail: perpetual.idehen@manchester.ac.uk

S. C. Ekpo (ed.), *The Second International Adaptive and Sustainable Science, Engineering and Technology Conference*, Signals and Communication Technology,
https://doi.org/10.1007/978-3-031-53935-0_21

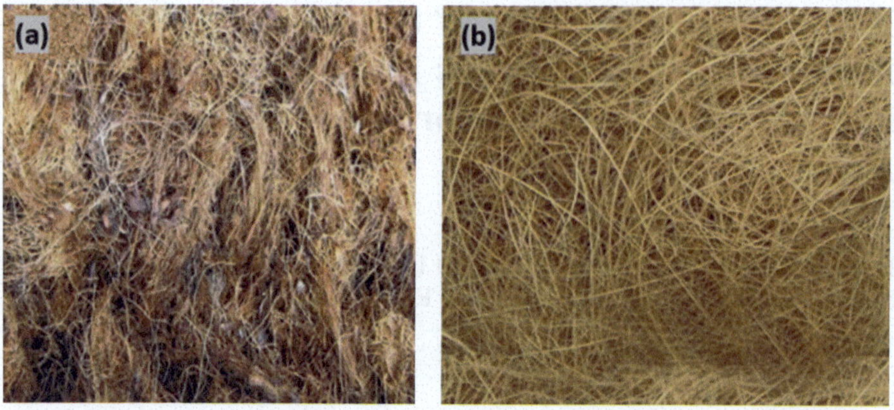

Fig. 1 (**a**) Palm fruit seed hairs and (**b**) coconut fibres

Palm fruit seed hair lignocellulosic composites are comprised of compactly arranged cellulose, hemicellulose, and lignins with 44.4% cellulose fraction [1], out of which as much as 60.6% can be turned into value-added biopolymer [2]. There have been several attempts to utilize the abundant and renewable resources of plant fibres in composite materials with a view to replace at least, partially, the use of expensive synthetic fibres and particulate fillers for a variety of applications [3].

Coir (also known as "kokos" or "coco") is a natural seed hair fibre obtained from the outer shell (endocarp) or husk of the coconut and palm fruit, respectively. The fruits *Cocos nucifera* L. and *Elaeis guineensis*, a tropical plant of Arecaceae (Palmae) family, are extensively grown in tropical countries. It poses high elastic limit and can even take up a permanent stretch and high resistance to microbial degradation. It comes as raw fibres, spun yarn, woven mats, mattings, rugs and carpets, coir rope, and several other materials [4, 5].

Therefore, the aim of this study is to determine the fibre properties together with the cellulose and hemicellulose components of coconut fruit fibres and palm fruit seed hairs, thereby converting agricultural wastes into useful manufacturing materials and chemical products for paper industry applications. Figure 1a, b show the samples of fibres used in this study.

Saka et al. [6] carried out a study on the chemical composition of palm fruit (*Elaeis guineensis*) fibre and its feasibility as cellulosic raw material for sugar production, whereas [5] studied extraction and characterization of coconut (*Cocos nucifera* L.) coir dust using water, acetone, acetone/water (70/30), (50/50). Akpakpan et al. [7] studied the evaluation of physiochemical properties and soda pulping of *Nypa fruticans* frond and petiole, where fibre dimensions and morphological indices were determined; meanwhile, [8] handled proximate and phytochemical studies of exudate of *Dacryodes edulis*. Saka et al. [6] undertook a study on chemical constituents of the different anatomical parts of oil palm (*Elaeis guineensis*) such as trunk, frond, shell, kernel cake, and empty fruit bunch (EFB) for sustainable utilization.

2 Materials and Methods

The materials used were coconut (*Cocos nucifera* L.) fruit fibres and palm fruit (*Elaeis guineensis*) seed hairs. The seed hairs were obtained from oil palm from Oron Market in Oron Local Government Area, and coconut fibres were procured from Use-Ofot Market in Uyo Local Government Area, all in Akwa Ibom State, Nigeria. The samples were washed with distilled water to remove the debris, air-dried, and then conveyed to the Chemistry Laboratory, University of Uyo, for further preparation and analysis.

2.1 Solubility in Water

Cold water solubility 2.0 g of the sample was placed in a beaker and 300 mL of distilled water was slowly added. The extraction was carried out at 23 °C with sequential stirring for 48 h. The materials were transferred to a tared filtering crucible which was previously dried to a constant weight at 105 °C, washed with 200 mL of cold distilled water, and dried to constant weight at 105 °C [9, 10].

Hot water solubility 2.0 g of sample was transferred to a 250 Erlenmeyer flask, and 100 mL of hot distilled water was added and placed in a boiling water bath. Reflux condenser was attached and digested for 3 h, making certain that the water level of the bath is held above the stock level in the flask. The contents of the flask were transferred to a tared filtering crucible which was previously dried to a constant weight at 105 °C, washed with 200 mL of hot water, and dried to constant weight at 105 °C [9, 10]:

$$\text{Water solubility } (\%) = \frac{W1 - W2}{W1} \times 100 \tag{1}$$

where W_1 = weight of sample before solubility test, g
W_2 = weight of sample after solubility test, g

2.2 Alkaline Solubility

Solubility in 1% NaOH 2.0 g of wood sample was put into a conical flask and 100 mL of 1% NaOH was added. The mixture was heated in a water bath for 1 h, after which it was filtered, washed with distilled water, and dried [9].

Solubility in 18% NaOH 2.0 g of wood sample was put into a conical flask and 100 mL of 18% NaOH was added. The mixture was treated as presented for solubility in 1% NaOH [9]. In each case, the loss in weight was determined as percentage solubility in alkali following Eq. (1).

2.3 Solubility in 1:2 Ethanol-Benzene Mixture

2.0 g of sample was wrapped with a filter paper and inserted into the thimble of the Soxhlet extractor; 50-ml ethanol and 100-mL benzene were poured into the 250-mL extractor flask, and the condenser of the extractor was connected to the cold water tap. The extractor flask was heated to 80 °C and maintained for 6 h. The crucible containing the extract was put in an oven maintained at 80 °C, then cooled in the desiccator, and weighed. The heating and cooling were repeated until a constant mass was obtained [11]. The soluble of the samples was calculated following Eq. (1).

2.4 Determination of Cellulose (The Kurschner-Hoffer Cellulose Method)

1.0 g of the air-dried sample was placed in a 250-mL beaker, 15 mL of 80% acetic acid and 5 mL of concentrated trioxonitrate (v) acid were added, and the mixture was heated for 20 min in a fume cupboard. After exactly 20 min, 20 ml of 95% cold ethanol was added, and then the mixture was cooled and filtered. The residue was then washed successively with hot benzene and hot ethanol, dried overnight in the oven to a constant weight, and weighed. It was then ashed in the muffle furnace at about 500 °C for 5 h [2, 12]:

$$\text{Cellulose }\% = \frac{W_o - W_r}{W_o} \times 100 \tag{2}$$

where W_o = weight of the ovendried residue
W_r = weight of the residue remaining after ignition

2.5 *Determination of Hemicellulose*

1.0 g of extracted dried sample was transferred into a 250-mL Erlenmeyer flask, and 150 mL of 500 mol/m^3 NaOH was added. The mixture was boiled for 4 h with distilled water, then filtered after cooling through vacuum filtration, and washed until neutral pH. The residue was dried to a constant weight at 1050 C in an oven [2, 7, 12]:

$$\text{Hemicellulose content }\%\frac{Ws - Wt}{Ws} \times 100 \tag{3}$$

where $Ws =$ weight of the ovendried sample, g, and $W_t =$ weight of the sample after the treatment, g.

2.6 *Determination of Fibre Characteristics*

The samples were put into test tubes and macerated with an equal volume (1:1) of glacial acetic acid and 30% hydrogen peroxide as adopted by [13, 14]. This was later heated in an oven for 4 h at 80 °C until bleached white and soft. The chemical mixture was decanted, and macerated fibres were rinsed several times with distilled water and defiberized to separate the fibres. The fibres were viewed under electron microscope to obtain its characteristics [13–15].

3 Results and Discussions

3.1 *Physiochemical Characterization of Coconut Fibre and Palm Fruit Seed Hairs*

Table 1 depicts the results of the percentage of the soluble contents and percentage chemical properties of coconut fibre and palm fruit seed hairs.

Table 1 Physiochemical characterization of coconut fibre (CF) and palm fruit seed hairs (PFSH)

Physiochemical parameters	CF	PFSH
Solubility in cold water (%)	16.0	19.0
Solubility in hot water (%)	22.0	26.5
Solubility in 1% NaOH (%)	26.0	30.0
Solubility in 18% NaOH (%)	36.0	42.0
Solubility in 2:1 ethanol-benzene solution (%)	27.5	31.0
Cellulose (%)	75.0	87.1
Hemicellulose (%)	54.0	62.0

The compositions of non-wood have tremendous variation in chemical and physical properties compared to wood fibres [4, 11]. Variations depended on non-wood species and the local conditions such as soil and climate [5].

3.2 Solubility in Cold Water

The components of wood that were soluble in cold water were extraneous components such as sugar, tannins, gums, and colouring matters present in the sample [15]. The solubility in the cold water had 16.0% and 19.0% for coconut fibre and palm fruit seed hairs, respectively, and were within the range for some non-woods [6, 7, 11]. It implies that the samples contained the stated properties, and these are good feedstock for the production of pulp for paper and polymer industry applications.

3.3 Solubility in Hot Water

The components that were soluble in hot water were the same with that of cold water but with high percentage of solubility. The coconut fibre had solubility of 22.0% in hot water while that of palm fruit seed hairs was 26.5%. The values obtained in both samples correlated with the results from coconut coir dust [6]. This follows that the test samples exhibited decomposition (extractive) components at high temperature [15].

3.4 Alkaline Solubility (1% NaOH and 18% NaOH)

Solubility in sodium hydroxide (NaOH) was used to determine the degree of fungus decay that can take place in a given wood or non-wood sample. The values of the solubility in 1% NaOH were 26.50% and 30.0% and in 18% NaOH 36.0% and 42.0% for coconut fibre and palm fruit seed hairs, respectively. The values obtained for both samples were within the range for those reported by [6] for some non-wood. This showed that the samples will undergo fungus degradation at higher rate due to the high pitch contents in the fibres, partial dissolution of waxes and resin, and extractive and partial degradation of bonds between lignin and cellulose [4]. The results increased with an increase in concentration of the alkali solution, as evident in increased fibre component breakdown into the solution [14].

3.5 Solubility in Ethanol and Benzene (1:2 v/v)

This indicated the measure of waxes, fats, resin, oils, tannin, and certain ether-insoluble components. The values obtained were 27.5% and 31.0% for coconut fibre and palm fruit seed hairs, respectively. These were higher compared to some other non-wood [1, 2, 8]. This implied that the samples have high prospects for paper and textile industries.

3.6 Cellulose/Hemicellulose Contents

Cellulose contents of coconut fibre and palm fruit seed hairs were 75.0% and 87.1%, respectively. High cellulose contents are desirable for high pulp yield. Both fibres recorded higher cellulose content compared to results obtained for some non-wood such as coconut coir dust (35.99%), fond (48.22%), and petiole (37.56%) [12]. [2] reported that materials with cellulose content ranging from 34% and above are considered as promising for pulp and paper. The percentage compositions of hemicellulose of coconut fibre and palm fruit seed hairs were 54.0% and 62.0%, respectively, higher than that of some non-wood, for example, sugarcane bagasse (19–33.28%), Siam weed (29.9%), and shea tree sawdust (20.3%) [12]. Result variations may be due to location of materials, method of analysis, biomass variety, solvent, and the part of plants used for analysis [14].

3.7 Fibre Properties of Coconut Fibre and Palm Fruit Seed Hairs

From Tables 2, 3, and 4, the fibre length for coconut fibre ranges from 0.886 mm to 0.708 mm with average fibre length of 0.88 mm which was lower than those of *Nypa*

Table 2 Variations in fibre characteristics of coconut fibre

Fibre length (mm)	Fibre diameter (μm)	Fibre width (μm)
0.8855	24.035	12.65
1.20175	31.625	25.3
0.56925	37.95	25.3
1.48005	36.685	25.3
1.1385	25.3	18.975
0.54395	25.3	12.65
0.6578	25.3	12.65
0.74635	22.77	12.65
0.8855	25.3	12.65
0.7084	31.625	21.505

Table 3 Variations in fibre characteristics of palm fruit seed hairs (PFSH)

Fibre length (mm)	Fibre diameter (μm)	Fibre width (μm)
0.3036	12.65	3.795
0.39215	11.385	3.795
0.3036	12.65	3.795
0.253	11.385	3.795
0.2783	12.65	3.795
0.26565	12.65	3.795
0.3289	18.975	12.65
0.44275	12.65	3.795
0.4301	18.975	12.65
0.3036	12.65	3.795

Table 4 Mean value for the fibre characteristic of coconut fibre and palm fruit feed hairs

Samples	CF	PFSH
FL (mm)	0.881705	0.330165
FD (μm)	28.589	13.662
FW (μm)	17.963	5.819
LW (μm)	14.295	6.831
CW (μm)	7.1473	3.4155
RR (μm)	0.5	0.5
FC (%)	50.0	50
SR	0.0308	0.0242

fruticans frond (1.59 mm) and petiole (1.06 mm) [13]. The average fibre length of palm fruit seed hairs was 0.330 mm, which was shorter than that of wood *Leucaena leucocephala* (1.131 mm) [14]. However, the average fibre lengths of coconut fibre was longer than of kenaf core fibre (0.6 mm), maize (0.75 mm), and raffia (0.588 mm) [6] whereas that of palm fruit seed hairs was lower than those obtained from some non-wood materials: fruit plants (0.931 mm) and *Luffa* (1.1928 mm). Hence, pulping raw materials with short fibre lengths are less suitable for paper making than long fibre [3, 13, 15].

The average fibre diameter of coconut was 28.589 μm. This was higher compared to that of core wood, middle wood, and other wood of *Vitex doniana* as reported by [13]. The average fibre diameter of palm fruit seed hairs was (13.66 μm) which was lower than that of *Luffa* (27.703 μm) and raffia (20.240 μm) but higher than pineapple (9.741 μm) which correlated with results obtained as reported for *Nypa fruticans* [13].

Fibre widths of both palm fruit seed hairs and coconut fibre were 5.819 μm and 17.963 μm, respectively. The values were lower than those of wood and other common non-wood fibres (approximately 18.0–30.0 μm and 32.0–43.0 μm for softwood and hardwoods, respectively) [15]. The fibre width of coconut fibre was comparable to those of wheat (3 μm), *Luffa* (18.722 μm), and rice straw (11 μm), while palm fruit seed hairs were comparable to those of fruit plants (5.313 μm), pineapple (4.807 μm), and maize (6.7045 μm) as reported for some non-wood materials [3].

The cell wall thickness determines the choice of fibrous raw materials for pulp and paper production. The values obtained were 7.147 μm and 3.416 μm for coconut fibre and palm fruit seed hairs, respectively. The lumen width of coconut fibre was 14.295 μm while that of palm fruit seed hairs was 6.381 μm. The values obtained were within the range for some hardwood and non-woods. For example, 3.25–8.70 μm and 15.38–13.73 μm, respectively, were recorded for *Vitex doniana* and *Leucaena leucocephala* [7, 13].

Runkel ratio is an important trait for pulp and paper properties in terms of conformity and pulp yield [15]. The Runkel ratio for both fibres was 0.5. This was in line with the results reported for *Anthonotha macrophylla* and Dalium guineensis (0.6–0.9 μm) [7]. [13] and [14] reported that a fibre Runkel ratio less than one (1) is an indication of suitability of the wood for paper making.

The flexibility coefficient for both samples was 50%, though this was lower than the range of 55–70% for most softwood as recorded for *Vitex doniana* [12]. It is expected that paper made from these fibres would have considerable interfibre bonding and hence great tensile strength. The slenderness ratio (SR) for the coconut fibre was 0.031 which was higher than that of palm fruit seed hairs with SR = 0.024. The results indicated that both samples are suitable for pulp and paper production since they fall within the accepted standard.

The novel research findings reported in this paper support the development of smart sustainable manufacturing infrastructure [16]. With advanced satellite-cellular convergence communication technologies [17–21], integrated Internet of Things sensors can be utilized to achieve precision pulp and paper production. The smart IoT sensors can provide the plant fibre's dielectric profiles at different stages of the production process from the agro-waste site to the final product. This would enable a holistic production, product quality, distribution, utilization, and recycling lifecycle analytics.

4 Conclusions

The results showed that coconut fibre and palm fruit seed hairs which are abundant environmental wastes in Nigeria contain cellulose, hemicellulose, and other fibre properties.

The results of physiochemical properties indicated that both fibres are suitable as pulping raw materials in paper industry having contained high cellulose and hemicellulose properties and their solubility is within the acceptable range for non-wood pulping raw materials. The values of fibre properties of coconut fibre and palm fruit seed hairs lie between that of hardwood and softwood; therefore, the fibres become good pulping intermediate material having had both properties for high strength paper production. It is also economically viable for developing countries since sophisticated equipment and expensive and hazardous chemicals are not required.

References

1. Ayeni, A.O., Omoleye, J.A., Mudliar, S., Hymore, F.K., Pandey, R.A.: Utilization of lignocellulosic water for ethanol production: enzymatic digestibility and fermentation of pretreated sheatree salidust. Korean J. Chem. Eng. **31**(7), 1180–1186 (2014)
2. Akpabio, U.D., Umoren, S.A., Umoudoh, A.J.: Conversion of agricultural waste into cellulose derivatives. Bull. Pure Appl. Sci. **23**(3), 9–13 (2004)
3. Hurter, A.M.: Utilization of annual plants and agricultural residues for the production of pulp and paper. In: Proceeding of Technical Association of Pulp and Paper Industry (TAPPI) Pulping Conference, pp. 124–129 (2011)
4. Bhat, J.V., Nambudiri, E.A.: The uniquity of coir retting. Microbial degradation substance. J. Sci. India. **5**(30), 17–28 (2005)
5. Thampan, P.K.: The Coconut Palm and Its Products, pp. 268–281. Green Villa Publishing House, Kerala (2006)
6. Saka, S., Munusamy, M.V., Shibata, M., Tono, Y., Miyafuji, H.: Chemical constituents of the different anatomical parts of the oil palm (*Elaeis guineensis*) for their sustainable utilization. Nat. Resour. Energy Environ. **3**(10), 606–8501 (2008)
7. Akpakpan, A.E., Akpabio, U.D., Obot, I.B.: Evaluation of physiochemical properties and soda pulping of *Nypa fruticans* frond and petiole. Elixir Appl. Chem. **45**, 7664–7668 (2012)
8. Udo, I.E., Obuzor, G.U., Horsfall Jnr, M.: Proximate and phytochemical studies of exudate of *Dacryodes edulis*. Res. J. Chem. Sci. **6**, 1–9 (2016)
9. Technical Association of the Pulp and Paper Industry (TAPPI). One Person Sodium Hydroxide Solubility of Wood and Pulp. T212 om 02: 1–5 (2002)
10. AOAC: Official Method of Analysis, 16th edn, pp. 327–332. Association of Official Analytical Chemist, Washington, DC (2002)
11. Israel, A.U., Ogali, R.E., Akaranta, O., Obot, I.B.: Extraction and characterization of coconut (*Cocos nucifera L.*) coir dust. Songklanakarin J. Sci. Technol. **33**(6), 717–772 (2011)
12. Gumuskaya, E., Kirci, H., Usta, M.: The effect of various pulping conditions on crystalline structure of cellulose in cotton linters. Polym. Degrad. Stab. **81**(3), 559–564 (2002)
13. Ojo, A.R., Alao, O.J., Onilearo, K.S., Asinwa, I.O., Badejo, B.Y., Bello, F.B.: Evaluation of selected fibre characteristics of the wood of *Leucaena leucocephala* (Lam) Dewit along the vertical axis. Obeche J. **27**(1), 34–37 (2008)
14. Ogunjobi, K.M., Adetogun, A.C., Omole, A.O.: Assessment of variation in the fibre characteristic of the wood of *Vitex doniana* sweet and its suitability for paper production. J. Res. For. Wildl. Environ. **6**(1), 2141–1778 (2013)
15. Anon, A.: Pulp and paper. West Afr. Technol. Revolut. **7**(3), 45–47 (2007)
16. Okon, B., Ekpo, S., Elhag, T.: A sustainable engineering infrastructure model for the 21st century. In: Proceedings of the World Congress on Engineering 2010, vol. II, pp. 1101–1105, London, UK, 30 June–2 July 2010 (2010)
17. Ekpo, S., George, D.: A deterministic multifunctional architecture design for highly adaptive small satellites. Int. J. Satell. Commun. Policy Manag. **1**(2/3), 174–194 (2012)
18. Ekpo, S., George, D.: A system engineering consideration for future-generations small satellites design. In: Proc. First IEEE European Satellite Telecoms Conference, pp. 1–4, Rome, Italy (2012)
19. Ekpo, S.C.: Thermal subsystem operational times analysis for ubiquitous small satellites relay in LEO. Int. Rev. Aerosp. Eng. J. **11**(2), 48–57 (2018)
20. Ekpo, S., Kettle, D.: Mm-wave LNAs design for adaptive small satellite applications. In: Proc. Joint 5th ESA Workshop on mmWave and 31st ESA Antenna Workshop, pp. 843–847, Noordwijk, Netherlands (2009)
21. Ekpo, S., George, D.: Reconfigurable cooperative intelligent control design for space missions. Recent Pat. Space Technol. **2**(1), 2–11 (2012)

Improving the Perception of Moronic Systems Whether Human or Self-Driving Vehicles Using Intelligent Agents: A Review

Fakhra Riaz, Faisal Riaz, Asma Jabeen, Umar Raza, William Holderbaum, Sidra Kouser, and Saeid Jamali

1 Introduction

Moronic systems allude to systems that are unequipped for wise decision production because of an absence of cutting-edge mental capacities. These systems are not up to date to meet the current standards and the requirements of machines that lack intelligent perception with these systems. Their perception capabilities can be enhanced by intelligent agents that can perceive and act based on their environment, so they are used to enhance the capacity for decision-making and overall performance [1]. According to Holstein et al. [2], intelligent agents can enable moronic systems to better perceive their environment and make more informed decisions by utilizing advanced sensing and cognitive capabilities.

F. Riaz (✉) · F. Riaz · A. Jabeen
Department of Computer Science and Information Technology, Mirpur University of Science and Technology, Mirpur, Pakistan
e-mail: fakhra.csit@must.edu.pk; faisal.riaz@must.edu.pk; asma.jabeen@must.edu.pk

U. Raza · W. Holderbaum
Department of Engineering, Manchester Metropolitan University, Manchester, UK
e-mail: u.raaza@mmu.ac.uk; w.holderbaum@mmu.ac.uk

S. Kouser
Medical Group of Health and Wellness, Barcelona, Spain

S. Jamali
Intellithing Lab Xyz Building, Level 1, Manchester, UK
e-mail: saeid@intellithing.tech

S. C. Ekpo (ed.), *The Second International Adaptive and Sustainable Science, Engineering and Technology Conference*, Signals and Communication Technology,
https://doi.org/10.1007/978-3-031-53935-0_22

The development of intelligent agents has changed the way we interact with technology in recent years. From voice assistants to self-driving vehicles, intelligent agents are progressively turning into a piece of our regular routines [3]. However, there is a rising worry regarding intelligent agents' capacity to effectively sense their environment as their use increases [4]. The safe and efficient functioning of intelligent agents, whether they are human-inspired or self-driving cars, depends on their ability to perceive accurately in the context of moronic systems [5].

To ensure the safe and reliable functioning of self-driving cars, it is imperative to change how people view them [6]. Intelligent agents can be used to improve the perception of self-driving cars by offering real-time data processing and decision-making skills that can help in recognizing possible risks and averting accidents [7, 8]. Like human-inspired moronic systems, intelligent agents can be used in gaming, mental health monitoring, and human-robot interaction to improve emotion recognition and support healthcare decision-making. Hassouneh et al. [9] discussed how the emotion recognition system was built with convolutional neural networks (CNN), and the architecture of the deep neural network was also discussed.

In this review paper, we explore the latest research and advancements in intelligent agents for improving the perception of moronic systems, including their application in self-driving vehicles and human-inspired systems.

The rest of the paper is structured as follows: In Sect. 2, the taxonomy of the paper is presented. Section 3 presents the literature review related to neural network techniques for object detection in autonomous vehicles (AVs) and for the diagnosis of diseases (human-inspired perception). Open research problems have been presented in Sect. 4. Furthermore, Sect. 5 concludes the paper.

2 Taxonomy

Taxonomy represents the neural network techniques used for human and autonomous vehicles. The first node represents the agent-based perception of moronic systems, which are divided into two nodes named self-driving vehicle perception and human-inspired perception. The self-driving vehicle's perception is further divided into neural network-based techniques. The neural network-based techniques are further subdivided into convolutional neural networks (CNN), region-based convolutional neural networks (R-CCN), faster R-CNN, single-shot detector (SDD), and You Only Look Once (YOLO). Neural network-based approaches are further split into human-inspired perception categories. Convolutional neural networks (CNNs) and recurrent neural networks (RNNs) are two categories of neural network-based approaches (Fig. 1).

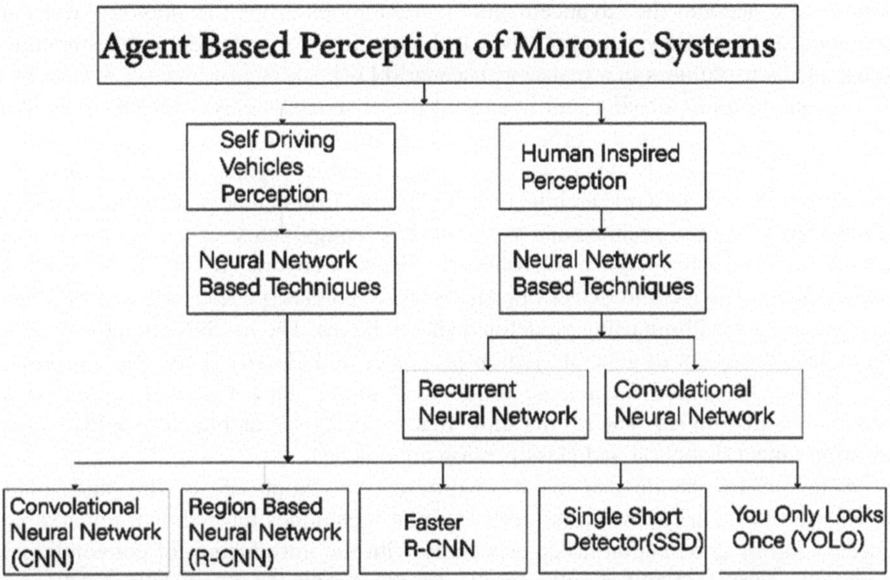

Fig. 1 Agent-based perception of moronic system used for AVs and humans

3 Agent-Based Perception of Moronic Systems

3.1 Neural Network Techniques for Object Detection in Autonomous Vehicles (AVs)

A broad range of strategies have been presented in the literature to address the well-researched computer vision topic of deep learning object detection. Convolutional neural networks (CNNs), regional convolutional neural networks (R-CNN), faster R-CNN, single-shot detector (SSD), and You Only Look Once (YOLO) are some of the most well-liked deep learning object recognition methods that have been developed in the past.

ANNs are used for a variety of tasks related to AV driving, ranging from environmental sensing and perception to decisions and control [10]. Adu-Gyamfi et al. [11] developed a solution for vehicle detection using convolutional neural networks (CNNs), which is a difficult problem to solve. The authors seek to develop an automated system that can identify automobiles with high efficiency and accuracy for a variety of uses, including security and surveillance. In order to teach CNN features that are helpful for vehicle recognition, the suggested method uses a huge collection of vehicle image examples. To improve performance, the authors tweak a pre-trained CNN using their dataset. On two benchmark datasets, the authors test their methodology and find that it performs better than others in terms of accuracy.

Taking into account the advancements in autonomous driving technology over the previous 20 years, they also apply their technique to a traffic camera video stream to show off its usefulness in a real-world scenario [12].

For autonomous driving and advanced driver assistance systems (ADAS), the challenge of perceiving the urban road environment is important. The perception system must accurately detect and recognize various objects. In [13], Lin et al. describe a neural network architecture for the environmental perception of urban roads. The proposed architecture integrates two components: an anchorless object detector and a region-based convolutional neural network (R-CNN). Anchorless object detector is used to detect objects such as vehicles, pedestrians, and bicycles in input images without using predefined anchor boxes. The R-CNN component uses the detected objects to generate region proposals and classify them. The proposed architecture is trained on a large urban road image dataset and evaluated on a separate validation set. The results show that the proposed architecture outperforms existing object detection and classification approaches.

Autonomous driving can make no difference without safe driving and smart navigation [14]. In recent years, deep learning techniques have shown remarkable success in object detection tasks, especially with the introduction of convolutional neural networks (CNNs) in autonomous driving. Faster R-CNN is one of the most popular CNN-based object detection methods, which achieves high accuracy and efficiency in object detection. Karthikeyan et al. [15] propose an automated object detection system using convolutional neural networks based on faster regions (faster R-CNN). The proposed system consists of three main stages: preprocessing, training, and testing. In the preprocessing step, the input images are resized and annotated using bounding boxes to indicate the locations of the mechanical fasteners. The faster R-CNN model is then trained on the annotated dataset to learn features and classify objects. During the testing phase, the trained model detects and classifies mechanical fixations in new input images. Drivers have a very limited amount of time to make a rapid and accurate judgment in order to avoid a road calamity [16].

Single-shot detectors (SSD), a common class of object detection techniques that can identify objects in real time, are widely used for this purpose. For object detection in temporal applications, Kumar et al. [17] provide an enhanced version of the single-shot multi-box detector (SSD) technique. The SSD algorithm is a well-known object identification system that predicts item locations and classifications using a single convolutional neural network (CNN). The suggested method is tested against several benchmark datasets, and the results demonstrate that it performs excellent object detection and classification efficiency and accuracy, particularly in real-time applications. The proposed method is also thoroughly analyzed by the authors, who also compare it to other cutting-edge object identification techniques to show how much more accurate and effective it is.

Identifying and finding numerous things, such as vehicles, people, and traffic signs in real time, are crucial tasks in autonomous driving. The YOLO (You Only Look Once) algorithm, a well-known deep learning-based object identification technique, can identify objects in real time. According to Abhishek Sarda et al. [18], an autonomous vehicle object identification system utilizing the YOLO (You

Table 1 Neural network techniques for object detection in autonomous vehicles

References	Contribution	Model	Type of detection	Dataset
Adu-Gyamfi et al. [11]	Automated vehicle recognition with deep CNN	CNN	Vehicle detection	Vehicle Image Database (VID), CompCars dataset
Lin et al. [13]	An anchor-free detector and R-CNN integrated neural network architecture for the environmental perception of urban roads	R-CNN	Vehicles, pedestrians, and bicycles	KITTI dataset, BDD100K dataset, and Waymo Open Dataset
Karthikeyan et al. [15]	Automated object detection of mechanical fasteners using faster region-based CNN	Faster R-CNN	Automated object detection	Pascal VOC 2012
Kumar et al. [17]	Object detection in real time based on an improved single-shot multi-box detector algorithm	SSD	Moving objects detection	Pascal VOC and common objects in context (COCO)
Sarda et al. [18]	Object detection for autonomous driving using YOLO	YOLO	Object detection	MS COCO

Only Look Once) method. For autonomous driving systems, object detection is essential because it enables the car to recognize and avoid obstacles, pedestrians, and other cars on the road. Preprocessing, training, and testing are the three primary stages of the suggested system. The input photos are scaled and annotated with bounding boxes to show where objects are during the preprocessing stage.

In general, neural network methods can be an effective tool for enhancing the perceptual capacities of autonomous cars. By using intelligent agents and deep learning algorithms, autonomous vehicles can better understand their environment and make informed decisions in real time (Table 1).

3.2 Neural Network Techniques for the Diagnosis of Autism Spectrum Disorder

The complicated neurobehavioral illness known as autism spectrum disorder (ASD) impacts linguistic and behavioral abilities, as well as communication, and starts in infancy and persists throughout adulthood. To achieve the greatest prediction outcomes for autism spectrum disorder, artificial intelligence (AI) approaches and machine learning (ML) algorithms played a crucial role in considerably supporting medical and healthcare professionals. Because it is challenging to diagnose autism in its early stages, experts have intensified their hunt for the most effective therapies.

ASD Detection

ASD is a neurodevelopmental disorder that has an impact on social interaction, communication, and behavior. ASD patients' outcomes can only be improved through early diagnosis and treatment. Zeinab et al. [19] explored the use of convolutional neural networks (CNNs) for the automated detection of autism spectrum disorder (ASD) based on facial images.

The researchers found several facial traits that were substantially related to ASD using a CNN to evaluate the facial images of the subjects. Also, the researchers discovered facial characteristics that were highly linked to ASD, including eye distance and brow shape.

Behavioral observations and evaluations are currently the basis for the diagnosis of autism spectrum disorder (ASD), which can be time-consuming and may result in a misdiagnosis or a delayed diagnosis. Utilizing machine learning, particularly neural networks, to aid in the diagnosis of ASD has received increasing interest. One promising way to deal with ASD findings is by utilizing convolutional neural networks.

CNN is a kind of deep learning system that can identify correlations and patterns in medical pictures. Rasha Kashef [20] put out a novel deep learning architecture. In order to extract meaningful features from the MRI scans, this architecture consists of two stages: a feature extraction stage that uses convolutional neural network (CNN) layers, and a classification stage that uses a fully connected layer to classify the extracted features as either ASD or non-ASD.

By using a type of genetic variation single-nucleotide polymorphisms, Soudeh Ghafouri-Fard et al. [21] investigated the use of artificial neural networks (ANNs) for diagnosing ASD (SNPs). A small sample of 31 individuals with ASD and 34 healthy controls participated in the study. To categorize people with ASD, the researchers employed SNP data from a genotyping array using ANN analysis. The researchers also discovered SNPs that were very significantly linked to ASD.

The use of deep neural networks (DNNs) for the classification of adults with autism spectrum disorder (ASD) based on their resting-state functional magnetic resonance imaging (fMRI) data was examined by the authors in this study [22]. In this study, 79 healthy controls and 76 individuals with ASD participated. To categorize people, the researchers employed a DNN to extract the functional connection patterns of each participant's brain regions. The researchers also pinpointed brain areas that had a high correlation with ASD.

ASD Treatments

Conventional ASD treatments like behavioral and speech therapy can be costly and time-consuming. In recent years, researchers have begun investigating whether video games may be used to treat ASD. Video games might replace traditional ASD therapies in an entertaining, affordable, and effective way. Video games have the potential to enhance the lives of those with ASD according to Jiménez-Muoz et al. [23]. The evaluation assesses how well commercial, serious, and virtual reality video games utilized in this research have improved the cognitive, social, and

Table 2 Neural network techniques for the diagnosis of autism spectrum disorder

References	Contribution	Model	Parameters	Dataset	Accuracy (%)
Zeinab et al. [19]	CNN for the automated detection of ASD based on facial images	CNN	Phenotypic information	ABIDE	70
Rasha Kashef [20]	Using structural MRI images, a revolutionary deep learning architecture is used	CNN	Brain activation patterns	ABIDE	80
Ghafouri-Fard et al. [21]	Using single-nucleotide polymorphisms, ANNs are used for diagnosing ASD	ANN	Genomic data	NA	73
Faiz Misman et al. [22]	The DNNs for the classification of adults with ASD are based on their resting state	DNN	24 features	UCI	96

behavioral skills of individuals with ASD. The research suggested that playing video games might improve the cognitive, social, and behavioral abilities of ASD patients, making them a useful therapeutic tool (Table 2).

4 Open Research Problems

In this section, we have discussed open research problems for assisting the researcher working in autonomous vehicles and health care:

- How can intelligent agents be designed to learn from human feedback and adapt their behavior accordingly to improve the perception of moronic systems?
- How can the design and implementation of intelligent agents be optimized to reduce the cognitive load on human operators and improve their overall performance when working with moronic systems in human or self-driving vehicles?
- What are the most effective methods for designing and implementing interfaces between human operators and intelligent agents, particularly in safety-critical scenarios and in human or self-driving vehicles with moronic systems?

5 Conclusion

Moronic systems lack the sophisticated cognitive abilities needed to make well-informed choices. These systems, on the other hand, can now make decisions based on more information for the development of intelligent agents. As self-driving cars become more prevalent in everyday life, accurate perception of intelligent agents is crucial for their safe and efficient operation. By providing capabilities for real-time data analysis and decision-making, intelligent agents can enhance the perception of these vehicles, identify potential dangers, and prevent accidents. In addition, intelligent agents can improve emotion recognition and support healthcare decision-making in a variety of applications for moronic systems based on humans. It has been demonstrated that the accuracy and performance of these systems are enhanced when they are constructed using machine learning and deep neural network methods.

References

1. Bredeweg, B., Kragten, M.: Requirements and challenges for hybrid intelligence: a case study in education. Front. Artif. Intell. **5**, 891630 (2022)
2. Holstein, K., Aleven, V., Rummel, N.: A conceptual framework for human–AI hybrid adaptivity in education. In: Lecture Notes in Computer Science, pp. 240–254. Springer, Cham (2020)
3. Anagnostou, M., Karvounidou, O., Katritzidaki, C., Kechagia, C., Melidou, K., Mpeza, E., Konstantinidis, I., Kapantai, E., Berberidis, C., Magnisalis, I., Peristeras, V.: Characteristics and challenges in the industries towards responsible AI: a systematic literature review. Ethics Inf. Technol. **24**(3), 1–18 (2022)
4. Kaplan, A., Haenlein, M.: Rulers of the world, unite! The challenges and opportunities of Artificial Intelligence. Bus. Horiz. **63**(1), 37–50 (2020)
5. Ma, Y., Wang, Z., Yang, H., Yang, L.: Artificial intelligence applications in the development of autonomous vehicles: a survey. IEEE/CAA J. Automat. Sin. **7**(2), 315–329 (2020). https://doi.org/10.1109/JAS.2020.1003021
6. Li, Z., Li, X., Jiang, B.: How people perceive the safety of self-driving buses: a quantitative analysis model of perceived safety. Transp. Res. Rec.: J. Transp. Res. Board. **2677**(1), 1356–1366 (2022)
7. Large, R., Harrington, K., Burnett, G., Luton, J., Thomas, P., Bennett, P.: To please in a pod. In: Proceedings of the 11th International Conference on Automotive User Interfaces and Interactive Vehicular Applications. Association for Computing Machinery, New York (2019)
8. Siegel, J., Pappas, G.: Morals, ethics, and the technology capabilities and limitations of automated and self-driving vehicles. AI & Soc. **38**(1), 213–226 (2021)
9. Hassouneh, A., Mutawa, A.M., Murugappan, M.: Development of a real-time emotion recognition system using facial expressions and EEG based on machine learning and deep neural network methods. Inform. Med. Unlocked. **20**, 100372 (2020)
10. Saleem, H., et al.: Evaluating the role of neural networks and cyber security for the development of next-generation autonomous vehicles: a survey. In: Proceedings of the 4th International Electrical Engineering Conference. IEP Karachi, Karachi (2019)
11. Adu-Gyamfi, Y.O., Asare, S.K., Sharma, A., Titus, T.: Automated vehicle recognition with deep convolutional neural networks. Transp. Res. Rec.: J. Transp. Res. Board. **2645**(1), 113–122 (2017)
12. Rasib, M., Butt, M.A., Khalid, S., Abid, S., Raiz, F., Jabbar, S., Han, K.: Are self-driving vehicles ready to launch? An insight into steering control in autonomous self-driving vehicles. Math. Probl. Eng. **2021**, 1–22 (2021)
13. Lin, C., Shi, Y., Zhang, J., Xie, C., Chen, W., Chen, Y.: An anchor-free detector and R-CNN integrated neural network architecture for the environmental perception of Urban Roads. Proc. Inst. Mech. Eng., Part D: J. Automob. Eng. **235**(12), 2964–2973 (2021)
14. Butt, M.A., Riaz, F.: Carl-D: a vision benchmark suite and large-scale dataset for vehicle detection and scene segmentation. Signal Process. Image Commun. **104**, 116667 (2022)
15. Karthikeyan, M., Subashini, T.S.: Automated object detection of mechanical fasteners using faster region-based convolutional neural networks. Int. J. Electr. Comput. Eng. **11**(6), 5430 (2021)

16. Riaz, F., Shah, S.I., Raees, M., Shafi, I., Iqbal, A.: Lateral pre-crash sensing and avoidance in emotion enabled cognitive agent-based vehicle-2-vehicle communication system. Int. J. Commun. Netw. Inf. Secur. **5**(2), 127–140 (2022)
17. Kumar, A., Zhang, Z.J., Lyu, H.: Object detection in real-time based on improved single shot multi-box detector algorithm. EURASIP J. Wirel. Commun. Netw. **2020**(1), 204 (2020)
18. Sarda, A., Dixit, S., Bhan, A.: Object detection for autonomous driving using Yolo [you only look once] algorithm. In: 2021 Third International Conference on Intelligent Communication Technologies and Virtual Mobile Networks (ICICV), Tirunelveli, India (2021)
19. Sherkatghanad, Z.: Automated detection of autism spectrum disorder using a convolutional neural network. Front. Neurosci. **13**, 1325 (2020). https://doi.org/10.3389/fnins.2019.01325
20. Kashef, R.: ECNN: enhanced convolutional neural network for efficient diagnosis of autism spectrum disorder. Cogn. Syst. Res. **71**, 41–49 (2022). https://doi.org/10.1016/j.cogsys.2021.10.002
21. Ghafouri-Fard, S., Taheri, M., Omrani, M.D., Daaee, A., Mohammad-Rahimi, H., Kazazi, H.: Application of single-nucleotide polymorphisms in the diagnosis of autism spectrum disorders: a preliminary study with artificial neural networks. J. Mol. Neurosci. **68**(4), 515–521 (2019). https://doi.org/10.1007/s12031-019-01311-1
22. Misman, M.F.: Classification of adults with autism spectrum disorder using deep neural network. In: 2019 1st International Conference on Artificial Intelligence and Data Sciences (AiDAS) (2019). https://doi.org/10.1109/aidas47888.2019.8970823
23. Jiménez-Muñoz, L., Peñuelas-Calvo, I., Calvo-Rivera, P., Díaz-Oliván, I., Moreno, M., Baca García, E., Porras-Segovia, A.: Video games for the treatment of autism spectrum disorder: a systematic review. J. Autism Dev. Disord. **52**(1), 169–188 (2021)

Design and Carbon Credit Evaluation of a Grid-Connected Solar Photovoltaic Power System for Campus Streetlight: Case Study of the University of Uyo Main Campus, Akwa Ibom State

Iniobong Edifon Abasi-obot, Eduediuyai Ekerette Dan, and Mfonobong Umoren

1 Introduction

In recent times, there has been an increasing response to the call for the adoption of green energy across Nigeria. Accordingly, many organizations in Nigeria are increasingly adopting solar power supply systems to meet some of their energy needs [1–3]. Some of the universities in Nigeria are using solar power for powering campus streetlights, as well as for powering laboratories and resources centers [4]. As such, this paper seeks to design and evaluate grid-connected solar photovoltaic (SPV) power system for a case study campus streetlight for the University of Uyo main campus.

The study focuses on the determination of the yearly energy yield, the monthly distribution of the energy yield, the system performance ratio, and the corresponding carbon credit associated with the energy yield of the SPV [5–7]. Specifically, analytical expressions used to determine the daily load demand and carbon credit of the SPV power system are presented, while the design of the grid-connected SPV, determination of the energy yield, and performance ratio are done using PVSyst simulation software [8, 9]. While most of the related works on street light solar

I. E. Abasi-obot (✉)
Department of Electrical and Electronic Engineering, Akwa Ibom State University, Ikot Akpaden, Akwa Ibom State, Nigeria
e-mail: iniobongabasiobot@aksu.edu.ng

E. E. Dan
Department of Computer Engineering, Federal Polytechnic, Ukana, Akwa Ibom State, Nigeria

M. Umoren
Department of Electrical/Electronic Engineering, University of Uyo, Uyo, Akwa Ibom State, Nigeria

S. C. Ekpo (ed.), *The Second International Adaptive and Sustainable Science, Engineering and Technology Conference*, Signals and Communication Technology,
https://doi.org/10.1007/978-3-031-53935-0_23

power system design focuses on using already defined lamp wattage to size the solar power, the approach used in this paper is to first determine the required lamp wattage based on the required illumination level along with the road and lamp parameters. This approach makes it easier to adapt the design to different street- and illumination-level requirements.

2 Methodology

The details of the design and carbon credit evaluation for a grid-connected solar photovoltaic (SPV) power system for a campus streetlight are presented. The streetlight is designed for the main campus of the University of Uyo. The Google Maps distance measurement tool is used to determine the streetlight path length, as shown in Fig. 1. According to Fig. 1, the campus streetlight has a total path length of 3.45 km. The analytical expressions used to determine the daily load demand of the campus streetlight are shown in Eqs. 1, 2, 3, 4, 5, and 6, while the parameter values used to determine the daily load demand are presented in Table 1.

The distance between the streetlight poles in m is denoted as DbtwPl and it is given as [10]:

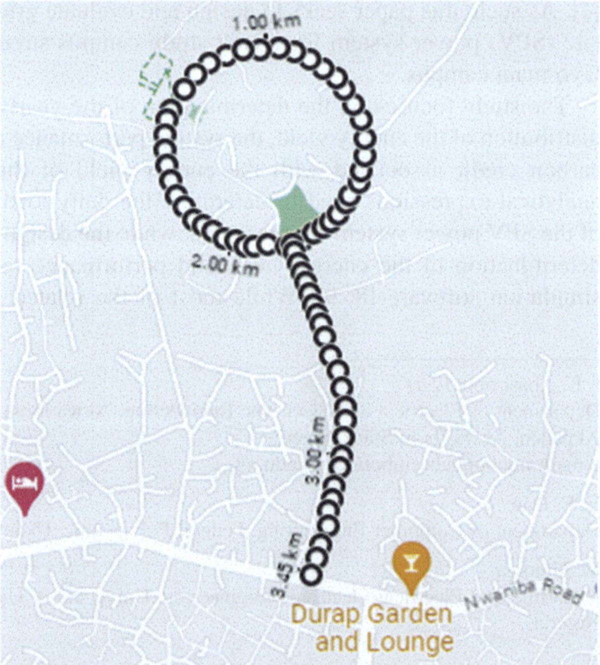

Fig. 1 The Google Maps distance measurement of the campus streetlight path length

Table 1 The design parameters of the campus streetlight

S/N	Parameter description	Parameter value
1	Pole height in m	8.3
2	Space-to-height ratio	2.8
3	Required illumination lux level for the street in lux/m^2 [10]	40
4	Road width in m	7
5	Coefficient of utilization [10]	0.77
6	Maintenance factor [10]	0.85
7	Luminous efficiency in lux/watt [1]	125
8	Distance between the streetlight poles in m	23.2
9	Average required lumen of the lamp in lumens	9942.2
10	Wattage of each streetlight per pole in watts	$79.5 \approx 80$
11	Streetlight path length (m) as shown in Fig. 1	3.5
12	Number of streetlight pole required	151
13	Total wattage of streetlight in kW	12.08
14	Number of operation hours per day	12
15	Daily energy demand of the streetlight in kWh	144.96

$$\mathrm{DbtwPl} = \mathrm{PHt}\ (\mathrm{PSpHtRt}) \tag{1}$$

where the pole height in m is denoted as PHt and the space-to-height ratio is denoted as PSpH tRt, the required illumination lux level for the street in lux/m^2 is denoted as RqLxLvl, the road width in m is denoted as RdW, the coefficient of utilization is denoted as CoU, the maintenance factor is denoted as Mf, and the average required lumen of the lamp in lumen is denoted as AL where:

$$\mathrm{AL} = \frac{\mathrm{RqLxLvl}\,(\mathrm{DbtwPl})\,(\mathrm{RdW})}{\mathrm{CoU}\,(\mathrm{Mf})} \tag{2}$$

The wattage of each streetlight in watts is denoted as WL and it is given as:

$$\mathrm{AL} = \frac{\mathrm{RAL}}{\mathrm{Leff}} \tag{3}$$

where the luminous efficiency in lux/watt is denoted as Leff. The total number of streetlight pole required is denoted as Np and it is given as:

$$\mathrm{Np} = \frac{\text{Street light path length in km}^{*}1000}{\mathrm{DbtwPl}} \tag{4}$$

The total wattage of streetlight in kW is denoted as WTTL and it is given as:

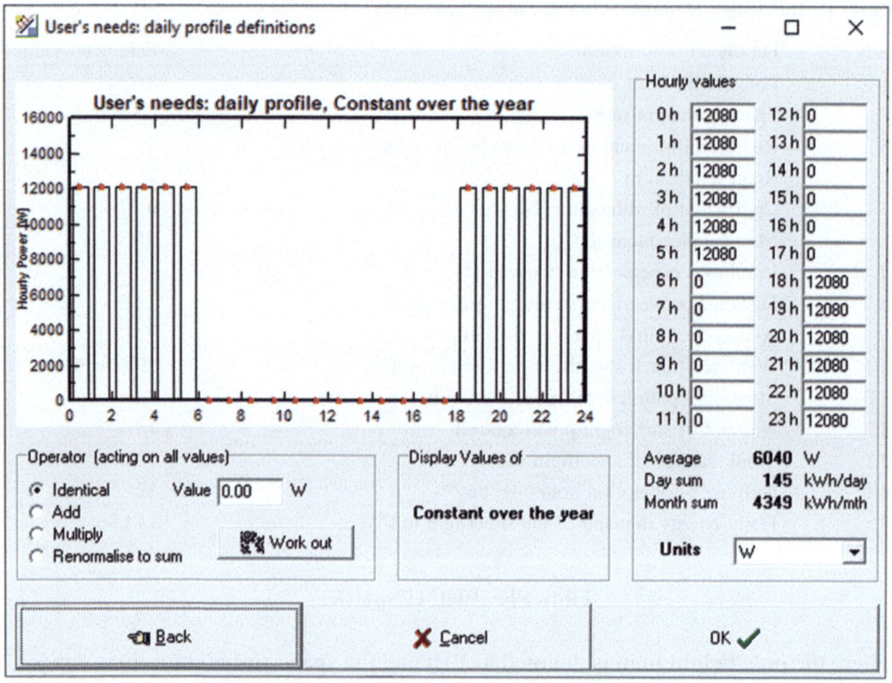

Fig. 2 The screenshot of the daily load profile of the campus streetlight as modeled in PVSyst simulation software

$$\text{WTTL} = \text{WL}(\text{Np}) \tag{5}$$

The daily energy demand of the streetlight in kWh is denoted as Edmdy and it is given as:

$$\text{Edmdy} = \text{WTTL}(\text{Nhpd}) \tag{6}$$

where Nhpd is the number of operation hours per day.

The case study of the University of Uyo main campus is located in Nsukara Offot in Uyo local government area of Akwa Ibom State with 1717.2 kWh/m^2 yearly solar irradiation on the horizontal plane and mean ambient temperature of 24.9 °C. The PVSyst simulation software is used to design the grid-connected SPV power from which the annual energy yield is determined. The screenshot of the daily load profile of the campus streetlight as modeled in PVSyst simulation software is presented in Fig. 2, while Fig. 3 shows the screenshot of the configurations of the PV modules and inverter used in the SPV system design.

The carbon credit is determined from the yearly energy yield of the SPV system. Specifically, SPV with annual energy yield is denoted as Ey_{yr} in MWh, CO_2 emission reduction factor as CORF with a value of 0.932 tons [11], CO_2 emission

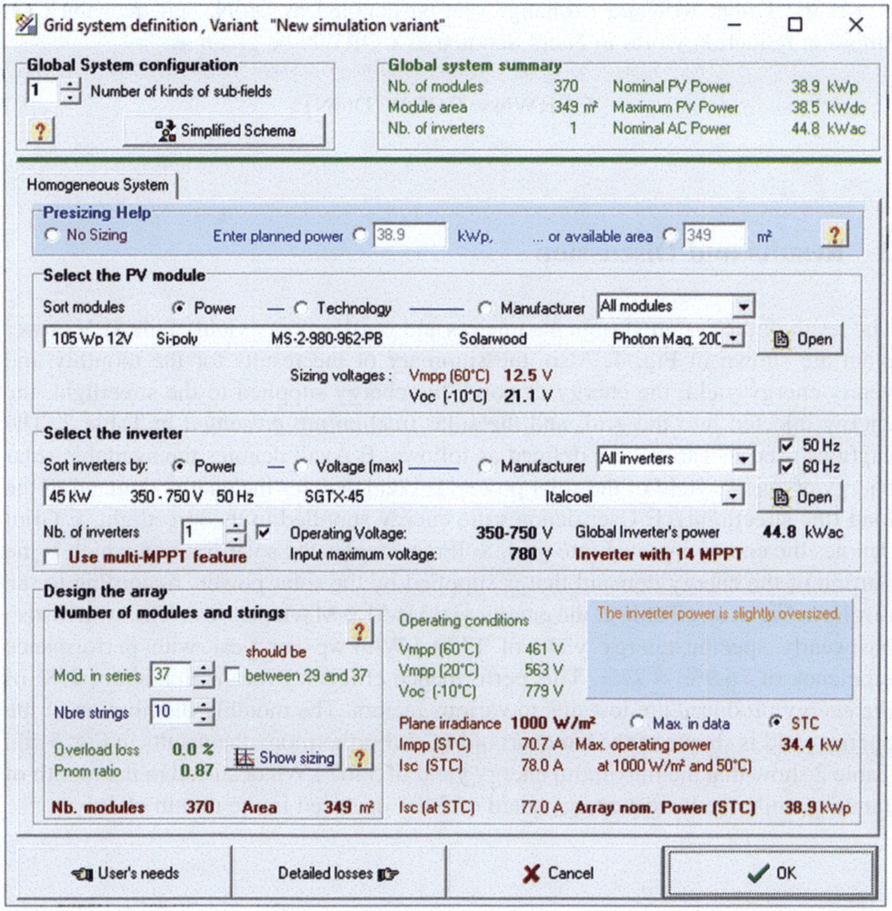

Fig. 3 The screenshot of the configurations of the PV modules and inverter used in the SPV system design

reduction trade value as COTRDV with a value of USD 27.5 [10–12], and CO_2 emission reduction value as CORV. The value of CORV is computed as follows [10, 11]:

$$\mathrm{CORV} = \mathrm{Ey}_{\mathrm{yr}}(\mathrm{CORF}) \tag{7}$$

Also, the CO_2 emission reduction worth denoted as CORW is given as:

$$\mathrm{CORW} = \mathrm{CORV}(\mathrm{COTV}) = \mathrm{Ey}_{\mathrm{yr}}(\mathrm{CORF})(\mathrm{COTV}) \tag{8}$$

Let the Dollar to Naira exchange rate be denoted as DtoN, and then the CO_2 emission reduction worth in Naira denoted as CORWN be given as:

$$\mathrm{CORWN} = \mathrm{CORW}(\mathrm{DtoN}) \tag{9}$$

3 Results and Discussion

The result of the system design parameters and yearly energy yield of the SPV power plant are shown in Fig. 4. Also, the summary of the results for the monthly and yearly energy yield, the energy demand, the energy supplied to the streetlight, the energy injected into the grid, and the solar fraction are presented in Table 2. The captions used in Table 2 are defined as follows: E Avail denotes the available solar energy or energy yield of the solar power. E Load denotes the energy demand of the load (the streetlight). E User denotes the energy supplied to the streetlight. E Gridl denotes the energy injected into grid. SolFrac denotes the solar fraction which is the portion of the energy demand that is supplied by the solar power. According to the results in Fig. 4 and Table 2, the energy yield is 51.6 MWh per year. The system also has yearly specific energy yield of 1329 kWh/kwp per year with performance efficiency of 76.9% ≈ 77%. This performance efficiency means that about 23% of the energy produced are lost due to various factors. The monthly distribution of the energy yield is shown in the bar chart of Fig. 5. Furthermore, the results in Fig. 5 and Table 2 show that the maximum energy yield of 5462 kWh occurred in the month of January while the lowest energy yield of 3401 occurred in the month of July.

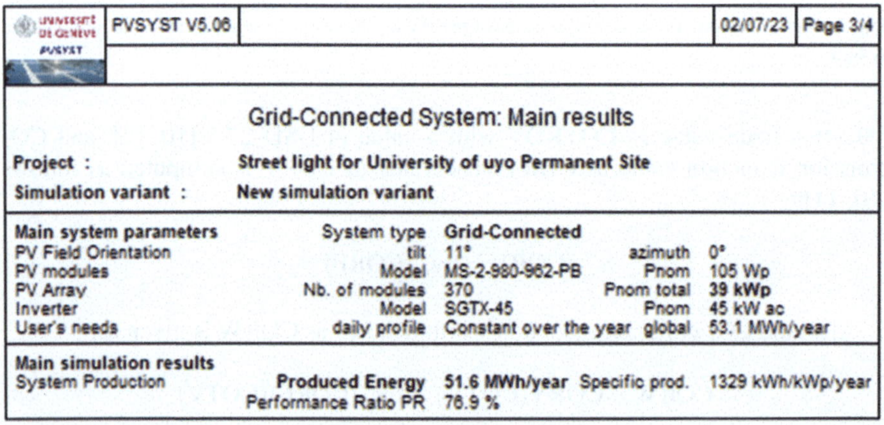
PVSYST V5.06 02/07/23 Page 3/4

Grid-Connected System: Main results

Project : Street light for University of uyo Permanent Site
Simulation variant : New simulation variant

Main system parameters	System type	**Grid-Connected**		
PV Field Orientation	tilt	11°	azimuth	0°
PV modules	Model	MS-2-980-962-PB	Pnom	105 Wp
PV Array	Nb. of modules	370	Pnom total	**39 kWp**
Inverter	Model	SGTX-45	Pnom	45 kW ac
User's needs	daily profile	Constant over the year	global	53.1 MWh/year
Main simulation results				
System Production	**Produced Energy**	**51.6 MWh/year**	Specific prod.	1329 kWh/kWp/year
	Performance Ratio PR	76.9 %		

Fig. 4 The cut section of the screenshot of the system design parameters and yearly energy yield of the SPV power plant

Table 2 The summary of the results for the monthly and yearly energy yield energy demand and energy supplied

Month	E Avail (kWh)	E Load (kWh)	E User (kWh)	E Grid (kWh)	SolFrac
January	5462	4494	0.038	5462	1.216
February	4796	4059	4.390	4792	1.182
March	4932	4494	2.093	4930	1.097
April	4401	4349	0.000	4401	1.012
May	4114	4494	0.000	4114	0.916
June	3661	4349	1.469	3660	0,842
July	3401	4494	2.022	3399	0.757
August	3428	4494	1.374	3427	0.763
September	3522	4349	0.000	3522	0.810
October	4049	4494	0.000	4049	0.901
November	4545	4349	0.000	4545	1.045
December	5305	4494	0.000	5305	1.181
Year	51,618	52,910	11.386	51,607	0.976

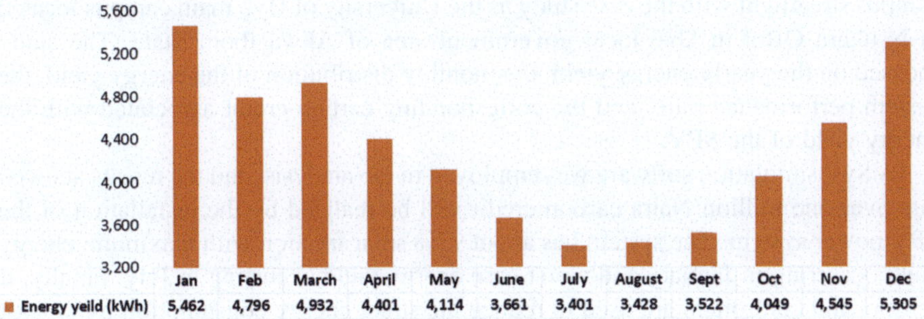

Fig. 5 The monthly distribution of the energy yield of the SPV power system

Table 3 The summary of the results for carbon credit analysis

S/N	Parameter description	Value
1	Yearly energy yield (MWh/year)	51.6
2	CO2 emission reduction factor (tons)	0.932
3	CO2 emission reduction value (tons/year)	48.0912
4	CO2 emission reduction trade value (USD/ton)	27.5
5	CO2 emission reduction worth (USD/year)	1322.51
6	Dollar to Naira exchange rate c($/Naira)	760
7	CO2 emission reduction worth (Naira/year)	1,005,106.08

The summary of the results for carbon credit analysis is presented in Table 3. According to the results in Table 3, the CO2 emission reduction value of 48.0912 tons per year is realized, and this amounts to carbon credit of 1322.51 USD, which is equivalent to 1,005,106.08 Naira. In essence, over one million Naira carbon credit is realized by the installation of the grid-connected SPV power system for the campus streetlight.

In any case, the analysis in this paper is based on the streetlight operating at the full illumination lux level. However, the use of proximity sensors and illumination-level control mechanisms can be used to significantly reduce the energy consumption of the streetlight. Notably, the worst-case energy yield has a solar fraction of 75.7% (as shown in Table 2). Now, available studies have shown that the use of sensors and light intensity controller on streetlight can reduce the energy consumption by 25–45% [13–15]. Hence, when the sensors and controllers are used to reduce the street energy consumption by at least 25%, the worst-case energy yield in July with 75.5% solar fraction is still adequate to satisfy the energy demand of the streetlight.

4 Conclusion

A combination of analytical and simulation approaches is employed in the design and evaluation of grid-connected solar photovoltaic (SPV) power system for a campus streetlight with the case study as the University of Uyo main campus located in Nsukara Offot in Uyo local government area of Akwa Ibom State. The study focused on the yearly energy yield, the monthly distribution of the energy yield, the system performance ratio, and the corresponding carbon credit associated with the energy yield of the SPV.

PVSyst simulation software was employed in the analysis, and the results showed that over one million Naira carbon credit will be realized by the installation of the SPV power system. The system has about 97% solar fraction with maximum energy yield occurring in January and worst-case energy yield occurring in July. Finally, if sensors and controllers are used to reduce the street energy consumption by at least 25%, the worst-case energy yield in July with 75.5% solar fraction is still adequate to satisfy the energy demand of the streetlight.

References

1. Gershon, O., Agbene, E.: Adopting hybrid energy technology for carbon emissions reduction in Nigeria's telecommunications industry. IOP Conf. Ser.: Earth Environ. Sci. **655**(1), 012048 (2021). IOP Publishing
2. Aluko, O.E., Fabunmi, T.E., Onibonoje, M.O., Dada, J.O.: A clean and renewable energy-utility solution in Nigeria. In: 2020 IEEE PES/IAS Power Africa, pp. 1–5. IEEE (2020)
3. Agbo, E.P., Edet, C.O., Magu, T.O., Njok, A.O., Ekpo, C.M., Louis, H.: Solar energy: a panacea for the electricity generation crisis in Nigeria. Heliyon. **7**(5), e07016 (2021)
4. Lawal, M.O., Bada, O.M., Ajewole, T.O., Ashaolu, S.O.: An experimental approach towards PV-based solar system sizing for an engineering laboratory. Tanz. J. Eng. Technol. **41**(2), 94–107 (2022)
5. Naskar, R., Ghosh, S., Mandal, R.: Analysis on carbon credit of a 5KWP solar photovoltaic power plant at JIS College of Engineering, Kalyani. Imp. J. Interdiscip. Res. **2**(10), 1506–1511 (2016)

6. Mukherjee, S., Ghosh, P.B.: Estimation of carbon credit and direct carbon footprint by solar photovoltaic cells in West Bengal, India. Int. J. Low Carbon Technol. **9**(1), 52–55 (2014)
7. Kumar, R., Rajoria, C.S., Sharma, A., Suhag, S.: Design and simulation of standalone solar PV system using PVsyst software: a case study. Mater. Today: Proc. **46**, 5322–5328 (2021)
8. Baqir, M., Channi, H.K.: Analysis and design of solar PV system using Pvsyst software. Mater. Today: Proc. **48**, 1332–1338 (2022)
9. Shrivastava, A., Sharma, R., Saxena, M.K., Shanmugasundaram, V., Rinawa, M.L.: Solar energy capacity assessment and performance evaluation of a standalone PV system using PVSYST. Mater. Today: Proc. **80**, 3385–3392 (2023)
10. George Allwyn, R., Al Abri, R., Malik, A., Al-Hinai, A.: Economic analysis of replacing HPS lamp with LED lamp and cost estimation to set up PV/battery system for street lighting in Oman. Energies. **14**(22), 7697 (2021)
11. Prabhakant, Tiwari, G.: Evaluation of carbon credits earned by a solar energy park in Indian conditions. Open Fuels Energy Sci. J. **1**, 57–66 (2008)
12. Archibong, E.I., Ozuomba, S., Ekott, E.E.: Life cycle cost and carbon credit analysis for solar photovoltaic powered internet of things-based smart street light in Uyo. Int. Multiling. J. Sci. Technol. **5**(1), 2793–2797 (2020)
13. Abdullah, A., Yusoff, S.H., Zaini, S.A., Midi, N.S., Mohamad, S.Y.: Energy efficient smart street light for smart city using sensors and controller. Bull. Electr. Eng. Inform. **8**(2), 558–568 (2019)
14. Escolar, S., Carretero, J., Marinescu, M.C., Chessa, S.: Estimating energy savings in smart street lighting by using an adaptive control system. Int. J. Distrib. Sens. Netw. **10**(5), 971587 (2014)
15. Elejoste, P., Angulo, I., Perallos, A., Chertudi, A., Zuazola, I.J.G., Moreno, A., et al.: An easy to deploy streetlight control system based on wireless communication and LED technology. Sensors. **13**(5), 6492–6523 (2013)

A Reconfigurable Load-Modulated Balanced 9-GHz Power Amplifier Design for Radar Applications

Joseph Christopher Skippings, Sunday Cookey Ekpo, Fanuel Elias, and Andy Gibson

1 Introduction

Radar systems deployed in various vertical industries and by the government agencies utilize detection systems that can determine the range, angle, or velocity of objects such as aircraft, ships, spacecraft, guided missiles, motor vehicles, weather formations, and terrain [1–5]. With the recent advancements in active and adaptive device technologies, radar systems can be designed to provide greater efficiency and enhance capabilities at a lower cost [1, 5–8]. New antenna and radar fabrication technologies including GaN HEMT-based power amplifiers (PAs), low-noise amplifiers (LNAs), beamformers, and active electronically steered array (ASEA) antenna have enabled highly efficient radar systems. This also includes advancements in MIMO radar systems and the use of ultra-wide bandwidth millimeter waves in radar.

The real-time multifunctional and multipurpose capabilities of future-generation radar transceiver front-end will depend upon the cognitive-aware architecture of the subsystems at the development and utilization phases. Smart user equipment performance requirements will heavily depend on how the physical layer subsystems performance metrics are validated. The massive multiple input multiple output (mMIMO) system is needed in capability-based reconfigurable radio frequency frontend (RFFE) ecosystem deployments. The key challenges of the mMIMO system implementation revolve around reduced energy consumption and dissipation, reduced weight/size, and increased capacity and coverage [3–6]. Addressing these challenges for the RFFE requires top-level design considerations around efficiency (thermals and power consumption), bandwidth (BW) (frequency range and input BW/output BW), and size (isolation and integration). Hence, the key

J. C. Skippings (✉) · S. C. Ekpo · F. Elias · A. Gibson
Manchester Metropolitan University, Manchester, UK
e-mail: joseph.skippings@stu.mmu.ac.uk

S. C. Ekpo (ed.), *The Second International Adaptive and Sustainable Science, Engineering and Technology Conference*, Signals and Communication Technology,
https://doi.org/10.1007/978-3-031-53935-0_24

design considerations of the receiver [9, 10] must be to reduce system size (high integration dual-channel integrated and high channel-to-channel isolation), expand capacity (wide bandwidth >400 MHz and lowest noise figure), and reduce thermals (low power consumption). Similarly, the design considerations for the pre-driver must aim to improve coverage (high-gain design to drive active semiconductor devices (e.g., LDMOS/GaN) amplifiers and output power), improve power consumption (excellent ACLR at low power), reduce design cycle time (pin-to-pin family to cover radar frequencies and gain adjustment to compensate for different losses/cable lengths), and improve capacity (high linearity and wide instantaneous BW >200 MHz) [6–8]. Since each MIMO system aims for a small weight and power and cost-effective implementation (SWaP-C), the key design considerations of the RFFE circulator must improve coverage and efficiency (lowest insertion loss and excellent power handling), improve capacity (high intermodulation performance), and reduce system size (smallest form factor: 7-mm, 10-mm, 12-mm packages) [9–14].

A PA is a critical component of radar design, serving as the main basis of the system's functionality and capabilities. There is a need for a reconfigurable satellite-terrestrial convergence surveillance ecosystem that is more advanced and efficient when compared to the predecessor [8, 12–16]. For example, British territories located in the Caribbean face the threat of illegal immigrants, drug smuggling, and human trafficking in the region. The current radar system utilizes a dish antenna with a limited detection range of 22 km or 12 nautical miles. A proposed solution to this is the adoption of a radar system that integrates a reconfigurable 9-GHz PA design with a phased array hybrid beamforming [17] that can detect a target at a range >44 km at an output of 34 Bm and gain compression of <3 dB. To meet this design goal, the WIN NP25 transistor is selected. The transistor is a 0.25-μm gate GaN on SiC, fabricated on a 4-inch silicon substrate that operates at a drain bias of 28 V. The transistor has a frequency range of DC to 18 GHz and a saturated output power of 5 W/mm. Its power gain is 19 dB at 10 GHz and power-added efficiency (PAE) of 65%. The main objective is to design and simulate a multifunctional and reconfigurable [18–20] PA architecture using the load modulation balanced amplifier methodology and analyze the device performance.

2 LMBA Design

The LMBA topology was first introduced in 2016 by Cripps as a way of improving the PA efficiency by adjusting the power and the phase of the control signal [3], deriving from the conventional balanced amplifier architecture consisting of two transistors connected to quadrature hybrid couplers at the input and output. Unlike the balanced amplifier, a control signal is injected in the isolate port of the output coupler instead of being terminated. This injected control signal is an active match for the two transistors, replacing the conventional passive output matching network. In [11] and [5] the authors design a reconfigurable X band power amplifier MMIC based on the load-modulated balanced amplifier technique using the gallium nitride (GaN) 0.25 μm gate length technology produced by WIN Semiconductors, which consisted of two 8 μm × 125 μm cell and Lange couplers. The LMBA design proposed achieved an output power of 14.1 W at 4 to 9.7 dB, with a PAE of 37% across 8 – 9 GHz, [5].

A DC-IV analysis determines conditions such as the operating point, initial conditions in transients, and transfer curves. The device size chosen for the design is 4 ∗ 200 μm, and by using Keysight ADS, the gate-source voltage VGS required to bias the transistor at VDS = 28V and IDS = 125 mA is determined through simulation, with VDS set at 28 VDC, VGS is swept, and IDS is recorded at each VGS value. The result is plotted and the corresponding VGS value that yields IDS = 81 mA is determined. The gate voltage of −2.6 V consumes 2.270 W and the expected output power of 34.204 dBm. The conduction angle is between 180 ° and 360° (at 220.253°) with a duty cycle of 61.181. The transistor S-parameters extracted from the transistor model in ADS features RF chokes (DC feeds) and DC blocks are ideal components required for simulation purposes. Input and output ports are terminated with 50 Ω. The simulation is run with frequency swept from 8 to 12 GHz. The transistor is unconditionally stable and broadband stable from 9 to 11 GHz. The load-pull simulation plots the transistor's power contours, and a point on the maximum output power contour is selected. From the load-pull analysis, at an input source of >33.4 dBm, the maximum gain of 10.9 dBm is achieved. The minimum gain compression is <4 dBm with a PAE of 54.75%. Load-pull analysis takeaways: input source = 22.5 dBm, input impedance = 5.767 − j * 14.738, and output impedance = 13.750 + j * 36.667. Therefore, transform the input impedance ZIN 5.767 − j * 14.738 Ω toward 50 Ω transmission line. The output matching network is designed to transform 50 Ω nominal load to the desired load impedance ZL = 13.750 + j * 36.667 Ω at which the transistor outputs maximum power.

3 LMBA-Based Radar Theory and Methodology

3.1 Radar Range Equations

The initial specification requirements for the proposed PA design are determined using the radar equation which is linked to the target properties, radar characteristics, distance between the target and the radar, as well as properties of the medium:

$$\mathrm{SNR} = \frac{P_\mathrm{s}}{P_\mathrm{n}} = \frac{P_T G_T G_R \lambda^2 \sigma}{(4\pi)^3 R^4 k T_0 B F_n L} \tag{1}$$

From (1) and the radar parameters mentioned in Table 1, the output of 34 dBm at an operating frequency of 9 GHz is required to achieve a range greater than 40 km.

Table 1 Radar parameters

Radar parameters	Values
Peak transmit power @ power tube	1.20E+06 W
Transmit losses, including feed and antenna	2 dB
Pulse width	4.00E−07 s
Antenna gain	34 dB
Operating frequency	9 GHz
Receive losses, including feed and antenna	3 dB
Noise figure	8 dB
Other losses	2 dB

Table 2 LMBA-based radar: PA requirements

Parameters	Values	Units
Frequency	9	GHz
Small signal gain	>10	dB
Output	>34	dBm
PAE at Psat	>50	%
Vdd	28	V
Gain compression	<4	dBm

The peak transmission power of 1.2 MW using power tube needs to satisfy the requirements of the PA and is an indication the device will fall under high-power amplifiers.

From this analysis of the radar range equation, the following parameter considerations for the power amplifier can be seen in Table 2.

Some of the main characteristics of the amplifier used for the design considerations are gain, output power, linearity, and efficiency. Operating at a frequency of 9 GHz, the PA should be capable of delivering an output power >34 dBm with a PAE >50% and a gain >10 dB.

3.2 LMBA PA Topology

The reconfigurability aspect of the power amplifier is achieved using the load-modulated balanced amplifier (LMBA) technique. This technique introduces a control signal power (CSP) at the isolated of the output 90-degree coupler which modulates a load of each of the balanced transistors [6]. The added feature of the CSP independently contributes to the total output power, therefore allowing for full recovery at the output, whereas with the Doherty PA method the auxiliary power recovery is dependent on the auxiliary output phase which is needed to modulate the load on the main transistor (Fig. 1).

The basic theory of LMBA is based on the following expression where b and c represent the drive levels of the balanced generators and the CSP, respectively, and R_0 the coupler impedance. Depending on the ratios between the drive levels, $\left(\frac{c}{b}\right)$

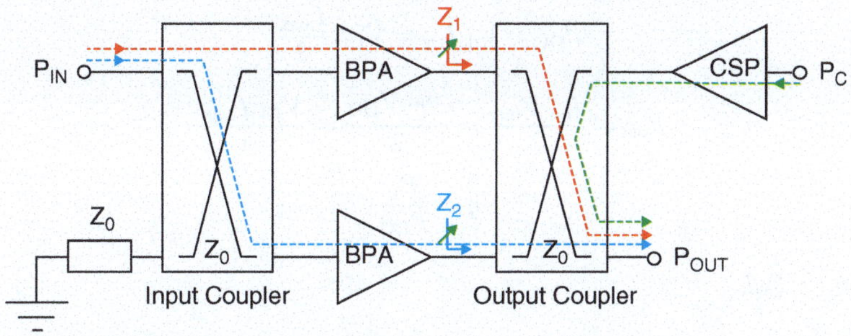

Fig. 1 LMBA topology

presented at the balanced amplifier when tuning the CSP generator's amplitude and phase. The total output power, P_{OUT}, of the LMBA is given (4) and (5):

$$\begin{cases} Z_2 = Z_4 = R_0\left(1+\dfrac{\sqrt{2}c}{b}\right) = Z_B \\ Z_3 = R_0 \end{cases} \tag{2}$$

$$P_{OUT} = P_2 + P_4 + P_3 = 2P_B + P_3 \tag{3}$$

$$= \frac{1}{8} R_0 I_M^2 \left| c + \sqrt{2}b \right|^2 \tag{4}$$

The DC power consumption, output, and PAE of the LMBA are calculated using the following:

$$P_{BA1} = \frac{V_{dc} I_{fund}}{2} \tag{5}$$

$$P_{BA2} = \frac{V_{dc} I_{fund}}{2} \tag{6}$$

$$P_{CA} = \frac{V_{dc} I_{fund}}{2} \tag{7}$$

$$P_{Total} = P_{BA1} + P_{BA2} + P_{CA} \tag{8}$$

$$\text{Gain} = \frac{P_{Total}}{(P_{RFin} + P_{CSP})} \tag{9}$$

$$\text{PAE} = 100 * \frac{P_{out} - P_{in} - P_{in.c}}{P_{DC}} \tag{10}$$

The following equations can be used to calculate the DC power consumption, output power, and drain efficiency:

$$P_{DC} = V_{dc} I_{Dc} = V_{DD} \frac{I_{max}}{2} \tag{11}$$

$$P_{out} = \frac{I_{out} V_{out}}{2} = \frac{\frac{I_{max}}{2} V_{DD}}{2} = \frac{V_{DD} I_{max}}{4} \tag{12}$$

$$\eta = \frac{P_1}{P_{DC}} \tag{13}$$

where $P_1 = \frac{V_{dc} I_1}{2}$.

4 Results

4.1 Single-Power Amplifier Design Results

The final analysis of the proposed single amplifier design for the LMBA circuit achieved a PAE of 49.08 % , output power of 33.15 dBm, and gain of 10.65 dB with a drain efficiency of 53.55%. The ACPRA analysis in Fig. 2 illustrates an input power of 45.599 dBm with an output power of 43.377 dBm. The lower ACPR was calculated to be −30.972 and the upper ACPR −17.545 (Fig. 3).

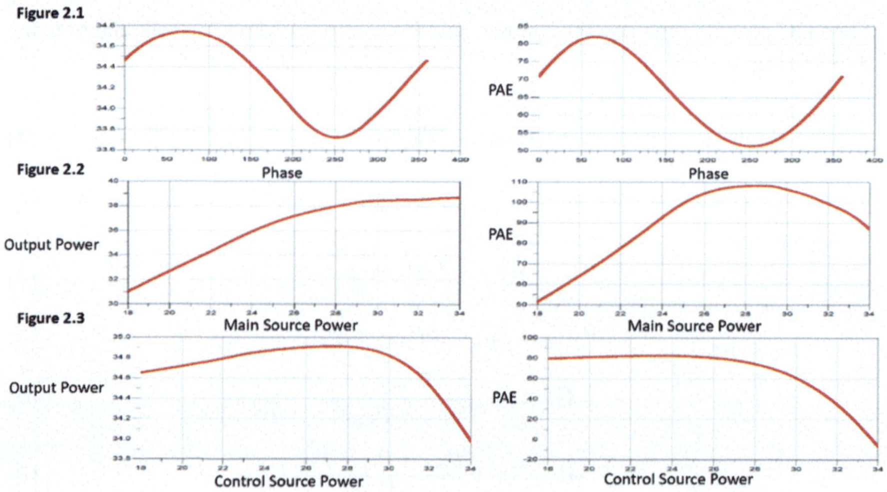

Fig. 2 Graph of PAE vs phase and output vs phase, graph of output power vs main source power and PAE vs main source power, graph of output power vs control source power, and PAE vs control source power

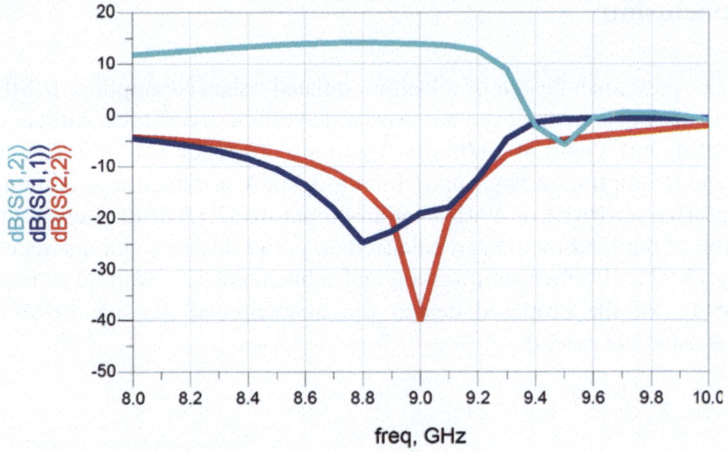

Fig. 3 LMBA S-parameter simulation results

4.2 *LMBA with Separate Power Source*

A PA design consists of two of the proposed power amplifiers design, the coupler, and the control amplifier is replaced with a separate control. Using the LMBA equations shown above, when the separate source power is set to the same of the parameters as the input RF source power, the LMBA achieved PAE of 68%, output power of 34.495 dBm with a gain of 10.06 dB, and drain efficiency of 78.662 % . If the phase of the separate source power is set to 50, it results in PAE of 82 % , gain of 9.099 dB, output power of 34.814 dBm, and drain efficiency of 93.62 % . If the input power RF source power is reduced by −3 dBm to 19.5 dBm, it produces a PAE of 80.96%, a gain of 12.421 dB, and an output power of 34.70 dBm. Therefore, when the control signal power phase is set to 50 and injected into the PA, higher efficiency is obtainable. As signal phase of increase, the power level increases which is due to the combining of both output power. The ratio of the control signal to the primary signal outpower combines with the output, which causes power to remain. To analyze the required phase and input sources for the main and separate source power required to arrive at a desired output, PAE, and gain, the verification test bench (VTB) setup is used to perform ACPR analysis. Both the main source power and the separate source are the same, resulting in an input power of 45.60 W and output power of 43.327 W. The lower ACPR was calculated to be −31.020 and upper ACPR −17.766.

5 Conclusion

This paper presents a design of a load-modulated balanced amplifier (LMBA). The 9-GHz LMBA design achieved a power-added efficiency of 68% with an output of 34.495 dBm and a gain of 10.06 dB. The drain efficiency was 78.662%. The base single-power amplifier design used for the LMBA achieved a maximum power-added efficiency (PAE) of 49.08% at an output of 33.15 dBm. Hence, the implementation of the load-modulated balanced amplifier design technique increased the PAE by 18.92%. Further optimization and refinements are required to improve the performance of the proposed design for an integrated massive MIMO 5G RF frontend radar use case(s).

References

1. Gigorro, E., Pascual, E., Martínez, J.J.S., Heras, M.L.G., Fernandez, B.V., Márquez, B.A., Grajal, J.: Design and characterization of a 6–18 GHz GaN on SiC high-power amplifier MMIC for electronic warfare. Int. J. Microw. Wirel. Technol. **11**(7), 625–634 (2019)
2. Ekpo, S.C.: Parametric system engineering analysis of capability-based small satellite missions. IEEE Syst. J. **13**(3), 3546–3555 (2019). https://doi.org/10.1109/JSYST.2019.2919526
3. Ekpo, S.C.: Thermal subsystem operational times analysis for ubiquitous small satellites relay in LEO. Int. Rev. Aerosp. Eng. J. **11**(2), 48–57 (2018). https://doi.org/10.15866/irease.v11i2.13663
4. Ekpo, S., George, D.: A system engineering analysis of highly adaptive small satellites. IEEE Syst. J. **7**(4), 642–648 (2013). https://doi.org/10.1109/JSYST.2012.2198138
5. Chen, R., Li, R., Zhou, S., Chen, S., Huang, J., Wang, Z.: An X-band 40 W power amplifier GaN MMIC design by using equivalent output impedance model. Electronics. **8**(99), 2–8 (2019)
6. Ekpo, S.C., Adebisi, B., George, D., Kharel, R., Uko, M.: A system-level multicriteria modelling of payload operational times for communication satellite missions in LEO. Recent Prog. Space Technol. **4**(1), 67–77 (2014). https://doi.org/10.2174/2210687104666140620221119
7. Shepphard, D.J., Powell, J., Cripps, S.C.: An efficient broadband reconfigurable power amplifier 92 using active load modulation. IEEE Microw. Wirel. Compon. Lett. **26**(6), 443–445 (2016)
8. Ekpo, S., George, D.: Reconfigurable cooperative intelligent control design for space missions. Recent Pat. Space Technol. **2**(1), 2–11 (2012). https://doi.org/10.2174/1877611611202010002
9. Ekpo, S., George, D.: Impact of noise figure on a satellite link performance. IEEE Commun. Lett. **15**(9), 977–979 (2011). https://doi.org/10.1109/LCOMM.2011.072011.111073
10. Ekpo, S., George, D.: 4–8 GHz LNA design for an adaptive small Satellite Transponder using InGaAs PHEMT technology. In: Proc. 11th IEEE Wireless & Microwave Conference, pp. 1–4, Melbourne, FL, USA (2010). https://doi.org/10.1109/WAMICON.2010.5461877
11. Powell, J.R., Shepphard, D.J., Quaglia, R., Cripps, S.C.: A power reconfigurable high-efficiency X-band power amplifier MMIC using the load modulated balanced amplifier technique. IEEE Microw. Wirel. Compon. Lett. **28**(6), 527–529 (2018)
12. Quaglia, R., Cripps, S.: A load modulated balanced amplifier for telecom applications. Trans. Microw. Theory Tech. **66**(3), 1328–1337 (2018)
13. Sowande, O., Idachaba, F., Ekpo, S., Faruk, N., Uko, M., Ogunmodimu, O.: Sub- 6 GHz 5G spectrum for satellite-cellular convergence broadband internet access in Nigeria. Int. Rev. Aerosp. Eng. J. **15**(2), 85–96 (2022). https://doi.org/10.1109/NIGERCON54645.2022.9803144

14. Zafar, M., Ekpo, S., George, J., Sheedy, P., Uko, M., Gibson, A.: Hybrid power divider and combiner for passive RFID tag wireless energy harvesting. IEEE Access J. **10**, 502–515 (2022). https://doi.org/10.1109/ACCESS.2021.3138070
15. Uko, M., Ekpo, S.: A 23–28 GHz pHEMT MMIC low-noise amplifier for satellite-cellular convergence applications. Int. Rev. Aerosp. Eng. J. **14**(5), 1–10 (2021). https://doi.org/10.15866/irease.v14i5.20361
16. Uko, M., Ekpo, S.: 8–12 GHz pHEMT MMIC low-noise amplifier for 5G and fiber-integrated satellite applications. Int. Rev. Aerosp. Eng. J. **13**(3), 99–107 (2020). https://doi.org/10.15866/irease.v13i3.17998
17. Ekpo, S., Adebisi, B., Wells, A.: Regulated-element Frost beamformer for vehicular multimedia sound enhancement and noise reduction applications. IEEE Access J. **5**, 27254–27262 (2017). https://doi.org/10.1109/ACCESS.2017.2775707
18. Uko, M.C., Ukommi, U., Ekpo, S.C., Kharel, R.: Effect of shadowing and multipath fading on the area spectral efficiency of a macro-femto heterogeneous network for cell-edge users. Appl. Comput. Electromagn. Soc. J. **35**, 1043–1046 (2016)
19. Ekpo, S., George, D.: A power budget model for highly adaptive small satellites. Recent Pat. Space Technol. **3**(2), 118–127 (2013). https://doi.org/10.2174/1877611611303020003
20. Ekpo, S., George, D.: A deterministic multifunctional architecture design for highly adaptive small satellites. Int. J. Satell. Commun. Policy Manage. **1**(2/3), 174–194 (2012). https://doi.org/10.1504/IJSCPM.2012.049543

UCoD: Ensemble BERT for Hierarchical Classification of the Urdu Disinformation Corpus

Umar Farooq, Omer Beg, Faisal Riaz, Saeid Jamali, William Holderbaum, and Umar Raza

1 Introduction

Disinformation, a potent mix of false content and propaganda, has been increasingly exploited on digital platforms, affecting critical events like the US presidential, Indian general, and Pakistan general assembly elections [1]. It has been instrumental in creating widespread fear and confusion, notably during the 2016 US election and the COVID-19 outbreak [2]. To counter this, we propose an approach using supervised machine learning and deep learning for the identification and classification of disinformation. The focus is on propaganda, which uses various manipulation techniques to sway public sentiment.

The prominence of Urdu, particularly on platforms like Twitter, has grown with the rise of multilingual online resources. However, despite its wide use, Urdu's complex orthographic features pose challenges for existing data preprocessing tools, limiting the effectiveness of automated propaganda and fake news detection [3]. Our study, thus, concentrates on developing a technique for the classification of propaganda and fake news in Urdu using deep learning architectures, sentiment analysis,

U. Farooq (✉) · O. Beg
National University of Science and Technology, Islamabad, Pakistan
e-mail: i191289@nu.edu.pk

F. Riaz
Mirpur University of Science and Technology, Mirpur, AK, Pakistan

S. Jamali
Intellithing Lab, Xyz Building, Manchester, UK

W. Holderbaum · U. Raza
Department of Engineering, Manchester Metropolitan University, Manchester, UK

S. C. Ekpo (ed.), *The Second International Adaptive and Sustainable Science, Engineering and Technology Conference*, Signals and Communication Technology,
https://doi.org/10.1007/978-3-031-53935-0_25

feature extraction techniques, and probabilistic models. We aim to detect and categorize disinformation at the sentence level in Urdu, marking a departure from previous efforts.

In this study, the focus is on developing a technique for the classification of propaganda and fake news in Urdu. Deep learning-based feature extraction techniques (such as Word2vec, GloVe, and FastText) and probabilistic models have shown promising results for Urdu language analysis. The proposed work differs from previous efforts, aiming to detect disinformation at the sentence level and determine the appropriate type for Urdu. In this study, we address the problem of detecting specific disinformation techniques in Urdu text. We also provide the Urdu Corpus of Disinformation (UCoD), consisting of over 19K tweets and news, to support research in this area. We ensure corpus quality through data annotation validation using inter-annotator agreement. We leverage various pre-trained models for transfer learning in the hierarchical classification task. We develop an ensemble of the best-performing BERT variants and demonstrate its superiority over other BERT-based baselines.

2 Literature Review

In today's digital landscape, the influx of data calls for advanced analysis and classification techniques, particularly to determine the authenticity of content on news and social media platforms. The increasing prevalence of disinformation witnessed during events like the COVID-19 pandemic [4] influences public mindset and manipulates narratives. This necessitates a robust model capable of discerning propaganda and fake news. Our proposed system aims to classify 18 types of disinformation by analyzing text data from news articles and tweets, using advanced text analysis methods to effectively detect misinformation. These disinformation types often exploit multimedia content, causing adverse effects and manipulation of public events [5, 6]. Detection of disinformation on social media, especially around emerging topics, is challenging due to the subtle use of psychological and rhetorical techniques like logical fallacies and emotional appeals employed by propagandists [7]. These tactics aim to influence audiences while avoiding detection.

2.1 Propaganda and Fake News Detection Systems

NLP research primarily revolves around textual data analysis. Studies have examined controversial speeches on social media to understand contentious events [8] and validate the authenticity of the news, crucial in the era of pervasive fake news [9]. Propaganda detection methods have been devised to categorize propagandistic content, mitigating disinformation's impact [10]. Despite significant progress, especially in English language studies, there remains a research gap in Urdu propaganda detection [9]. Propaganda, often emotionally appealing, employs techniques like loaded language, flag-waving, appeal to authority, slogans, clichés, and logical fallacies, presenting alternatives, whataboutism, and irrelevant data. Scholars vary in their classification of these techniques [11]. Some tried to detect propagandistic fragments and identify different propaganda types using models [12] and classification approaches at the sentence level and fragment level [9]. However, annotating these fragments requires significant time and expertise [13, 14]. The rise of fake news, ranging from simple misinformation to intentional societal deception [15], affects both industry and academia. Many users, desktop and mobile, regularly access fake news sites along with legitimate ones [16]. Fake news varies from real news in writing style, syntax, semantics, discourse, and lexicon usage and can be further classified into hoaxes, satire, and fabricated news [17, 18].

2.1.1 Limitations

The lack of research and structured data in Urdu poses challenges in disinformation detection. Existing techniques and models are predominantly designed for English, and there is a scarcity of annotated corpora for low-resource languages [19, 20]. A comprehensive study on disinformation detection in Urdu is currently lacking. Determining the most suitable mechanism for generating accurate predictions in Urdu is difficult due to limited data. Further investigation is required to explore the effectiveness of transfer learning with transformer models for the Urdu corpus. The challenge lies in identifying these discrepancies and establishing a robust detection system.

3 Proposed Approach

We aim to determine the best-performing classifier for disinformation detection in Urdu by evaluating various machine learning classifiers and seq-2-seq models, including BERT and its variants, as well as convolutional neural networks.

The proposed system diagram shown in Fig. 1 illustrates the overall structure, encompassing data collection, cleansing, preprocessing, rule-based techniques, and training of ML/DL algorithms.

3.1 ML Models

For disinformation detection, we consider a range of machine learning classifiers such as MNB, BNB, SVM, LR, RF, DT, AB, and NB. This approach helps us find the best-performing classifier for our corpus. Baseline methods like LR, SVMs, and

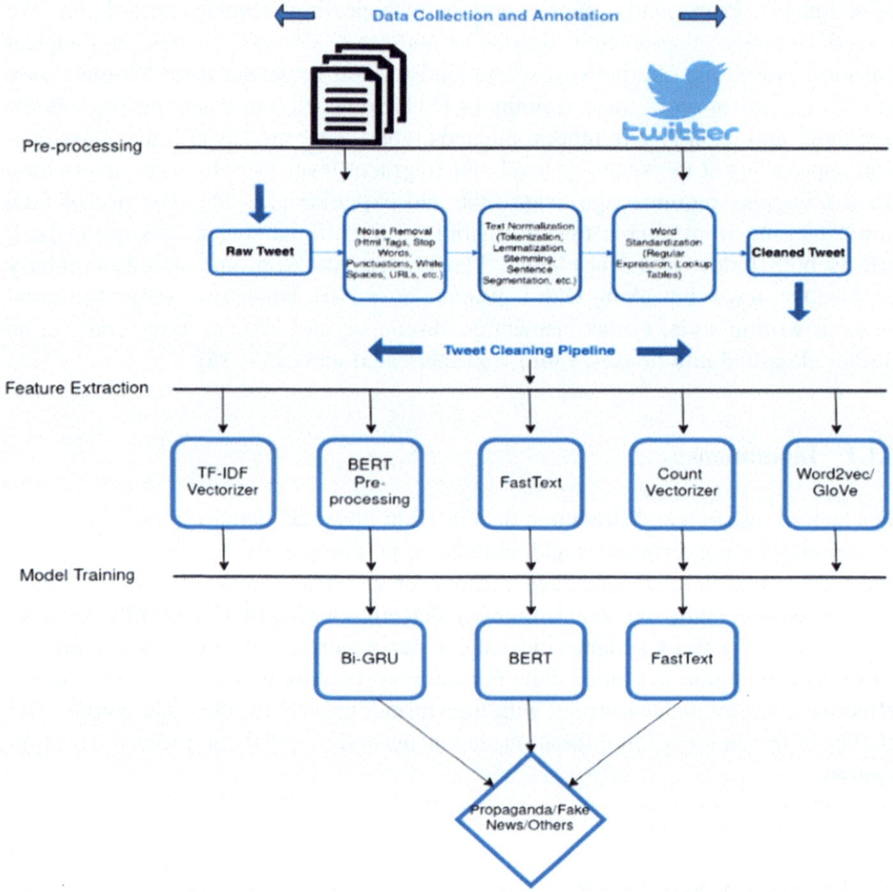

Fig. 1 The proposed system diagram from data collection to final prediction

RF are commonly used in the literature to demonstrate results. We employ a three-fold approach using TF-IDF, CountVectorizer, and Word2Vec with different machine learning models.

3.2 DL Models

In terms of vector embeddings, we explore pre-trained FastText embeddings on our corpus using FastText-like and BiGRU models for classification. Additionally, we utilize two variants of BERT: one with a feed-forward network and another with transfer learning on our corpus. DL methods, including CNNs, LSTMs, transformers, and GRUs, are widely used for text classification. Attention mechanisms in transformers capture contextual information effectively.

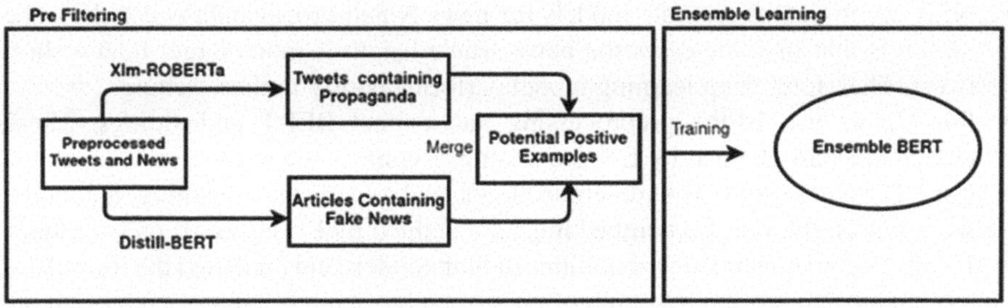

Fig. 2 The ensemble of distill BERT and XLM-RoBERTa for disinformation detection

3.3 Ensembles

Ensemble models are known for their competitiveness and have been shown to outperform baselines in various fields. We train separate models for tweets and news, as individual models may struggle to extract information from different types of corpora. Figure 2 illustrates the ensemble classifiers used to predict the technique employed in misinformation.

3.4 Loss Function

To develop better machine/deep learning models for new ML/DL tasks, it is crucial to determine a suitable loss function for the proposed problem [21]. Cross-entropy is widely used for multi-level classification tasks and has proven effective in distinguishing numerous classes [22, 23].

The cross-entropy loss function is defined as follows:

$H(p, q) = -\Sigma p(x)\log(q(x))$, where $p(x) \in R3$ is the label vector and Nc is the number of classes in the corpus. The cost function based on multi-class log loss for a dataset of size N is given by:

$J = -\Sigma\ (yi \log(y^{i}))/N$, where yi and y^{i} represent the true and predicted values, respectively, and the sum is taken over all N samples.

3.5 Algorithm

- Fake news (Amjad et al. [24]) with an added class of satire are defined by $N \in n1$, $n2$, ...nm.
- After that, we make use of commonly used words that are used in propaganda and events influenced by propaganda.
- These propaganda tweets are defined by PT $\in$ Pt1, Pt2, ...Ptm.

- Next, we train two separate models for news N and propaganda tweets PT. The reason is that in some cases the news article length is much larger than a short tweet. Therefore, deep learning model performs poorly in these settings.
- For N, we use BERT preprocessing and extract BERT embeddings. These embeddings are defined by EN $\in$ en1, en2, ...enm.
- For PT, we use FastText Urdu embeddings, and as a machine learning, the model we use is KNN. FastText embeddings are defined by PE U $\in$ eu1, eu2, ...eum.
- Finally, we averaged the probabilities of both models and predicted the final class.

4 Result and Discussions

In this section, we discuss the implementation details of our experiments using various baseline models and the best-performing model for disinformation detection and analysis. We explore different embedding techniques and approaches used in the experiments.

To improve classification accuracy and broaden the scope of prediction, we conducted several experiments to validate the effectiveness of our methods. Local contextual information has a limited impact in hierarchical classification tasks, so leveraging different word embedding techniques becomes crucial to capture contextual information more effectively [25]. In addition to traditional techniques like Word2Vec, TF-IDF vectorizer, and CountVectorizer used in the baseline models, we incorporated FastText and BERT multilingual embeddings. For tokenization, we utilized the Hugging Face library tokenizer for both FastText and BERT models.

4.1 Baselines

For baseline results, we used various machine learning classifiers and embedding techniques. Baseline models included K-NN, random forests, support vector machine, AdaBoost, simple perceptron, Gaussian process classifier, and QDA. We used TF-IDF, CountVectorizer, and Word2Vec embeddings. Performance details can be found in Table 1.

The support vector machine (SVM) classifier and random forest classifier with CountVectorizer achieved the best performance, with a weighted $F1$-score of 0.52. Overall, the random forest classifier performed better than other models when using TF-IDF, CountVectorizer, and Word2Vec features. SVM showed better performance compared to other baselines. AdaBoost, simple perceptron, and GaussianNB had similar performance with slight variations across different features. GaussianNB with CountVectorizer had a low $F1$-score of 0.35 and an accuracy of 0.29. Handling overlapping classes that can be used interchangeably posed a significant challenge in most baseline cases. Our focus is to develop a method that can effectively handle these overlapping classes.

Table 1 Baseline classifiers and their results in terms of precision (*P*), recall (*R*), *F*1-score (*F*1), and accuracy (Acc)

Classifier	Features	*P*	*R*	*F*1	Acc
KNN	TF-IDF	0.41	0.48	**0.44**	0.48
	CV	0.40	0.47	0.43	0.47
	W2V	0.45	0.49	0.47	0.49
SVM	TF-IDF	0.49	0.52	0.46	0.52
	CV	**0.50**	**0.55**	**0.52**	**0.55**
	W2V	0.48	0.54	0.51	0.54
RF	TF-IDF	**0.50**	**0.55**	0.51	**0.55**
	CV	**0.50**	0.54	**0.52**	0.54
	W2V	0.48	0.50	0.49	0.50
AdaBoost	TF-IDF	0.42	0.44	0.37	0.44
	CV	0.39	0.41	0.40	0.41
	W2V	0.32	0.39	0.35	0.39
Simple perceptron	TF-IDF	0.49	0.52	0.50	0.52
	CV	**0.50**	0.53	0.51	0.53
	W2V	0.49	0.52	0.50	0.52
GuassianNB	TF-IDF	0.49	0.28	0.36	0.28
	CV	0.45	0.29	0.35	0.29
	W2V	0.47	0.36	0.41	0.36
QDA	TF-IDF	0.32	0.43	0.37	0.43
	CV	0.35	0.44	0.39	0.44
	W2V	0.47	0.48	0.48	0.48

Table 2 Results of various embeddings on UCoD

	Acc	*F*1	*P*	*R*
FastText	0.52	0.50	0.52	0.49
FastText + BiGRU	0.55	0.53	0.56	0.50
Bert	0.59	0.54	0.56	0.52
DistilBert	**0.60**	0.56	0.59	0.51
XLM-RoBERTa	0.59	**0.57**	0.59	0.54

4.2 Results of Deep Learning Approaches

Our experiments with deep learning models on the tweets and news corpus showed comparable or slightly improved results compared to the baselines, as shown in Table 2. However, pre-trained models like BERT and its variants did not achieve a higher $F1$-score than the baseline models. Initially, we trained FastText embeddings with a maximum vector length of 500d, obtaining an accuracy of 0.52 and a weighted $F1$-score of 0.50. We also employed bidirectional gated recurrent units (Bi-GRU) with 300d FastText word vectors, achieving an accuracy score of 0.55 and a weighted $F1$-score of 0.53.

By fine-tuning BERT multilingual embeddings on our corpus using the standard BERT model, we obtained an accuracy score of 0.59 and an overall macro-$F1$-score of 0.54. We also experimented with DistilBERT, a smaller and faster version of

BERT, which significantly reduced the training time. DistilBERT yielded an accuracy of 0.60 and an $F1$-score of 0.56. Finally, we explored XLM-RoBERTa, which demonstrated similar results to DistilBERT with accuracy and $F1$-score of 0.59 and 0.57, respectively. Both DistilBERT and XLM-RoBERTa were trained for a higher number of epochs than BERT and exhibited satisfactory performance on UCoD.

4.3 Ensemble Results

Our corpus consists of tweets and news blogs, which we split into two separate parts for training and evaluation. This division resulted in improved results, as shown in Tables 3 and 4. By separating long and short texts, the models performed better on each part of the corpus.

Significant improvements can be observed in the corpus compared to using the entire dataset. The same trend can be seen for long texts or fake news, where substantial improvements are observed.

Among the various embeddings used, FastText, FastText + BiGRU, BERT, DistilBert, and XLM-RoBERTa were evaluated on propaganda tweets and fake news corpus. The results are summarized in Tables 3 and 4, showcasing accuracy, $F1$-score, precision, and recall for each model.

5 Conclusion

In conclusion, we propose a new approach to studying disinformation in news media, focusing on identifying specific disinformation techniques. This fine-grained analysis allows for more accurate and reliable systems, providing explanations for

Table 3 Results of various embeddings on propaganda tweets

	Acc	*F*1	*P*	*R*
FastText	0.64	0.60	**0.67**	0.54
FastText + BiGRU	0.62	0.58	0.58	0.60
Bert	0.65	0.60	0.60	0.59
DistilBert	0.66	0.60	0.59	0.61
XLM-RoBERTa	**0.66**	**0.62**	0.60	0.64

Table 4 Results of various embeddings on fake news

	Acc	*F*1	*P*	*R*
FastText	0.62	0.57	0.61	0.56
FastText + BiGRU	0.62	0.58	0.64	0.53
Bert	0.65	0.65	0.63	0.66
DistilBert	**0.69**	**0.66**	0.68	0.66
XLM-RoBERTa	0.67	0.64	0.60	0.61

why sentences are classified as propaganda and news as biased, satirical, or authentic. We developed an annotation schema for 22 disinformation techniques and annotated a substantial corpus of tweets and news articles accordingly. We have experimented with various BERT-based models, which effectively learn contextual information and accurately classify disinformation into subclasses. Furthermore, we are collaborating with other researchers to expand the corpus and extend the task to electronic media and speech.

References

1. Agarwal, N., Balasubramanian, V.N., Jawahar, C.: Improving multi-class classification by deep networks using dagsvm and triplet loss. Pattern Recogn. Lett. **112**, 184–190 (2018)
2. Amjad, M., Sidorov, G., Zhila, A.: Naive Bayes data augmentation using machine translation for fake news detection in the Urdu language. In: Proceedings of the 12th Language Resources and Evaluation Conference, pp. 2537–2542. European Language Resources Association, Marseille (2020)
3. Shu, K., Sliva, A., Wang, S., Tang, J., Liu, H.: Fake news detection on social media: a data mining perspective. ACM SIGKDD Explor. Newsl. **19**(1), 22–36 (2017)
4. Nelson, T., Nicole K., Claire C., Alan H., Albert H.: The danger of misinformation in the COVID-19 crisis. Missouri Medicine **117**(6), 510 (2020)
5. Allcott, H., Gentzkow, M.: Social media and fake news in the 2016 election. J. Econ. Perspect. **31**(2), 211–236 (2017)
6. Lazer, D.M., Baum, M.A., Benkler, Y., Berinsky, A.J., Greenhill, K.M., Menczer, F., et al.: The science of fake news. Science. **359**(6380), 1094–1096 (2018)
7. Da San Martino, G., Barrón-Cedeno, A., Rosso, P.: Automatic detection and classification of propaganda techniques in news articles. In: Proceedings of the 2019 Conference on Empirical Methods in Natural Language Processing and the 9th International Joint Conference on Natural Language Processing (EMNLP-IJCNLP). Association for Computational Linguistics, Hong Kong (2019)
8. Mustafa, R.U., Nawaz, M.S., Lali, M., Shahzad, B.: Early detection of controversial urdu speeches from social media. Data Sci. Pattern Recognit. **1**(2), 26–42 (2017)
9. Barrón-Cedeño, Da San Martino, Jaradat, and Nakov, Proppy: Organizing News Coverage on the Basis of Their Propagandistic Content, Information Processing and Management, 2019
10. Conroy, N.K., Rubin, V.L., Chen, Y.: Automatic deception detection: methods for finding fake news. Proc. Assoc. Inf. Sci. Technol. **52**, 1–4 (2015)
11. Torok, R.: Symbiotic radicalisation strategies: Propaganda tools and neuro linguistic programming (2015)
12. Barrón-Cedeno, A., Giovanni Da San M., Israa J., Preslav N.: Proppy: A system to unmask propaganda in online news. In Proceedings of the AAAI Conference on Artificial Intelligence, **33**(01), 9847–9848 (2019)
13. Rashkin, H., Choi, E., Jang, J.Y., Volkova, S., Choi, Y.: Truth of varying shades: analyzing language in fake news and political fact-checking. In: Proceedings of the 2017 Conference on Empirical Methods in Natural Language Processing, pp. 2931–2937. Association for Computational Linguistics, Copenhagen (2017)
14. Habernal, I., Hannemann, R., Pollak, C., Klamm, C., Pauli, P., Gurevych, I.: Argotario: computational argumentation meets serious games. arXiv preprint arXiv:1700.06002 2017

15. Wang, W.Y.: "Liar, liar pants on fire": a new benchmark dataset for fake news detection. In: Proceedings of the 55th Annual Meeting of the Association for Computational Linguistics (Volume 2: Short Papers), pp. 422–426. Association for Computational Linguistics, Vancouver, Canada (2017)
16. Nelson, J.L., Taneja, H.: The small, disloyal fake news audience: the role of audience availability in fake news consumption. New Media Soc. **20**, 3720–3737 (2018)
17. Rubin, V.L., Chen, Y., Conroy, N.K.: Deception detection for news: three types of fakes. Proc. Assoc. Inf. Sci. Technol. **52**(1), 1–4 (2015)
18. Zhou, X., Jain, A., Phoha, V.V., Zafarani, R.: Fake news early detection: a theory-driven model. Digital Threats: Res. Pract. **1**, 1–25 (2020)
19. Daud, A., Wahab K., Dunren C.: Urdu language processing: a survey. Artificial Intelligence Review 47, 279–311 (2017)
20. Devlin, J., Chang, M.W., Lee, K., Toutanova, K.: Bert: pre-training of deep bidirectional transformers for language understanding. In: Proceedings of the 2019 Conference of the North American Chapter of the Association for Computational Linguistics: Human Language Technologies, Volume 1 (Long and Short Papers), pp. 4171–4186. Minneapolis, Association for Computational Linguistics (2019)
21. Demirkaya, A., Chen, J., Oymak, S.: Exploring the role of loss functions in multiclass classification. In: 2020 54th Annual Conference on Information Sciences and Systems (CISS), pp. 1–5. IEEE (2020)
22. Ho, Y., Wookey, S.: The real-world-weight cross-entropy loss function: modeling the costs of mislabeling. IEEE Access. **8**, 4806–4813 (2019)
23. Semenov, A., Boginski, V., Pasiliao, E.L.: Neural networks with multi-dimensional cross-entropy loss functions. In: International Conference on Computational Data and Social Networks, pp. 57–62. Springer (2019)
24. Amjad, M., Grigori S., Alisa Z., Helena Gómez-Adorno, Ilia V., Alexander G.: Bend the truth: Benchmark dataset for fake news detection in Urdu language and its evaluation. Journal of Intelligent & Fuzzy Systems **39**(2), 2457–2469 (2020)
25. Benito, D., Araque, O., Iglesias, C.A.: Gsi-upm at semeval-2019 task 5: semantic similarity and word embeddings for multilingual detection of hate speech against immigrants and women on twitter. In: Proceedings of the 13th International Workshop on Semantic Evaluation, pp. 396–403. Association for Computational Linguistics, Minneapolis, Minnesota, USA (2019)

Human Health Risk Assessment of Trace Metals Due to Dietary Intake of Edible Fishes Obtained from Qua Iboe River, Ibeno Local Government Area, Akwa Ibom State

Uwem U. Ubong, Akanimo D. Akpan, and Ifiok O. Ekwere

1 Introduction

Aquatic organisms are desirable and nutritious source of high-quality protein, minerals, and vitamins. However, marine pollution is a serious environmental problem that poses health issues all over the world. In the aquatic food chain, fishes are good biomarkers for long-term monitoring of metal pollution due to their accumulation of high concentrations of metals in tissues [1]. Using fishes as biological indicators can help detect the temporal and spatial changes in trace metals and help understand pollution trends in water bodies [2]. Trace metal accumulation by edible fish species in polluted aquatic environment depends on ecological requirements, metabolisms, and other factors such as salinity, water pollution level, food, and sediment. [3] stated that increasing human population intensifies the anthropogenic threat exerted on the environment. According to [4], depending on their lifetime, the particulates observed at a location can be both of local origin and the product of the transport over distances of hundreds to thousands kilometers. Metals entering the aquatic ecosystem can be stored in fish and other aquatic organisms through the process of bioaccumulation and biological magnification via the food chain, becoming potentially toxic when accumulation reaches a substantially high level [5]. According to [6], water, sediments, and biota are generally metal reservoirs in aquatic environment. The level of harmful and toxic substances is usually higher in river sediments and biological tissues than in the water itself.

U. U. Ubong · A. D. Akpan (✉) · I. O. Ekwere
Department of Chemistry, Akwa Ibom State University, Mkpat Enin, Akwa Ibom State, Nigeria

S. C. Ekpo (ed.), *The Second International Adaptive and Sustainable Science, Engineering and Technology Conference*, Signals and Communication Technology,
https://doi.org/10.1007/978-3-031-53935-0_26

Apart from soil environment and aquatic ecosystem, atmospheric inorganic contaminants of natural origin or anthropogenic sources with heavy metals at high concentrations could lead to serious ecological consequences and pose human health risks [7]. Sediments are regarded as any particulate matter that can be transported by fluid flow and which eventually is deposited as a layer of solid particles on the bed or bottom of body of water or other liquid. Lead is transportation-related arising from fuel combustion and rainfall, which washes particles out of the air and stops re-entrainment of particles [8]. The increased use of metal-based fertilizer in agricultural revolution of government could result in continued rise in concentration of metal pollution in fresh water reservoir due to water run-off [9]. Heavy metals are potentially hazardous to humans. Various ecological sources of metal pollution in the environment includes: weathering, industrial processing of ores, leaching of metals from garbage and solid waste dumps, animal and human excretions. Heavy metals in trace amounts are normal constituents of marine organisms [10], but at high concentrations, they sufficiently become toxic to these organisms. According to [11], nickel, cadmium and lead have no function in biological system and are potentially toxic, even at trace concentration. In general, they are not biodegraded and they bioaccumulate in fish, oyster, crabs, mussels, sediments. Feeding habit plays significant roles in the accumulation of metals in organism tissues because metals have the tendency to be bio-magnified through food chain. [12] stated that the key sources of metal pollution are domestic/industrial sewages; industrial effluents; oil and chemical spills; combustion emission, mining operations, metallurgical activities and non-hazardous landfill sites. Often times, these anthropogenic actions usually lead to accelerated release of other chemical pollutants into the aquatic environment. This poses serious and harsh threats to aquatic life and man, due to their proven toxicity, persistence, bioaccumulation and biomagnifications in the food chain.

According to [3], human health risk assessment (HHRA) of consuming heavy metal-contaminated fish is key to assessing the risk posed by the presence of toxic compounds in the environment. HHRA is made up of four steps, namely, hazard identification, dose-response assessment, exposure assessment, and risk characterization. Risk assessment is a quick method of evaluating the impact of the hazards on human health and also in determining the level of treatment to solve the environmental problems occurring in daily life. In addition, estimated daily intake (EDI), hazard quotient (HQ), and hazard index (HI) are used to assess human health risks posed by these trace metals. Accordingly, the objective of this study is to compare the level of hazardous heavy metals in the muscles and gills of some edible fish species, water, and sediment obtained from Qua Iboe River, Ibeno local government area, Akwa Ibom State, as well as assess the human health risk associated with the consumption of the fishes.

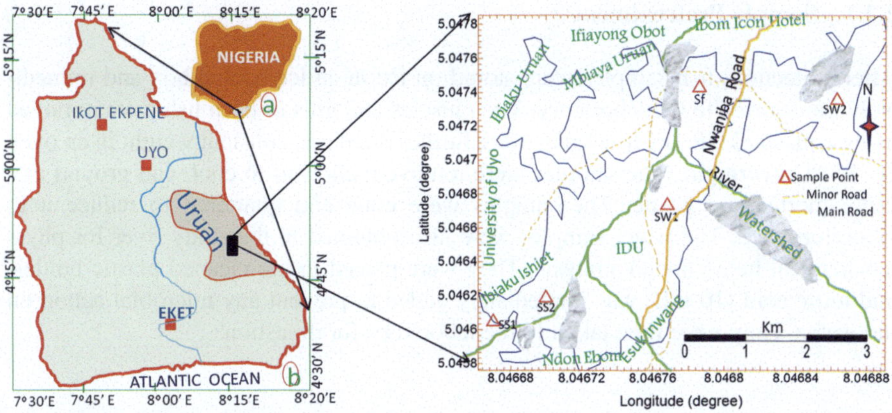

Fig. 1 Map of the study area showing sampling stations compiled using open street map database and fieldwork

2 Materials and Methods

2.1 *Description of the Study Area*

The study was carried out in Qua Iboe River, Ibeno, Akwa Ibom State, Nigeria. It is strategically located at the Qua Iboe River estuary in Ibeno, Akwa Ibom State. Three landing area (Iwokpom, Iwuchang, and Upenekang) were established along the spatial grid of Stubbs Creek. Ibeno is located in the south of Nigeria and is a local government area of Akwa Ibom State. The Global Positioning System (GPS) coordinates for the sampling sites and station code are Sf (N 4′3456704; E 7°2′ 48.834″), Ss1 (N 4°34′345012, E 7°2′47.946″), Ss2 (N 4°34′0036; E 7°2′48.174″), Sw1 (N 4°34′12933; E 7°2′48.63″), and Sw2 (N 4.56781; E 7°2′48.762″) (Fig. 1).

2.2 *Sample Collection and Preparation*

Water samples, fish, and soil sediments were collected from the sample sites in January 2023. Water samples (1 L) were collected, and 5 mL of concentrated nitric acid was added; sediment samples were collected at 30 cm depth using Ekman grab, placed in polyethene container, and air-dried in the laboratory. Five different fish species were also collected from the sampling sites and transported to the laboratory in an ice cooled box. The fish samples were bisected and dried in an oven prior to analysis.

A total of five sample specimens, comprising of different species of fish, were sampled. The species collected were *Oreochromis niloticus* (ON) ($n = 5$), *Pseudotolithus typus* (PT) ($n = 5$), *Caranx hippos* (CH) ($n = 5$), *Ethmalosa fimbriata* (EF) ($n = 5$), and *Polydactylus quadrifilis* (PQ) ($n = 5$).

2.2.1 Sample Pretreatment

After collection, fish samples were stored on ice in an ice cooled box and immediately transported to the laboratory. The muscles and gills of each fish were removed using a plastic knife. Fish samples were further dried to a constant weight in an oven 80–85 °C overnight. The samples were removed, allowed to cool, and ground in a clean mortar and pestle. The samples were cone and quartered to reduce it to laboratory size. The water samples were also obtained in the study river for physicochemical heavy metal analysis. They were placed in pre-cleaned plastic bottles, and nitric acid (10 mL) was immediately added to prevent any microbial action on the water. They were then taken to the laboratory for digestion.

2.2.2 Digestion and Analysis

All samples were digested following the American Society for testing and materials method [13], with a mixture of HNO_3/HCl in 3:1 ratio on a hot plate at 80 °C (gradually increased) until all the materials were dissolved. Finally, volumes of digests were made up to 100 ml with distilled water. All the digested samples were analyzed for Cu, Zn, Mn, Cr, and Pb using atomic absorption spectrophotometer (AAS, Perkin Elmer-2280, USA). The analysis of the standard, reagent blanks and replicates were all run in the same way, and concentrations were determined using standard solutions prepared in the same acid matrix.

2.3 *Risk Assessment Procedure*

HHRA is considered as the characterization of potential adverse effects on humans as a result of exposures to environmental hazards [14]. This process employs the tools of science and statistics to identify and measure a hazard, determine possible routes of exposure, and finally use that information to calculate a numerical value to represent the potential risk [3, 15]. In this study, hazard identification was accomplished through the collection of fish samples from Qua Iboe River and subsequent analysis to quantify the contamination level of trace metals. In order to estimate the noncarcinogenic risks, the dose-response assessment was carried out using standard reference dose (RfD) values for the trace metals considered in this study, as outlined below.

2.3.1 Exposure Assessment

The exposure assessment is the determination or estimation (qualitative or quantitative) of the magnitude, frequency, duration, and route of exposure with respect to the chemical contaminant in question [16]. According to [17], it is important to analyze contaminant release, identify populations and potential pathways of exposure, as well as estimate intakes for specific pathways.

Table 1 Summary of risk assessment parameters

Symbol	Description	Unit	Value(s)
CF	Metal concentration in fish	mg/kg	Presented in Table 4.3
IR	Food ingestion rate	kg/day	*Adults = 0.024, children = 0.019
FI	Fraction ingested	Unitless	1, assuming that the whole fillet is consumed
EF	Exposure frequency	Days/year	365
ED	Exposure duration	Years	*Adults = 70, children = 10
BW	Body weight	kg	*Adults = 70, children = 10
AT	Average time	Days	ED × EF
RfD	Reference dose	mg/kg/day	*Cr = 1.50, Cu = 0.04, Mn = 0.14, Zn = 0.30 and Pb = 0.004

*Derived from a similar study conducted in Nigeria on consumption of muscles of cow meat [3, 18, 19] and having considered that the ingestion rate proposed by [16] for fish and meat consumption is essentially the same; [20], a [21]; b [22]

2.3.2 Estimated Daily Intake (EDI)

Health risk was estimated considering the average concentrations of all fish tissues and daily heavy metal intake (EDI) following Eq. (1), as reported by [3, 18]:

$$\mathrm{EDI} = \frac{\mathrm{CF} \times \mathrm{IR} \times \mathrm{FI} \times \mathrm{ED} \times \mathrm{EF}}{\mathrm{BW} \times \mathrm{AT}} \tag{1}$$

where CF is the chemical concentration in fish, IR is ingestion rate, FI is the fraction ingested, EF is the exposure frequency, ED is exposure duration, BW is body weight, and AT is the average. Specific values for each of the exposure parameters are provided in Table 1:

2.3.3 Evaluation of Noncarcinogenic (Toxic) Risk

Toxic risks can also be called noncarcinogenic effects that would arise as a result of the exposure to contaminant [17]. The magnitude of the effects is estimated in terms of hazard quotients (HQ) [3]. This may be evaluated by employing the equation below:

$$\mathrm{HQ} = \frac{\mathrm{EDI}}{\mathrm{RfD}} \tag{2}$$

The values for the reference dose (RfD) of trace metal considered in this study are listed alongside the risk assessment parameters in Table 1. HQ value <1 implies that the level of exposure is not likely to cause any obvious adverse effects [14]. According to [16, 17], in a situation where there are multiple toxicants and/or multiple exposure pathways, it is important that their possible interaction be considered. The assumption is that the toxic risk due to potentially hazardous chemicals in the same medium is cumulative. The HQs would then be added up to obtain overall toxic risk, the hazard index (HI), as indicated in the expression below:

$$HI = \sum_{i=1}^{n} THQ \tag{3}$$

where n is the number of trace metals.

This computation was carried out for all the five species of fish. If the calculated HI < 1, then the noncarcinogenic adverse effect due to the exposure pathway or toxicant will be assumed to be negligible.

3 Results and Discussion

3.1 Levels of Trace Metals in Fish Species

The mean concentrations of Cr in the muscles and gills of fish species ranged from 0.00 to 8.57 mg/kg and 0.45 to 47.35 mg/kg, respectively. The highest amount of Cr was found in the muscle of PQM (8.57 mg/kg) and gills of CHG (47.35 ± 0.20 mg/kg). The lowest amount of Cr concentrations was observed in the muscle of CHM (0.00 mg/kg) and gills in ONG with a value of 0.45 ± 0.05 (Table 2). The mean concentrations of Cr in the gills tissue of all fish species considered in this study were higher than the concentrations reported by [24] except of ONG. The result revealed that chromium accumulated more in the gills than in the muscles.

The lead concentration in the different species followed the trend CHG > PTG > EFG > ONG > PQG (Table 1). The study revealed that the level of lead in the muscles of all the species are saved for consumption, whereas the gills are hazardous as the observed levels of lead in all the species were above FAO/WHO maximum permissible limit of 1.30 mg/kg except PQG. The levels of lead in the fish species muscles considered in this study were below the values reported in some selected literature, while a contrast was observed for the gills except PQG.

The mean concentrations of Cu in fish species ranged from 7.70 mg/kg to 13.10 mg/kg in muscles and 7.70 to 35.81 mg/kg in gills (Table 2). Cu concentrations in the gills in the different species followed the trend PQG > CHG > EFG > ONG > PTG. In this study, Zn concentrations in the examined fish muscles ranged from 3.44 (PQM) to 507.0 mg/kg (PTM) dry weight (Tables 2 and 3). With the exception of PQM, the levels of Zn recorded in the muscles of examined fish in this study were higher than the permissible limit of 50.00 mg/kg [23] and values reported in some selected literature (Table 4). The mean Zn concentrations in the gills of fish species ranged from 1.03 mg/kg (EFG) to 20.70 mg/kg (ONG). The levels of Zn recorded in the gills of the examined fish in this study were comparable to those reported in some selected literature (Table 4). Manganese accumulated high concentrations in muscles and gills in most of the fish

Table 2 Mean concentration of the trace metals in the fish organs in mg/kg and FAO/WHO permissible limits

Fish species	Common names	CODE	Cu	Zn	Pb	Cr	Mn
Oreochromis niloticus	Tilapia	ONM	12.20 ± 0.20	350.30 ± 0.10	BDL	1.50 ± 0.05	15.45 ± 0.05
Pseudotolithus typus	Croaker	PTM	13.10 ± 0.10	507.00 ± 1.00	BDL	6.35 ± 0.05	10.80 ± 0.10
Caranx hippos	Threadfin	CHM	7.80 ± 0.10	327.40 ± 0.20	0.00 ± 0.00	0.00 ± 0.00	40.55 ± 0.05
Ethmalosa fimbriata	Bonga	EFM	7.70 ± 0.20	328.20 ± 0.10	0.00 ± 0.00	0.35 ± 0.05	12.80 ± 0.10
Polydactylus quadrifilis	Shiny nose fish	PQM	8.15 ± 0.15	3.44 ± 0.02	0.00 ± 0.00	8.57 ± 0.29	15.08 ± 0.10
Oreochromis niloticus	Tilapia	ONG	15.65 ± 0.15	20.70 ± 0.10	4.22 ± 0.01	0.45 ± 0.05	46.07 ± 0.08
Pseudotolithus typus	Croaker	PTG	7.70 ± 0.10	1.05 ± 0.02	10.61 ± 0.01	3.74 ± 0.31	18.37 ± 0.08
Caranx hippos	Threadfin	CHG	31.42 ± 0.40	7.77 ± 0.02	426.93 ± 288.85	47.35 ± 0.20	15.92 ± 0.21
Ethmalosa fimbriata	Bonga	EFG	22.60 ± 0.22	1.03 ± 0.03	9.47 ± 0.01	4.13 ± 4.95	21.07 ± 0.29
Polydactylus quadrifilis	Shiny nose fish	PQG	35.81 ± 0.47	6.22 ± 0.02	0.02 ± 0.01	13.95 ± 0.36	47.67 ± 0.23
[23]			30.00	50.00	1.30	2.00	1.00

Guideline values obtained from [3, 18]
ONM Oreochromis niloticus muscles, *PTM* Pseudotolithus typus muscles, *CHM* Caranx hippos muscles, *EFM* Ethmalosa fimbriata muscles, *PQM* Polydactylus quadrifilis muscles, *ONG* Oreochromis niloticus gills, *PTG* Pseudotolithus typus gills, *CHG* Caranx hippos gills, *EFG* Ethmalosa fimbriata gills, *PQG* Polydactylus quadrifilis gills

Table 3 Mean value ± standard deviation concentration of the trace metals in the water (mg/L) and sediments in mg/kg

Sites	Trace metals				
	Cu	Zn	Pb	Cr	Mn
Water (AWA)	35.81 ± 0.47	5.56 ± 0.02	0.00 ± 0.00	1.25 ± 0.05	0.30 ± 0.10
Water (AWB)	13.20 ± 13.20	1.03 ± 0.01	0.02 ± 0.01	0.00 ± 0.00	10.25 ± 0.05
Sediment (ASA)	0.09 ± 16.55	1.71 ± 0.02	0.02 ± 0.01	9.20 ± 0.10	35.10 ± 0.05
Sediment (ASB)	15.65 ± 0.15	2.04 ± 0.02	0.00 ± 0.00	0.15 ± 0.05	1.40 ± 0.05
Sediment WHO (2011)	50.00	300.00	50.00	100. 00	2000
Water [25]	2.00	3.00	0.01	0.1	0.4

Water (AWA) = the middle of the river; water (AWB) = bank of the river
Sediment (ASA) = the middle of the river; sediment (ASB) = bank of the river

Table 4 Levels of trace metals in the fish organs reported in the literature (mean values in mg/kg dry weight)

Sample location/references	Fish species	Organs	Cu	Zn	Pb	Cr	Mn
Kolkata wetland, India [26]	*Labeo rohita*	Muscles	3.93	24.26	0.16	0.63	0.47
		Gills	3.39	19.73	0.26	0.59	1.51
Shirinsu wetland, Iran [27]	*Cyprinus carpio*	Muscles	0.01	0.007	0.015		
		Gills	0.26	0.53	0.009		
Enyong Creek, Itu [18]	*Xenomystus nigri*	Muscles	0.03	0.32	0.04	0.18	0.74
Baghdad [24]	*B. sharpeyi*	Muscles	1.10	1.05	1.50	2.20	
		Gills	0.50	0.90	2.05	2.50	

species considered in this study than other trace metals. Mn concentrations in the muscles of the fish species followed the trend CHM > ONM > PQM > EFM > PTM (Table 2). The observed values were found to be higher than the recommended level of 1.0 mg/kg and as such may be dangerous to health.

3.2 *Human Health Risk Assessment*

3.2.1 Exposure Assessment for Fish Consumption

The evaluation of human exposure to trace metals in the muscles and gills of the fish species, sediments, and water samples was obtained for adults and children, and the estimated dietary intake of trace metals in the edible muscles and gills of the fish species, sediments, and water samples collected from Qua Iboe River is summarized in Table 5. The results revealed that intake values for trace metals due to consumption of the investigated muscles of the fish species were of the trend Zn > Mn > Cu > Cr > Pb for both adults and children while in the gills of the fish species were of the trend Mn > Cu > Cr > Zn > Pb for both adults and children.

3.2.2 Evaluation of Toxic (Noncarcinogenic) Risks

In this study, it was assumed that the toxic risk due to potentially hazardous chemicals in the same medium was cumulative; therefore, the summation of the HQs gave the overall toxic risk, which is the hazard index. On a general note, all the computed HI values for adults in the muscles, gills, water, and sediment were below unity except the gill of CHG that accumulated 32.307. For children, HI values for EFM, AWB ASA, and ASB were below unity indicating that there are no harmful effects for both fish and human (Table 6), while every other studied species were higher than unity implying harmful effect for them.

Table 5 Estimated dietary intake (mg/kg/day) of trace metals in fish species, sediments, and water for children and adults

Code	Cu		Pb		Zn		Cr		Mn	
	Adult	Children	Adult	Children	Adult	Children	Adult	Children	Adult	Children
ONM	0.0037	0.0232	0	0	0.1051	0.6656	0.0005	0.0029	0.0046	0.0294
PTM	0.0039	0.0249	0	0	0.1521	0.9633	0.0019	0.0121	0.0032	0.0205
CHM	0.00234	0.0148	0	0	0.0982	0.6221	0	0	0.0122	0.0770
PQM	0.0023	0.0146	0	0	0.0985	0.6236	0.0001	0.0007	0.0038	0.0243
EFM	0.0024	0.0155	0	0	0.0010	0.0065	0.0026	0.0163	0.0045	0.0287
ONG	0.0047	0.0297	0.0013	0.0080	0.0062	0.0393	0	0	0.0138	0.0875
PTG	0.0023	0.0146	0.0032	0.0202	0.0003	0.0020	0.0011	0.0071	0.0055	0.0349
CHG	0.0094	0.0597	0.1281	0.8112	0.0023	0.0148	0.0142	0.0900	0.0048	0.0302
EFG	0.0068	0.0429	0.0028	0.0180	0.0003	0.0020	0.0012	0.0078	0.0063	0.0400
PQG	0.0107	0.0680	0.0000	0	0.0019	0.0118	0.0043	0.0272	0.0143	0.0906
AWA	0.0107	0.0680	0	0	0.0017	0.0106	0.0004	0.0024	0.0001	0.0006
AWB	0.0040	0.0251	0.0000	0	0.0003	0.0020	0	0	0.0031	0.0195
ASA	0.0000	0.0002	0.0000	0	0.0005	0.0033	0.0028	0.0175	0.0105	0.0667
ASB	0.0047	0.0297	0	0	0.0006	0.0039	0.00005	0.0003	0.0004	0.0028

Table 6 Hazard quotient and hazard index of trace metals in fish species, sediments, and water for children and adults

	Cu		Zn		Pb		Cr		Mn		HI	
Code	Children	Adult	Children	Adult	Children	Adult	Children	Adult	Children	Adult	Children	Adult
ONM	0.580	0.092	2.219	0.350	0	0	0.002	0.000	0.210	0.033	3.010	0.475
PTM	0.622	0.098	3.211	0.507	0	0	0.008	0.001	0.147	0.023	3.988	0.630
CHM	0.371	0.059	2.074	0.327	0	0	0	0	0.550	0.087	2.994	0.473
PQM	0.366	0.058	2.079	0.328	0	0	0.000	0.000	0.174	0.027	2.619	0.413
EFM	0.387	0.061	0.022	0.003	0	0	0.010	0.002	0.205	0.032	0.624	0.099
ONG	0.743	0.117	0.131	0.021	2.005	0.317	0	0	0.625	0.099	3.504	0.553
PTG	0.366	0.058	0.007	0.001	5.040	0.796	0.005	0.001	0.249	0.039	5.666	0.895
CHG	1.493	0.236	0.050	0.008	202.79	32.02	0.060	0.010	0.216	0.034	204.6	32.31
EFG	1.073	0.170	0.007	0.001	4.499	0.710	0.005	0.001	0.286	0.045	5.870	0.927
PQG	1.701	0.269	0.040	0.006	0.007	0.001	0.018	0.003	0.647	0.102	2.413	0.381
AWA	1.701	0.269	0.035	0.006	0	0	0.002	0.000	0.004	0.001	1.742	0.275
AWB	0.627	0.099	0.007	0.001	0.007	0.001	0	0	0.139	0.022	0.780	0.123
ASA	0.0041	0.001	0.011	0.002	0.007	0.001	0.012	0.002	0.476	0.075	0.510	0.081
ASB	0.7434	0.118	0.0139	0.002	0	0	0.000	0.000	0.020	0.003	0.776	0.123

4 Conclusion

The computed HI values for adults' consumption of fish (muscles and gills), water, and sediment were below unity except CHG (32.307). For children, HI values for EFM, AWB, ASA, and ASB were below unity indicating that there are no harmful effects for both fish and human (Table 6), while every other studied species were higher than unity implying harmful effect for them. Furthermore, the HQ values calculated for adults consuming both the muscles and gills analyzed were all below unity indicating that low or no toxic risks are associated with them except for zinc in CHG. Computed HQ for children consuming the muscles of all the species were all below unity indicating low or no toxic risks except Zn in EMF, which implies possible risk.

References

1. Briaudeau, T., Zorita, I., Cuevas, N., Franco, J., Marigomez, I., Izagirre, U.: Multi-annual survey of health status disturbance in the Bilbao estuary (Bay of Biscay) based on sediment chemistry and juvenile sole (*Solea spp.*) histopathology. Mar. Pollut. Bull. **145**, 126–137 (2019)
2. Azizi, G., Layachi, M., Akodad, M., Yanez, R., Martin, I., Baghour, M.: Seasonal variations of heavy metals content in mussels *(Mytilus galloprovincialis)* from Cala Iris offshore (Northern Morocco). Mar. Pollut. Bull. **137**, 688–694 (2018)
3. Ubong, U., Ekwere, I., Ekanem, A., Ite, A.: Human health risk assessment of cadmium (Cd), lead (Pb) and mercury (Hg) levels in organs of fish obtained from Iko River, Eastern Obolo Local Government Area (LGA), Akwa Ibom State, Nigeria. J. Mater. Environ. Sci. **14**(3), 384–394 (2023)
4. Ubong, I., Ubong, U., Ubong, E., Ukonta, R., Ishmael, D.: Distribution of particulate matter in Cawthorne Channels air basin in Nigeria. Environ. Pollut. **4**(3), 19 (2015)
5. Huang, J., Yu, H., Guan, X., Wang, G., Guo, R.: Accelerated dryland expansion under climate change. Nat. Clim. Chang. **6**(2), 166–171 (2016)
6. Ubong, U., Ekwere, I., Ikpe, E.: Risk and toxicity assessments of heavy metals in *Tympanotonus fuscatus* and sediments from Iko river, Akwa Ibom state, Nigeria. Int. J. Environ. Clim. Chang. **10**(3), 38–47 (2020). https://doi.org/10.9734/ijecc/2020/v10i330186
7. Ite, A.E., Ubong, U.U., Etesin, U.M., Nsi, E.W., Ukpong, E.J., Ekanem, A.N., Udo, A.I.: Heavy metals in epiphytic lichens and mosses of oil-producing communities of Eket and Ibeno, Akwa Ibom State-Nigeria. Am. J. Environ. Prot. **4**(2), 38–47 (2016). https://doi.org/10.12691/env-4-2-1
8. Ubong, U., Okoye, P., Ubong, I.: Temporal variations of suspended particulate matter in Port Harcourt metropolis. Afr. J. Environ. Pollut. Health. **9**(1), 1–6 (2011)
9. Ubong, U.U., Ubong, I.U., Ubong, E.U., Etukudo, O.U.: Assessment of quality of potable water sources in eket local government area of Akwa Ibom State, Nigeria. Int. J. Adv. Innov. Res. **4**(8), 2278–7844 (2015)
10. Ubong, U.U., Horsfall, M., Ubong, I.U.: Distribution of heavy metals in tissues of *Callinectus latimanus* from the New Calabar River Nigeria. Afr. J. Environ. Pollut. Health. **9**(1), 7–13 (2011)

11. Ubong, U., Ekwere, I.: Distribution and human health risk assessment of heavy metals in tissues of *Callinectes sapidus* from Iko River, Akwa Ibom State, Nigeria. Asian J. Environ. Ecol. **19**(3), 1–8 (2022). https://doi.org/10.9734/ajee/2022/v19i3408
12. Ikpe, E., Ubong, U., Archibong, U.: Proximate analysis, heavy metals and total hydrocarbon content of *Callinectes sapidus* obtained from Ibaka River, Akwa Ibom State, Nigeria. Res. J. Sci. Technol. **2**(2), 32–43 (2022)
13. American Society for Testing Materials (ASTM): Standard Test Methods for Tension Testing of Metallic Materials. ASTM International, West Conshohocken (2009)
14. Wang, X., Sato, T., Xing, B.: Health risks of heavy metals to the general public in Tianjin, China via consumption of vegetables and fish. Sci. Total Environ. **350**(1–3), 28–37 (2005). https://doi.org/10.1016/j.scitotenv.2004.09.044
15. Lushenko, M.A.: A Risk Assessment for Ingestion of Toxic Chemicals in Fish from IMPERIAL Beach, California. San Diego State University (2010)
16. Means, B.: Risk-Assessment Guidance for Superfund. Volume 1. Human Health Evaluation Manual. Part A. Interim Report (Final). Environmental Protection Agency. Office of Solid Waste, Washington, DC (1989)
17. Lee, S., Oh, B., Kim, J.: Effect of various amendments on heavy mineral oil bioremediation and soil microbial activity. Bioresour. Technol. **99**(7), 2578–2587 (2008). https://doi.org/10.1016/j.biortech.2007.04.039
18. Udosen, E., Offiong, N., Alade, I.: Human health risk assessment of trace metals due to dietary intake of some edible fish species collected from Enyong Creek, Itu–Nigeria. In: A Paper Presented at the 37th International Conference of Chemical Society of Nigeria held at Uyo Nigeria (2014)
19. Ihedioha, J., Okoye, C.: Dietary intake and health risk assessment of lead and cadmium via consumption of cow meat for an urban population in Enugu State, Nigeria. Ecotoxicol. Environ. Saf. **93**, 101–106 (2013). https://doi.org/10.1016/j.ecoenv.2013.04.010. PMID: 23664087
20. Urban, J., Tachovsky, J., Haws, L., Staskal, D.W., Harris, M.: Assessment of human health risks posed by consumption of fish from the Lower Passaic River, New Jersey. Sci. Total Environ. **408**(2), 209–224 (2009). https://doi.org/10.1016/j.scitotenv.2009.03.004
21. Inyang, H.: Beyond contaminant concentration measurements: use of data in contaminant migration analyses and environmental risk assessment. An invited lecture by a Duke Energy Distinguished Professor of Environmental Engineering and Science, and Professor of Earth Science, University of North Carolina, Charlotte, USA at the University of Uyo, Uyo, on 3rd January, 2013
22. Barlow, S., Schlatter, J.: Risk assessment of carcinogens in food. Toxicol. Appl. Pharmacol. **243**(2), 180–190 (2010). https://doi.org/10.1016/j.taap.2009.11.004
23. Council, N.R: Recommended Dietary Allowances. Food and Nutrition Board, Commission on Life Sciences, National Research Council. National Academic, Washington (1989)
24. Mensoor, M., Said, A.: Determination of heavy metals in freshwater fishes of the Tigris River in Baghdad. Aust. Fish. **3**(2), 23 (2018). https://doi.org/10.3390/fishes3020023
25. WHO: Guidelines for Drinking-Water Quality, 4th edn. © World Health Organization, Geneva (2011)
26. Kumar, N., Chandan, K., Bhushan, S., Singh, D., Satish, K.: Health risk assessment and metal contamination in fish, water and soil sediments in the East Kolkata Wetlands, India, Ramsar site. Sci. Rep. **13**, 1546 (2023). https://doi.org/10.1038/s41598-023-28801-y
27. Ardakani, S.S., Jafari, S.M.: Assessment of heavy metals (Cu, Pb and Zn) in different tissues of common carp (*Cyprinus carpio*) caught from Shirinsu Wetland, Western Iran. J. Chem. Health Risks. **4**(2), 47–54 (2018)

Evaluation of Standalone Solar Photovoltaic Power System with Load Scheduling for Automated Teller Machine Gallery

A. Umoren Mfonobong, Eduediuyai Ekerette Dan, and Nseobong Okpura

1 Introduction

Over the years, Nigerians have suffered perennial power shortage from the national grid [1–4]. As such, power consumers within Nigeria have resorted to alternative power supply solutions. Among the several options, solar photovoltaic (SPV) power system has attracted many users because it is sustainable and environmentally friendly and also there is a steady drop in the overall cost of acquisition and maintenance of the SPV [5–7].

In addition, in recent times, the Nigerian banking and financial service providers are increasingly adopting the use of self-service banking tools like the automated teller machine (ATM) to reduce the workload on their staff [8–11]. As such, presently, several banks across Nigeria have installed several automated teller machine (ATM) galleries. However, ensuring steady power supply for the ATM gallery to avoid any downtime due to power supply has become a running challenge to the banks, especially in the locations without access to the national grid [12–14]. In such case, standalone solar photovoltaic (SSPV) can be a viable option [15–17]. Accordingly, in this paper, PVsyst simulation-based evaluation of standalone solar photovoltaic (SSPV) power system with load scheduling for automated teller machine gallery is presented. The study considered the loss of load performance of the SSPV and introduced the use of load scheduling to avert total loss for the ATM gallery.

A. Umoren Mfonobong (✉) · N. Okpura
Department of Electrical/Electronic Engineering, University of Uyo, Uyo, Akwa Ibom State, Nigeria
e-mail: nseobongokpura@uniuyo.edu.ng

E. E. Dan
Department of Computer Engineering, Federal Polytechnic, Ukana, Akwa Ibom State, Nigeria

S. C. Ekpo (ed.), *The Second International Adaptive and Sustainable Science, Engineering and Technology Conference*, Signals and Communication Technology,
https://doi.org/10.1007/978-3-031-53935-0_27

2 Methodology

The study considered an automated teller machine (ATM) gallery consisting of ten units of ATMs with daily energy demand of 608.4 kWh, shown in Table 1. The ATM gallery site has a latitude of 4.621393 and a longitude of 7.763904 with mean daily average global irradiation on the horizontal plane of 5.05 kWh/m^2/day (as shown in Fig. 1) and mean daily ambient temperature of 26.13 °C (as shown in Fig. 2). The optimal tilt angle of 13% was selected based on PVsyst software that [18, 19] generated optimal transposition factor of 1.01% (as shown in Fig. 3). The PVsyst system configurations for the sizing of the PV array and battery bank are presented in Fig. 4. According to the screenshot in Fig. 4, the PV power system is designed with maximum loss of load probability of 3% and 4-day autonomy.

Table 1 The daily energy demand of the automated teller machine (ATM) gallery

S/N	Equipment description	No. of ATM units	QTY per ATM	Power rating (kW)	Duration (h)	Electrical load (kW)	Energy demand per day (kWh)
1	ATM	10	1	0.8	24	8	192
2	AIR Conditioner	10	1	1.6	24	16	384
3	Lightings	10	1	0.09	24	0.9	21.6
4	Hub		1	0.45	24	0.45	10.8
Total energy consumed						25.35	608.4
Equivalent of 25.41 kW running for 24 hours a day to give energy demand of 609.84 kWh per day							

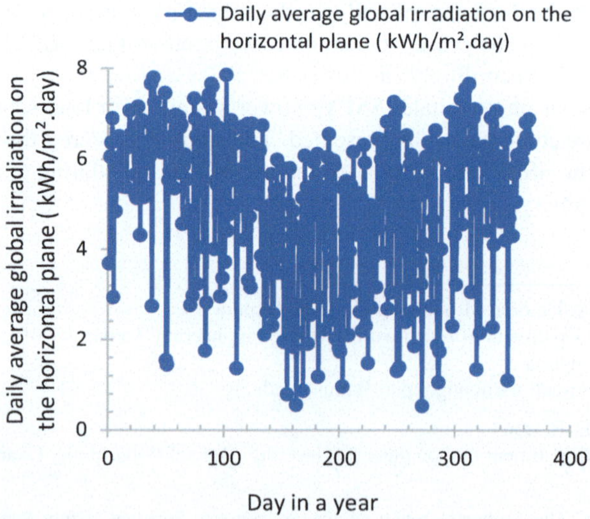

Fig. 1 Daily average global irradiation on the horizontal plane (kWh/m^2.day)

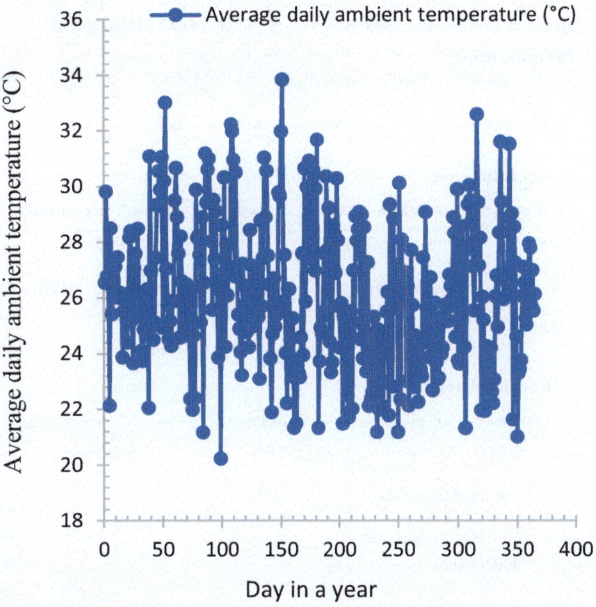

Fig. 2 Mean daily ambient temperature (°C)

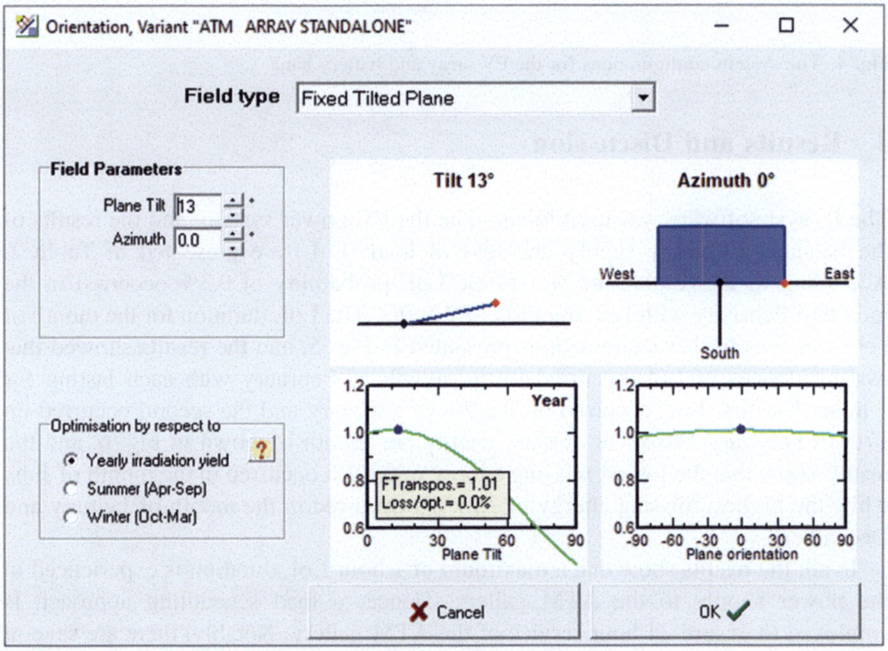

Fig. 3 The optimal tilt angle and transposition ratio for the PV array

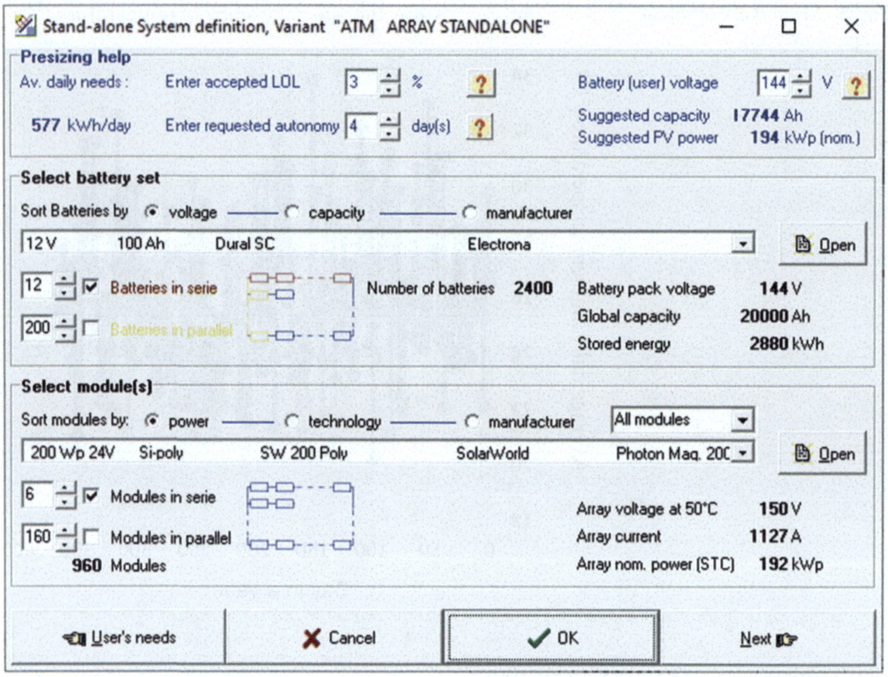

Fig. 4 The system configurations for the PV array and battery bank

3 Results and Discussion

The PVsyst software was used to simulate the PV power system, and the results of the balance of energy supply and loss of load (LoL) are presented in Table 2. According to the results, the worst-case LoL probability of 0.3% occurred in the month of February with LoL duration of 2 hours. The LoL duration for the month of February was further examined, as presented in Fig. 5, and the results showed that two incidences of LoL occurred in the month of February with each lasting for 1 hour. The first LoL occurred on the 7th of February and the second occurred on 17th of February. Also, the missing energy per month is shown in Fig. 6, and the results show that the lowest missing energy of 0 kWh occurred in the month of June while the highest missing energy of 25 kWh occurred in the month of January and December.

In all, the results show that a maximum of 1 hour LoL duration is experienced in the power supply to the ATM gallery. Hence, a load scheduling approach is employed to ensure 24-hour service of the ATM gallery. Notably, there are several scheduling options, but the one adopted in this work is to run five of the ATM machines for 24 hours per day while the remaining five ATM machines are scheduled to run from 5 am till 11 pm (which amounts to 18 hours of operation per day).

Table 2 The results of the balance of energy supply and loss of load

Month	Energy supplied to the user (kWh) without load scheduling	Energy need of the user (kWh) without load scheduling	Missing energy (kWh) without load scheduling	Solar fraction	Duration of loss of load (hour)	Loss of load probability (%)
January	18,889	18,864	25	0.999	1	0.13
February	17,061	17,039	22	0.999	2	0.3
March	18,889	18,866	23	0.999	1	0.13
April	18,279	18,256	23	0.999	1	0.14
May	18,889	18,872	17	0.999	0	0
June	18,279	18,279	0	1	0	0
July	18,889	18,887	2	1	0	0
August	18,889	18,877	11	0.999	0	0
September	18,279	18,262	18	0.999	1	0.14
October	18,889	18,869	20	0.999	1	0.13
November	18,279	18,257	23	0.999	1	0.14
December	18,889	18,864	25	0.999	1	0.13
Year	222,400	222,192	208	0.999167	9	0.103333

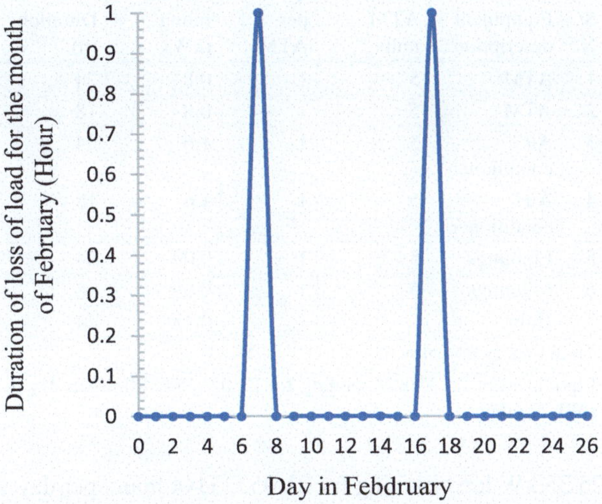

Fig. 5 Duration of loss of load for the month of September (hour)

The period from 11 pm to 5 am is considered off-peak period. As such, five out of the ten ATM machines can serve the few users that may need ATM services in such night hours. Accordingly, the daily energy demand of the ATM gallery with the load scheduling is given in Table 3. In the light of the daily load demand shown in Table 1, the daily energy demand of the ATM array in Table 3 is equivalent to

Fig. 6 The missing energy per month without load scheduling

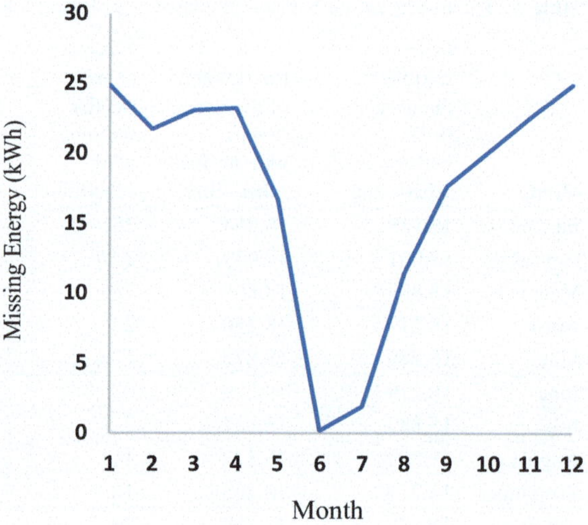

Table 3 The daily energy demand of the automated teller machine (ATM) gallery consisting of ten ATM machines with five running for 24 hours per day and the remaining five running for 18 hours per day

S/N	Equipment description	No. of ATM units	QTY per ATM	Power rating (kW)	Duration (h)	Electrical load (kW)	Energy demand per day (kWh)
1	ATM	5	1	0.8	24	4	96
2	ATM	5	1	0.8	18	4	72
3	Air conditioner	5	1	1.6	24	8	192
4	Air conditioner	5	1	1.6	18	8	144
5	Lightings	5	1	0.09	24	0.45	10.8
6	Lightings	5	1	0.09	18	0.45	8.1
7	Hub		1	0.45	24	0.45	10.8
Total energy consumed						25.35	533.7
Equivalent of 25.35 kW running for 21.05325444 hours a day to give energy demand of 533.7 kWh per day							

25.35 kW load running for 21.05325444 hours per day which gives energy demand of 533.7 kWh per day. The results of the energy available to the user (kWh) and the energy need of the user (kWh) with load scheduling are shown in Fig. 7. It can be seen that with the load scheduling, the available energy delivered to the load will meet the energy demand of the ATM gallery without any loss of load or missing energy.

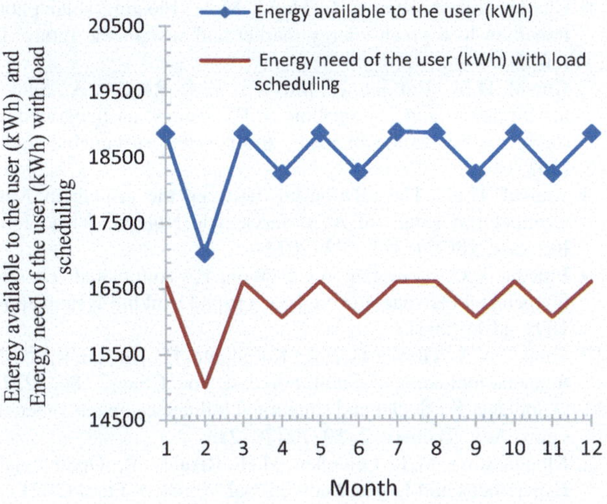

Fig. 7 The graph of the energy available to the user (kWh) and the energy need of the user (kWh) with load scheduling

4 Conclusion

A simulated evaluation of the solar power supply for automated teller machine (ATM) gallery is presented. The study examined the energy demand of the ATM gallery with full load and also with load scheduling whereby a given number of ATMs are always operational 24 hours while the remaining ATMs are shut down in the off-peak periods which in this case are night hours from 11 pm to 5 am. With the load scheduling, a 24-hour ATM service is guaranteed all through the year. The ideas presented in this study are useful for the design of solar power system for users with dispatchable load.

References

1. Nnamani, R.G., Chukwu, Q.C., Ekekwe, E.: Exploring the development of infrastructure in the oil producing areas of Nigeria, 2009–2017. EPRA Int. J. Res. Dev. **8**(4), 114–122 (2023)
2. Jack, J.T.B., Jack, B.: Nigeria's energy crisis and the sustainability question. Soc. Health Sci. **20**(1 and 2), 19–24 (2022)
3. Tukur, M.A.: Issues and challenges of power sector reforms in Nigeria: the political economy approach. Am. J. Arts Hum. Sci. **1**(2), 128–137 (2022)
4. Akinbomi, J.G., Ezeilo, L.C., Fagbuyi, O., Onyebueke, L.: Optimal microgrid power supply system for Nigerian detached communities: environmental impact and energy cost criteria. Niger. J. Technol. **40**(3), 491–500 (2021)
5. Batool, K., Zhao, Z.Y., Irfan, M., Ullah, S., Işik, C.: Assessing the competitiveness of Indian solar power industry using the extended Five Forces Model: a green innovation perspective. Environ. Sci. Pollut. Res. **30**(34), 82045–82067 (2023)

6. Charles Rajesh Kumar, J., Majid, M.A.: Floating solar photovoltaic plants in India–a rapid transition to a green energy market and sustainable future. Energy Environ. **34**(2), 304–358 (2023)
7. Khalid, H.M., Rafique, Z., Muyeen, S.M., Raqeeb, A., Said, Z., Saidur, R., Sopian, K.: Dust accumulation and aggregation on PV panels: an integrated survey on impacts, mathematical models, cleaning mechanisms, and possible sustainable solution. Sol. Energy. **251**, 261–285 (2023)
8. Ahmad, U.H.: The relationship between the growth in Automated Teller Machine (ATM) terminal and values of ATM services in Nigeria. World Journal of Advanced Research and Reviews, **18**(03), 511–523 (2023)
9. Ehiedu, V.C., Onuorah, A.C., Okoh, E.: Automated teller machine (ATM) penetration and financial inclusiveness in Nigeria: a tripod banking system approach. Indian J. Econ. Bus. **20**(3), 1093–1104 (2021)
10. Shaikh, A.A., Glavee-Geo, R., Karjaluoto, H., Hinson, R.E.: Mobile money as a driver of digital financial inclusion. Technol. Forecast. Soc. Chang. **186**, 122158 (2023)
11. Galdolage, S.: Social and situational influences on customer shift towards self-service technologies. Adv. Technol. **2**, 50–70 (2022)
12. Khiaonarong, M.T., Leinonen, M.H., Rizaldy, R.: Operational Resilience in Digital Payments: Experiences and Issues. International Monetary Fund (2021)
13. Baba, M., Dogonyaro, N.M., Yisa, V.L., Olatunde, M.O.: Evaluation of Mobile Banking Services Usage in Minna, Niger State. i-manager's Publication (2019)
14. Baca, M., Schenkman, B., Hightower, M.: Use of advanced microgrids to support community resilience. Nat. Hazards Rev. **22**(4), 05021012 (2021)
15. Nag, A.K., Sarkar, S.: Techno-economic analysis of a micro-hydropower plant consists of hydrokinetic turbines arranged in different array formations for rural power supply. Renew. Energy. **179**, 475–487 (2021)
16. Ebhota, W.S., Jen, T.C.: Fossil fuels environmental challenges and the role of solar photovoltaic technology advances in fast tracking hybrid renewable energy system. Int. J. Precis. Eng. Manuf.-Green Technol. **7**, 97–117 (2020)
17. Li, D., Guo, J., Zhang, J., Zhan, L., Alizadeh, M.: Numerical assessment of a hybrid energy generation process and energy storage system based on alkaline fuel cell, solar energy and Stirling engine. J. Energy Storage. **39**, 102631 (2021)
18. Kumar, R., Rajoria, C.S., Sharma, A., Suhag, S.: Design and simulation of standalone solar PV system using PVsyst software: a case study. Mater. Today: Proc. **46**, 5322–5328 (2021)
19. Belmahdi, B., El Bouardi, A.: Solar potential assessment using PVsyst software in the northern zone of Morocco. Proc. Manuf. **46**, 738–745 (2020)

Index

S. C. Ekpo (ed.), *The Second International Adaptive and Sustainable Science, Engineering and Technology Conference*, Signals and Communication Technology,
https://doi.org/10.1007/978-3-031-53935-0

The manufacturer's authorised representative in the EU is Springer Nature Customer Service Centre GmbH, Europaplatz 3, 69115 Heidelberg, Germany. If you have any concerns regarding our products, please contact ProductSafety@springernature.com

Printed and bound by CPI Group (UK) Ltd, Croydon, CR0 4YY
01/10/2026
02225896-0012